Combinatorial Library Methods and Protocols

METHODS IN MOLECULAR BIOLOGY™

John M. Walker, SERIES EDITOR

221. **Generation of cDNA Libraries:** *Methods and Protocols,* edited by *Shao-Yao Ying, 2003*
220. **Cancer Cytogenetics:** *Methods and Protocols,* edited by *John Swansbury, 2003*
219. **Cardiac Cell and Gene Transfer:** *Principles, Protocols, and Applications,* edited by *Joseph M. Metzger, 2003*
218. **Cancer Cell Signaling:** *Methods and Protocols,* edited by *David M. Terrian, 2003*
217. **Neurogenetics:** *Methods and Protocols,* edited by *Nicholas T. Potter, 2003*
216. **PCR Detection of Microbial Pathogens:** *Methods and Protocols,* edited by *Konrad Sachse and Joachim Frey, 2003*
215. **Cytokines and Colony Stimulating Factors:** *Methods and Protocols,* edited by *Dieter Körholz and Wieland Kiess, 2003*
214. **Superantigen Protocols,** edited by *Teresa Krakauer, 2003*
213. **Capillary Electrophoresis of Carbohydrates,** edited by *Pierre Thibault and Susumu Honda, 2003*
212. **Single Nucleotide Polymorphisms:** *Methods and Protocols,* edited by *Piu-Yan Kwok, 2003*
211. **Protein Sequencing Protocols, 2nd ed.,** edited by *Bryan John Smith, 2003*
210. **MHC Protocols,** edited by *Stephen H. Powis and Robert W. Vaughan, 2003*
209. **Transgenic Mouse Methods and Protocols,** edited by *Marten Hofker and Jan van Deursen, 2002*
208. **Peptide Nucleic Acids:** *Methods and Protocols,* edited by *Peter E. Nielsen, 2002*
207. **Recombinant Antibodies for Cancer Therapy:** *Methods and Protocols.* edited by *Martin Welschof and Jürgen Krauss, 2002*
206. **Endothelin Protocols,** edited by *Janet J. Maguire and Anthony P. Davenport, 2002*
205. ***E. coli* Gene Expression Protocols,** edited by *Peter E. Vaillancourt, 2002*
204. **Molecular Cytogenetics:** *Protocols and Applications,* edited by *Yao-Shan Fan, 2002*
203. ***In Situ* Detection of DNA Damage:** *Methods and Protocols,* edited by *Vladimir V. Didenko, 2002*
202. **Thyroid Hormone Receptors:** *Methods and Protocols,* edited by *Aria Baniahmad, 2002*
201. **Combinatorial Library Methods and Protocols,** edited by *Lisa B. English, 2002*
200. **DNA Methylation Protocols,** edited by *Ken I. Mills and Bernie H, Ramsahoye, 2002*
199. **Liposome Methods and Protocols,** edited by *Subhash C. Basu and Manju Basu, 2002*
198. **Neural Stem Cells:** *Methods and Protocols,* edited by *Tanja Zigova, Juan R. Sanchez-Ramos, and Paul R. Sanberg, 2002*
197. **Mitochondrial DNA:** *Methods and Protocols,* edited by *William C. Copeland, 2002*
196. **Oxidants and Antioxidants:** *Ultrastructure and Molecular Biology Protocols,* edited by *Donald Armstrong, 2002*
195. **Quantitative Trait Loci:** *Methods and Protocols,* edited by *Nicola J. Camp and Angela Cox, 2002*
194. **Posttranslational Modifications of Proteins:** *Tools for Functional Proteomics,* edited by *Christoph Kannicht, 2002*
193. **RT-PCR Protocols,** edited by *Joe O'Connell, 2002*
192. **PCR Cloning Protocols, 2nd ed.,** edited by *Bing-Yuan Chen and Harry W. Janes, 2002*
191. **Telomeres and Telomerase:** *Methods and Protocols,* edited by *John A. Double and Michael J. Thompson, 2002*
190. **High Throughput Screening:** *Methods and Protocols,* edited by *William P. Janzen, 2002*
189. **GTPase Protocols:** *The RAS Superfamily,* edited by *Edward J. Manser and Thomas Leung, 2002*
188. **Epithelial Cell Culture Protocols,** edited by *Clare Wise, 2002*
187. **PCR Mutation Detection Protocols,** edited by *Bimal D. M. Theophilus and Ralph Rapley, 2002*
186. **Oxidative Stress Biomarkers and Antioxidant Protocols,** edited by *Donald Armstrong, 2002*
185. **Embryonic Stem Cells:** *Methods and Protocols,* edited by *Kursad Turksen, 2002*
184. **Biostatistical Methods,** edited by *Stephen W. Looney, 2002*
183. **Green Fluorescent Protein:** *Applications and Protocols,* edited by *Barry W. Hicks, 2002*
182. **In Vitro Mutagenesis Protocols, 2nd ed.,** edited by *Jeff Braman, 2002*
181. **Genomic Imprinting:** *Methods and Protocols,* edited by *Andrew Ward, 2002*
180. **Transgenesis Techniques, 2nd ed.:** *Principles and Protocols,* edited by *Alan R. Clarke, 2002*
179. **Gene Probes:** *Principles and Protocols,* edited by *Marilena Aquino de Muro and Ralph Rapley, 2002*
178. **Antibody Phage Display:** *Methods and Protocols,* edited by *Philippa M. O'Brien and Robert Aitken, 2001*
177. **Two-Hybrid Systems:** *Methods and Protocols,* edited by *Paul N. MacDonald, 2001*
176. **Steroid Receptor Methods:** *Protocols and Assays,* edited by *Benjamin A. Lieberman, 2001*
175. **Genomics Protocols,** edited by *Michael P. Starkey and Ramnath Elaswarapu, 2001*
174. **Epstein-Barr Virus Protocols,** edited by *Joanna B. Wilson and Gerhard H. W. May, 2001*
173. **Calcium-Binding Protein Protocols, Volume 2:** *Methods and Techniques,* edited by *Hans J. Vogel, 2001*
172. **Calcium-Binding Protein Protocols, Volume 1:** *Reviews and Case Histories,* edited by *Hans J. Vogel, 2001*
171. **Proteoglycan Protocols,** edited by *Renato V. Iozzo, 2001*
170. **DNA Arrays:** *Methods and Protocols,* edited by *Jang B. Rampal, 2001*
169. **Neurotrophin Protocols,** edited by *Robert A. Rush, 2001*
168. **Protein Structure, Stability, and Folding,** edited by *Kenneth P. Murphy, 2001*
167. **DNA Sequencing Protocols,** *Second Edition,* edited by *Colin A. Graham and Alison J. M. Hill, 2001*
166. **Immunotoxin Methods and Protocols,** edited by *Walter A. Hall, 2001*
165. **SV40 Protocols,** edited by *Leda Raptis, 2001*
164. **Kinesin Protocols,** edited by *Isabelle Vernos, 2001*
163. **Capillary Electrophoresis of Nucleic Acids, Volume 2:** *Practical Applications of Capillary Electrophoresis,* edited by *Keith R. Mitchelson and Jing Cheng, 2001*
162. **Capillary Electrophoresis of Nucleic Acids, Volume 1:** *Introduction to the Capillary Electrophoresis of Nucleic Acids,* edited by *Keith R. Mitchelson and Jing Cheng, 2001*

METHODS IN MOLECULAR BIOLOGY

Combinatorial Library

Methods and Protocols

Edited by

Lisa Bellavance English

BD Biosciences, San Jose, California

Springer Science+Business Media, LLC

Originally published by Humana Press Inc. in 2000
Softcover reprint of the hardcover 1st edition 2000

www.humanapress.com

This publication is printed on acid-free paper. ∞
ANSI Z39.48-1984 (American Standards Institute)

Permanence of Paper for Printed Library Materials.

Production Editor: Adrienne Howell
Cover design by Patricia F. Cleary.
Cover illustration: (A) Total ion image showing two selected areas (*A1* and *A2*) each corresponding to one bead. (B) Negative S-SIMS image of Boc-Pro-Phe-Leu (carboxylate ion at *m/z* 474); (C) Negative S-SIMS image of the deleted sequence Boc-Pro-Leu (carboxylate ion at *m/z* 327). *See* full caption and discussion on page 18 (Fig. 2 A, B, and C in Chapter 2).

Library of Congress Cataloging-in-Publication Data

Combinatorial library methods and protocols / edited by Lisa B. English.
p. cm. -- (Methods in molecular biology ; v. 201)
ISBN 978-1-4899-3899-2 ISBN 978-1-59259-285-2 (eBook)
DOI 10.1007/978-1-59259-285-2
1. Combinatorial chemistry--Laboratory manuals. I. English, Lisa B. II. Series.

RS419 .C6568 2002
615'.19--dc21

2001051653

Preface

The continued successes of large- and small-scale genome sequencing projects are increasing the number of genomic targets available for drug discovery at an exponential rate. In addition, a better understanding of molecular mechanisms—such as apoptosis, signal transduction, telomere control of chromosomes, cytoskeletal development, modulation of stress-related proteins, and cell surface display of antigens by the major histocompatibility complex molecules—has improved the probability of identifying the most promising genomic targets to counteract disease. As a result, developing and optimizing lead candidates for these targets and rapidly moving them into clinical trials is now a critical juncture in pharmaceutical research. Recent advances in combinatorial library synthesis, purification, and analysis techniques are not only increasing the numbers of compounds that can be tested against each specific genomic target, but are also speeding and improving the overall processes of lead discovery and optimization.

There are two main approaches to combinatorial library production: parallel chemical synthesis and split-and-mix chemical synthesis. These approaches can utilize solid- or solution-based synthetic methods, alone or in combination, although the majority of combinatorial library synthesis is still done on solid support. In a parallel synthesis, all the products are assembled separately in their own reaction vessels or microtiter plates. The array of rows and columns enables researchers to organize the building blocks to be combined, and provides an easy way to identify compounds in a particular well. In contrast, the split-and-mix approach relies largely on solid-based synthetic methods, and produces a mixture of related compounds in the same reaction vessel. Although most combinatorial synthesis is done on solid support, solution-based synthetic methods offer some advantages. For example, solution-based synthesis offers the flexibility to use a larger number of chemical reactions; however, one classic problem of this approach is keeping track of which building blocks are added to which reaction vessel or microtiter plate well. In addition, because the compounds are not attached to a solid support, it is difficult to isolate them. Chapters 1–12 of *Combinatorial Library Methods and Protocols* discuss a variety of strategies for combinatorial library synthesis and quality control.

A combinatorial library only brings value when screened. The way library members are screened for activity depends on the form in which they were synthesized. For solid-based methods, the compounds are usually cleaved from the solid support on which they were made and eluted into microtiter plates with one or more compounds per well. For solution-based methods, the compounds of interest must be isolated, purified, and then distributed to microtiter plates. The exact method used to determine the activity of individual compounds is dependent on the screening assay used. Assays often involve displacement of another ligand, or release of a reporter element to give a readout signal. Most commonly, screening assays involve measuring radioactivity, fluorescence, or absorbance in each reaction well and comparing those to measurements on positive and negative controls. Chapters 13–16 of *Combinatorial Library Methods and Protocols* discuss purification and screening of combinatorial libraries.

The design, production, characterization, tracking, and screening of many combinatorial libraries in multiple biological assays presents an enormous computational and information management challenge. There is a need for integrated library specification, design, synthesis, screening, and analysis with the ability to feed back information from completed experiments iteratively during the entire process. Such integration requires a combination of computational informatics and analysis solutions. Chapters 17–21 of *Combinatorial Library Methods and Protocols* discuss a range of computational approaches to combinatorial library design.

Combinatorial chemistry has rapidly evolved from its early focus on the generation of large numbers of molecules to a powerful combinatorial design technology for the generation and optimization of pharmaceutical leads to produce drug candidates. Developing trends in combinatorial chemistry that promise to further improve drug design include the integration of combinatorial approaches with a range of design strategies, including structure-based design, physiochemical parameters, and combinatorial methods to optimize natural products. Because only a very small number of biologically active compounds have been sampled from all possible chemicals, the potential to discover new pharmaceuticals by applying combinatorial techniques has opened a new frontier in biology and medicine.

Lisa B. English

Contents

Contributors

FAHAD AL-OBEIDI • *Aventis Combinatorial Technologies Center, Aventis Pharmaceuticals, Tucson, AZ*
ALEX M. ARONOV • *Vertex Pharmaceuticals Inc., Cambridge, MA*
JEAN-LOUIS AUBAGNAC • *Université de Montpellier II, Montpellier, France*
RICHARD E. AUSTIN • *Aventis Combinatorial Technologies Center, Aventis Pharmaceuticals, Tucson, AZ*
LOUIS H. BLUHM • *Department of Chemistry, Vanderbilt University Nashville, TN*
DAN R. S. BOND • *Aventis Combinatorial Technologies Center, Aventis Pharmaceuticals, Tucson, AZ*
FRANK R. BURDEN • *SciMetrics, Harrow Enterprises Ltd, Victoria, Australia*
ANDREW BURRITT • *ChemRx, Tucson, AZ*
A. RICHARD CHAMBERLIN • *Department of Chemistry, University of California-Irvine, Irvine, CA*
SEEMA CHOUDHARY • *Chemistry Department, State University of New York at Buffalo, Amherst, NY*
ROBERT COMBARIEU • *Ecole des Mines de Paris, Sophia-Antipolis, France*
CHRISTINE ENJALBAL • *Université de Montpellier II, Montpellier, France*
LILING FANG • *ChemRx Advanced Technologies, South San Francisco, CA*
GIANINE M. FIGLIOZZI • *ChemRx Advanced Technologie, South San Francisco, CA*
TAO GUO • *Pharmacopeia Inc., Princeton, NJ*
STEFAN GÜSSREGEN • *Tripos Receptor Research Ltd., United Kingdom*
GREG HARLOW • *Aventis Combinatorial Technologies Center, Aventis Pharmaceuticals, Tucson, AZ*
DAVID S. HARTSOUGH • *ArQule, Inc., Woburn, MA*
CHARLES HENSON • *The Procter & Gamble Company, Mason, OH*
DOUG W. HOBBS • *Pharmacopeia Inc., Princeton, NJ*
MARK M. IRVING • *ChemRx Advanced Technologies, South San Francisco, CA*
ERIC A. JAMOIS • *Accelrys, San Diego, CA*
VIKTOR KRCHŇÁK • *Encore International Corporation, Tucson, AZ*
CLINTON A. KRUEGER • *ChemRx Advanced Technologies, South San Francisco, CA*

PETER O. KRUTZIK • *Department of Chemistry, University of California-Irvine, Irvine, CA*
AMY LEW • *Exelixis, Inc., South San Francisco, CA*
TINGYU LI • *Department of Chemistry, Vanderbilt University, Nashville, TN*
BRIAN R. LINTON • *Department of Chemistry, Bowdoin College, Brunswick, ME*
DIANA LIU • *ChemRx Advanced Technologies, South San Francisco, CA*
JOHN I. MANCHESTER • *Infection Chemistry, AstraZeneca R&D Boston, Waltham, MA*
JEAN MARTINEZ • *Université de Montpellier I, Montpellier, France*
JANET R. MORROW • *Chemistry Department, State University of New York at Buffalo, Amherst, NY*
PETER NEIDIG • *Bruker Analytik GmbH, Silberstreifen, Rheinstetten, Germany*
JOHN F. OKONYA • *Aventis Combinatorial Technologies Center, Aventis Pharmaceuticals, Tucson, AZ*
SEAN X. PENG • *The Procter and Gamble Company, Mason, OH*
ANNA ROBINSON • *Aventis Combinatorial Technologies Center, Aventis Pharmaceuticals, Tucson, AZ*
GÉRARD ROSSÉ • *Aventis Combinatorial Technologies Center, Aventis Pharmaceuticals, Tucson, AZ*
HARALD SCHRÖDER • *F. Hoffman-La Roche Ltd., Basel, Switzerland*
CARSTEN SPANKA • *Novartis Pharma AG, Switzerland*
CHUNG-MING SUN • *Department of Chemistry, National Dong Hwa University, Taiwan*
DAVID S. THORPE • *Aventis Combinatorial Technologies Center, Aventis Pharmaceuticals, Tucson, AZ*
YAN WANG • *Department of Chemistry, Vanderbilt University, Nashville, TN*
MARK WARNE • *Tripos Receptor Research Ltd., United Kingdom*
BERND WENDT • *Tripos Receptor Research Ltd., United Kingdom*
PAUL WENTWORTH, JR. • *The Scripps Research Institute, La Jolla, CA*
KENNETH F. WERTMAN • *Aventis Combinatorial Technologies Center, Aventis Pharmaceuticals, Tucson, AZ*
PATTI WILLSON • *Aventis Combinatorial Technologies Center, Aventis Pharmaceuticals, Tucson, AZ*
SYDNEY WILSON • *Aventis Combinatorial Technologies Center, Aventis Pharmaceuticals, Tucson, AZ*
DAVID A. WINKLER • *CSIRO Molecular Science, Clayton, Australia*

FRANK WOOLARD • *ChemRx Advanced Technologies, South San Francisco, CA*
BING YAN • *ChemRx Advanced Technologies, South San Francisco, CA*
HELEN YEOMAN • *Aventis Combinatorial Technologies Center, Aventis Pharmaceuticals, Tucson, AZ*
JIANG ZHAO • *ChemRx Advanced Technologies, South San Francisco, CA*

I

Library Synthesis and Quality Control

1

Using a Noncovalent Protection Strategy to Enhance Solid-Phase Synthesis

Fahad Al-Obeidi, John F. Okonya, Richard E. Austin, and Dan R. S. Bond

1. Introduction

Since the introduction of solid-phase peptide synthesis by Merrifield *(1)* nearly forty years ago, solid-phase techniques have been applied to the construction of a variety of biopolymers and extended into the field of small molecule synthesis. The last decade has seen the emergence of solid-phase synthesis as the leading technique in the development and production of combinatorial libraries of diverse compounds of varying sizes and properties. Combinatorial libraries can be classified as biopolymer based (e.g., peptides, peptidomimetics, polyureas, and others *[2,3]*) or small molecule based (e.g., heterocycles *[4]*, natural product derivatives *[5]*, and inorganic complexes *[6,7]*). Libraries synthesized by solid-phase techniques mainly use polystyrene-divinylbenzene (PS) derived solid supports. Owing to physical and chemical limitations of PS-derived resins, other resins have been developed *(8,9)*. Most of these resins are prepared from PS by functionalizing the resin beads with oligomers to improve solvent compatibility and physical stability *(8,9)*.

Solid-phase synthesis offers several attractive features over solution-phase synthesis: (1) Molecules are synthesized while covalently linked to the solid support, facilitating the removal of excess reagents and solvents. (2) The solid-supported reaction can be driven to completion through the use of excess, soluble reagents. (3) Mechanical losses are minimized as the compound–polymer beads remain in single-reaction vessels throughout the synthesis. (4) Physical manipulations are easy, rapid, and amenable to automation. (5) The physical separation of the reaction centers on resin furnishes a "pseudo-dilution" (physi-

From: *Methods in Molecular Biology, Combinatorial Library Methods and Protocols*
Edited by: L. B. English © Humana Press Inc., Totowa, NJ

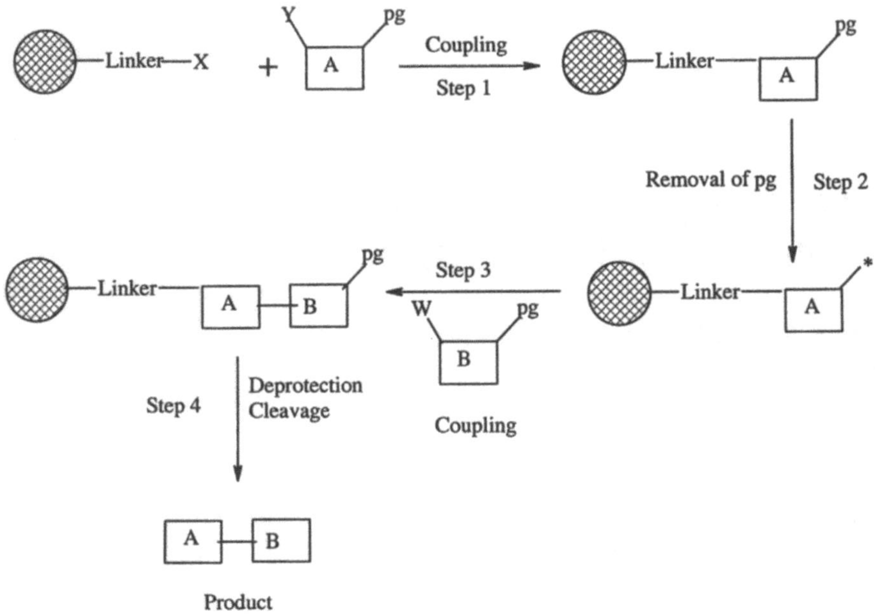

Fig. 1. Linear solid-phase synthesis of biopolymer-like peptides and polynucleotides.

cal separation in space minimizes or eliminates contact between resin-bound reacting sites), which makes certain transformations more successful when compared to solution-phase synthesis. A general schematic representation of the steps involved in a linear synthesis of compounds on solid phase is outlined in **Fig. 1**.

In linear solid-phase synthesis, the building blocks (i.e., A and B in **Fig. 1**) are covalently attached to the solid support via a linker ***(10)***. In the case of peptide synthesis, the building blocks are protected amino acids. Usually the N^{α}-group is protected by an acid-sensitive *tert*-butyloxycarbonyl (Boc) group, a base-sensitive 9-fluorenylmethyloxycarbonyl (Fmoc) group, or Pd(0)-sensitive allyloxycarbonyl (Alloc) group. The use of protecting groups (pg in **Fig. 1**) prevents side reactions and complications arising from the incorporation of multiple building blocks in the desired product. The presence of a protecting group requires additional chemical step(s) for deprotection and exposure of the functional group (in the present example, an amino group). Only then can further coupling with other amino acids be performed. Similar strategies are used in the construction of peptide nucleic acid oligomers using Boc or Fmoc protection ***(11,12)***.

It was envisaged that instead of using covalently linked protecting groups that require chemical synthesis and removal, a transient protection scheme could be used to facilitate the same overall chemical transformation. Noncovalent protection was first used in peptide synthesis under solution- and solid-phase protocols *(13–17)* to prevent double coupling and other side reactions. One approach is based on the fact that crown ethers can form stable complexes with ammonium ions *(18–20)*. Because crown ethers selectively sequester potassium ions, solutions containing potassium salts can be used to remove the crown ether from the ammonium group. Similarly, it was found that the noncovalent nature of the protection afforded by the crown ether entity allowed its mild and rapid removal from resin-bound peptides by treatment with 1% *N,N*-diisopropylethylamine (DIEA) solutions *(16)*.

1.1. Noncovalent Protection in Solid-Phase Peptide Synthesis

The use of crown ethers for protection of the amino group of amino acids offers, in principle, several advantages over the more commonly used protecting groups *tert*-Boc and Fmoc. The noncovalent nature of the interaction between crown ethers and ammonium ions, coupled with the high affinity of crown ethers for inorganic ions *(21)*, provides the basis for a rapid but mild protection and deprotection scheme. The crown ether protection of N^{α}-amino acids in solution *(13–15)* and solid-phase syntheses *(16,17)* has been extensively studied.

Mascagni and co-workers *(13–17,22)* have investigated conditions under which peptide synthesis by the fragment condensation approach in the solid phase can be carried out using crown ethers as noncovalent protecting groups for the N^{α}-amino group. As a model system, the syntheses of tripeptides was performed by coupling the 18-crown-6 complex of the dipeptide Gly-Gly-OH (III and IV, **Fig. 2**) with either resin-bound Tyr or Pro amino acids while varying the solvent choice between *N,N*-dimethylformamide (DMF) and dichloromethane (DCM). Each coupling was carried out with a fourfold excess of the activated dipeptide–crown ether complex using 1,3-dicyclohexylcarbodiimide (DCC, **Fig. 2**) and 1-hydroxybenzotriazole (HOBt, **Fig. 2**) as activating reagents. The couplings were run for 30–45 min at room temperature. In these experiments the goal was to evaluate the effect of solvent, counter ion, the nature of the carboxy-(C)-terminal amino acid, and the viability of noncovalent protection in fragment condensation. Synthetic performance of the syntheses was judged by the level of the desired peptides vs the presence of double-coupled side products (**Table 1**). It should be noted that preliminary experiments found that a polyacrylamide-based support performed poorly in comparison to a PS support (i.e., Wang resin). The ability to control the reaction was found to vary as a function of solvent and the C-terminal amino acid. The identity of the counter ion appeared to have no effect. The best results were obtained

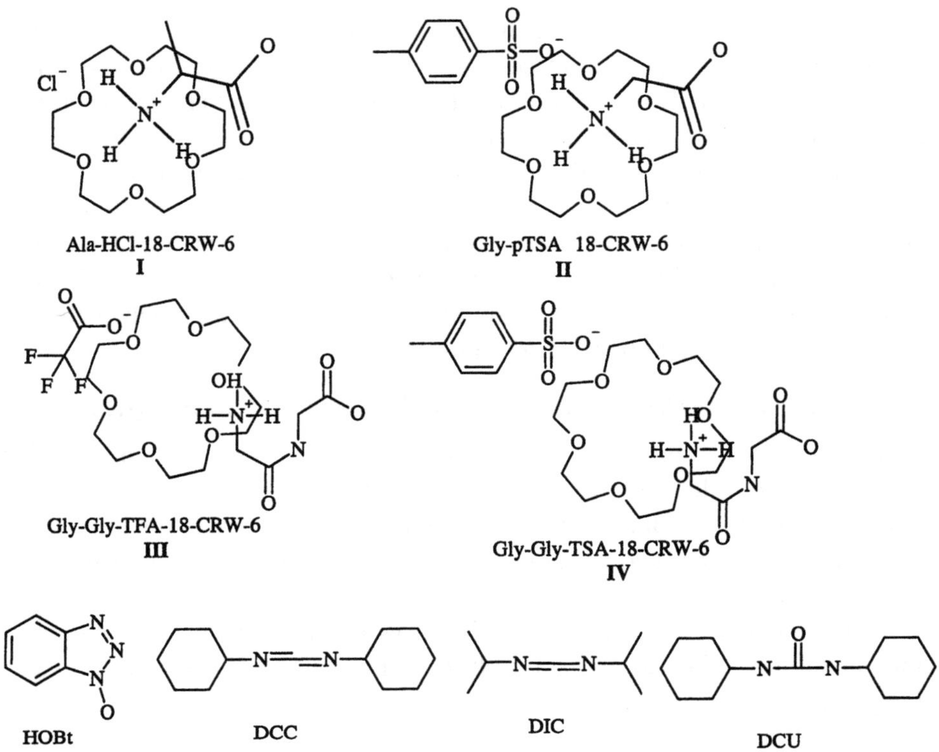

Fig. 2. Chemical structures of reagents and building blocks for peptide synthesis using noncovalent protection.

using Wang resin functionalized with Pro and DCM as a solvent. Interestingly, reactions involving Tyr as the C-terminal amino acid tended not to go to completion. Detailed studies established that the crown ether protection was transferred from the terminal Gly of the activated dipeptide to the resin-bound amino- (N)-terminus, a likely cause for the observation of double-coupled products and unreacted, resin-bound amines. That Pro was not affected by this same circumstance is in accord with the observation that 18-crown-6 selectively forms a complex with primary ammonium salts in preference to secondary ammonium salts. The use of a secondary amine as the C-terminal group in noncovalent protection was investigated as well *(16)*. The observed solvent effect is believed to be related to the greater solvating ability of DMF for the ammonium salt relative to DCM. It is postulated that a competition is established between DMF and the crown ether for solvation of the ammonium ion. The authors also found that this protection scheme is not applicable to single amino acid condensation, as polymerization results immediately after activation *(22)*.

Table 1
Peptide Sequences Synthesized by Non-Covalent Protection on a Solid Phase *(16)*

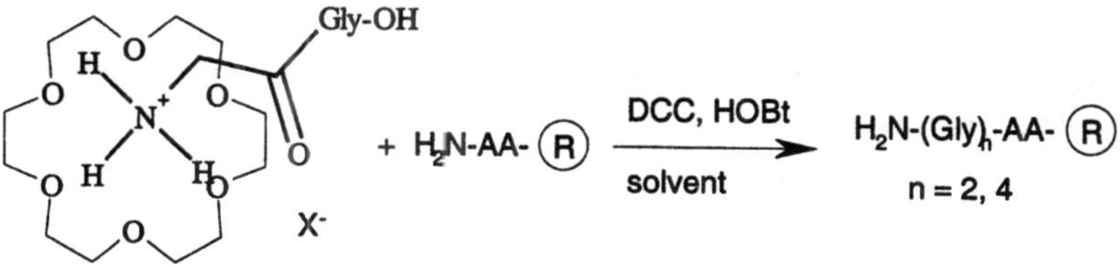

X^- = 4-$MePhSO_3^-$, $CF_3CO_2^-$

Entry	C-Terminal amino acid	Solvent	Product ratio ($n = 2{:}n = 4$)
1	Tyr	DMF	1:1
2	Tyr	DCM	5:2
3	Pro	DCM	96:4

The use of crown ethers for noncovalent protection of N^{α}-amino acids and for protection of side chains of Lys or Arg residues has found the most successful utility in the fragment condensation approach to solid- and solution-phase peptide synthesis *(15–17)*.

1.2. Noncovalent Protection in Solid-Phase Rhodamine-Labeled Peptide Nucleic Acid Synthesis

Another investigation employing noncovalent protection was the labeling of peptide nucleic acids (PNAs) with fluorophores as probes for characterizing nucleic acid sequences by *in situ* hybridization *(23)*. Cellular uptake of PNAs was monitored using fluorescent microscopy *(24)*. Non-bonded interactions between the lipophilic resin backbone and the fluorophore reagent carboxytetramethylrhodium succinimidyl ester (CTRSE) hindered full incorporation of the fluorophore on the PNAs *(25)*. To improve efficiency, noncovalent protection was employed by addition of an analog (sulforhodamine sodium [CTRS]) of the intended fluorophore prior to the coupling of CTRSE to the resin-bound PNAs. CTRS served to noncovalently block the interfering lipophilic sites on the resin. The incorporation of CTRSE was improved by more than fivefold relative to the reaction in the absence of CTRS. The result was that a cheap reagent was used to improve efficiency and reduce the amount needed of a more expensive building block (e.g., CTRSE).

Based on these findings on noncovalent protections, similar approaches could be proposed in cases where either temporary protection is needed for chemical transformation or where resin–reagent compatibility is an issue *(8,9)*.

The potential of noncovalent protection schemes to address these kinds of issues has not been fully explored.

2. Materials

2.1. Preparation of 18-Crown-6 Ether Complexes of Peptides and Amino Acids

1. Solvents: *N,N*-Dimethylformamide (DMF), dichloromethane (DCM).
2. Fmoc-Tyr(OtBu)-Wang (0.59 mmol/g) from Calbiochem-Novabiochem (San Diego, CA).
3. Coupling reagents: *N*-Hydroxybenzotriazole (HOBt), dicyclohexylcarbodiimide (DCC), and diisopropylcarbodiimide (DIC) from Aldrich (Wisconsin).
4. Gly-Gly-OH dipeptide from Sigma Biochemicals (St. Louis, MO).
5. 18-Crown-6 from Aldrich.
6. Trifluoroacetic acid (TFA) and piperidine from Aldrich Chemical.

2.2. Preparation of Fluorescein-Labeled PNAs on a Solid Support

1. Fmoc-PNA monomers (**Fig. 3**) protected nucleic acid bases from Applied Biosystems (http://www.appliedbiosystems.com/ds/pna/) *(26)* (*see* **Note 1**).
2. Dry DMF (Sigma, St. Louis, MO) (*see* **Note 2**).
3. Fluorescein tags (**Fig. 3**) Carboxytetramethylrhodamine succinimidyl ester from Molecular Probes (Eugene, OR and Leiden, The Netherlands) and sulforhodamine from Sigma-Aldrich, St. Louis, MO.
4. Coupling reagent HATU ([*O*-(7-aza-benzo-triazol-1-yl)-1,1,3,3-tetramethyluronium hexafluorophosphate]) (**Fig. 3**) from PerSeptive Biosystem (Framingham, MA).
5. PEG-PS resin functionalized with XAL linker (9-Fmoc-aminoxanthen-3-yloxymethyl) (**Fig. 3**) from Applied Biosystem (Foster City, CA) (*see* **Note 3**).
6. PE (Perkin-Elmer) Biosystems Expedite 8909 automated synthesizer.

3. Methods

3.1. Preparation of Amino Acid and Peptide Complexes with 18-Crown-6 (see Note 4)

1. Alanine hydrochloride-18-crown-6 complex: Dissolve alanine (1 Eq) in aqueous hydrochloric acid (1.1 Eq) and lyophilize to dryness to give alanine hydrochloride in quantitative yield. Suspend alanine hydrochloride (1 Eq) with 1 Eq of 18-crown-6 in chloroform and stir the mixture at room temperature to give a clear solution. Evaporate chloroform to dryness to give the title compound as a powder (*see* **Note 5**).
2. Alanine tosylate-18-crown-6 complex: Lyophilize alanine (1 Eq) from 5 mL of water containing p-toluenesulfonic acid monohydrate (1.1 Eq). The alanine–tosylate salt is added to a chloroform solution of 18-crown-6 (1 Eq) and the mixture stirred until homogeneous. Evaporation of chloroform and crystallization of the residue from methanol–ethyl acetate (*see* **Note 6**) yields the solid alanine–crown ether complex with a melting point of 123–125°C.

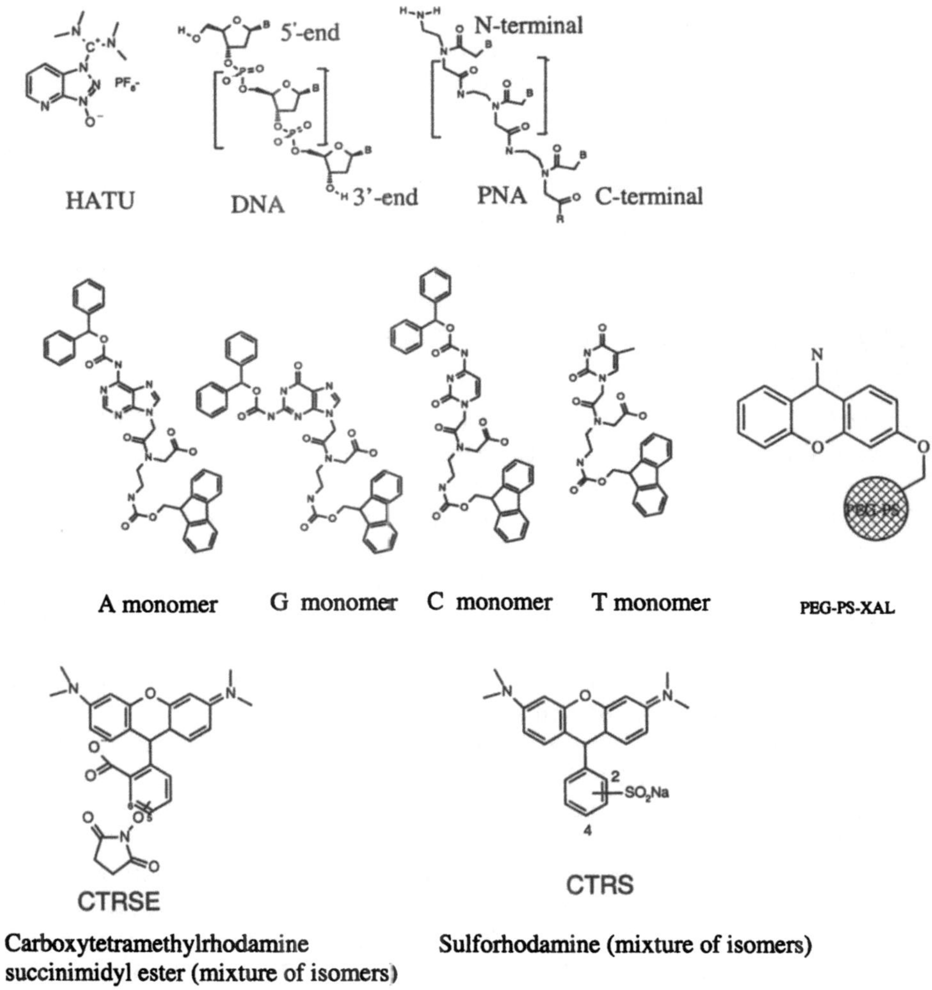

Fig. 3. Chemical structures of reagents and building blocks for synthesis of rhodamine-labeled PNA oligomers.

3. Gly-Gly trifluoroacetate crown ether complex (**III** in **Fig. 2**): To a solution of Gly-Gly trifluoroacetate in water (1 Eq) is added 18-crown-6 (1 Eq) with stirring. Lyophilize the reaction solution. Dissolve in water, and lyophilize again. This process is repeated until all traces of acid are eliminated (monitored by pH paper). The complex is used without further purification.
4. Gly-Gly tosylate crown ether complex (**IV** in **Fig. 2**): Gly-Gly (5 g, 38 mmol) is added to a solution of *p*-toluenesulfonic acid (7.2 g, 38 mmol) in water–ethanol (50 mL, 1:1). Stir the reaction mixture at room temperature for 1–2 h and then evaporate to dryness. Suspend the residual dipeptide salt in 50 mL of ethanol (*see* **Note 7**) and add 18-crown-6 (10 g, 38 mmol). Stir the reaction mixture with

warming to give a clear solution. Cool the solution to room temperature and add dry ethyl acetate dropwise until the solution becomes turbid. Leave the suspension at room temperature for 6–8 h and filter the precipitated crystals to give 20 g (93%) of compound **IV** (**Fig. 2**).

5. Gly-Gly hydrochloride crown ether complex: Prepare as described in **step 4**. Use similar equivalents as in the synthesis of **IV**. The yield is 80% of glycylglycine hydrochloride–18-crown-6 complex.

3.2. Solid-Phase Synthesis of NH_2-Phe-Gly-Gly-Pro-Asp-Leu-Tyr-OH Heptapeptide by the Fragment Condensation Approach Using Noncovalent Protection of Dipeptide Glycylglycine (IV, Fig. 2, see Note 8)

1. Add 1.5 mL of 50% piperidine in DMF to 100 mg of Fmoc-Tyr(OtBu)-Wang resin (loading 0.52 mmol/g). Agitate the resin for 1 h at room temperature. Filter the resin and wash with DMF (1.5 mL ×6).
2. Add a solution of Fmoc-Leu (73.5 mg, 208 µmol), HOBt (28.1 mg, 208 µmol), and DIC (26.2 mg, 208 µmol) in 1 mL of dry DMF to the resin from the above step. Agitate the suspension at room temperature for 45 min. Monitor the completion of coupling with the ninhydrin test. Wash the fully coupled resin with DMF (1.5 mL ×6). Remove the protecting group by adding 1.5 mL of 50% piperidine in DMF and shaking at room temperature for 10 min. Wash the resin with DMF (1.5 mL ×8) and use in the next step.
3. Repeat **step 2** using Fmoc-Asp(OtBu) (85.6 mg, 208 µmol) with equivalent amounts of DIC and HOBt in 1.5 mL of DMF. Continue coupling for 45 min at room temperature. Treat the resin as in **step 2** and use in the next step.
4. Repeat **step 2** using Fmoc-Pro (70.1 mg, 208 µmol). After completion of the coupling, remove the protecting group with 50% piperidine in DMF and wash with DMF (1.5 mL ×8), DCM (1.5 mL ×6). Suspend the product in DCM.
5. In a separate vial dissolve 106 mg (208 µmol) of Gly-Gly trifluoroacetate–crown ether complex (prepared as described in **Subheading 3.1., step 3**, compound **III** in **Fig. 2**), in 2 mL of dry DCM (*see* **Note 9**). To the solution add sequentially 28 mg of HOBt (208 µmol) and 42.6 mg of DCC (208 µmol). Stir the mixture at room temperature for 12 min and then filter the precipitated DCU (*see* **Fig. 2**). Transfer the clear solution to the reactor containing the filtered tetrapeptide Pro-Asp (OtBu)-Leu-Tyr (OtBu)-Wang resin from **step 4** (*see* **Note 10**). Add more DCM to facilitate the suspension of the resin (about 300 µL) and agitate the reaction mixture for 45 min (*see* **Note 11**). Test for completion of coupling by placing a few resin beads into a small test tube and running the ninhydrin test. On completion of the coupling, filter the resin and wash with DCM (3×), DMF (2×), and then treat with 1% DIEA in DMF 2× (3 min each) to remove the crown ether protecting group.
6. Suspend the resin from step 5 in DMF (1.7 mL) and add Fmoc-Phe-Pfp activated ester (115.1 mg, 208 µmol). Agitate the suspended resin at room temperature for 1 h and monitor for completion of the coupling by ninhydrin

analysis. Filter the reagents and solvent, wash the resin with DMF (2 mL ×4), and then suspend in 2 mL of 50% piperidine in DMF for 20 min to remove the Fmoc protecting group. Wash the deprotected resin with DMF (2 mL ×8) and DCM (2 mL ×8). Dry the finished resin in a desiccator over anhydrous potassium carbonate for 2 h.

7. Transfer the dried resin from **step 6** to a glass vial with a screw cap and add 2 mL of a trifluoroacetic acid–water mixture (95% TFA, 5% H_2O). Close the vial and allow the cleavage reaction to proceed at room temperature for 1 h. Filter the cleavage mixture, wash the resin with additional TFA–water, and combine the filtrates. Evaporate TFA at room temperature using a rotary evaporator or acid-resistant centrifugal vacuum system. Triturate the residual product with anhydrous ether and separate the white solid product by decantation or centrifugation. Dry the crude peptide over potassium hydroxide pellets under vacuum for 1 h.
8. Take a sample of the dried, crude peptide made in **step 7** (0.05–0.1 mg) and dissolve in a water–methanol mixture. Add acetonitrile until the solution clears. Analyze by high-performance liquid chromatography (HPLC) and liquid chromatography–mass spectrometry (LC–MS) to verify the purity and identity of the synthesized peptide. For Phe-Gly-Gly-Pro-Asp-Leu-Tyr, MS: Expected 768.8 or 769 for M+1 by electrospray mass spectrometry.

3.3. Solid-Phase Synthesis of Rhodamine Labeled Peptide Nucleic Acids using Noncovalent Protection

1. Fmoc-Gly-CCCTAACCCTTACCCTAA-Lys(Boc)-RAM-PS: Synthesis of the protected PNA on a small scale (0.05 mmol) can be achieved by the Fmoc strategy ***(12,25,27)*** on PE Biosystems Expedite 8909 automated synthesizer using the protocol supplied by the manufacturer (http://www.appliedbiosystems.com/ds/pna/) (*see* **Notes 12–14**)
2. Suspend the resin-bound, protected PNA synthesized in **step 1** in DMF containing 20% piperidine in a reaction tube (500 µL). Agitate the resin for 20 min, filter the reagent and the solvent, and wash the resin with DMF (500 µL ×8).
3. Connect the reaction tube containing the resin from **step 2** to two 1-mL syringes. Dissolve 70 m*M* of sulforhodamine in 300 µL of 1:30 mixture of DIEA–DMF in one syringe. Keep the other syringe empty. Pass the sulforhodamine solution over the PNA resin in the reaction tube for 20 min using the two syringes. Wash the resin with DMF–DCM (1:1) 8×.
4. Connect the reaction tube of the resin from **step 3** with two 1-mL syringes. In one syringe load 300 µL of a 10 m*M* solution of tetramethylrhodamine succinimydyl ester in DIEA–DMF (1:30) and pass the solution over the resin using the dual syringes for 20 min. Wash the resin with DMF (0.5 mL ×8), DCM (0.5 mL ×8), and dry under vacuum for 2 h.
5. Suspend the dry resin made in **step 4** in 1 mL of TFA containing 25% *m*-cresol for 45 min at room temperature (*see* **Note 15**). Filter the cleavage mixture, wash the resin with the same cleavage solution and combine the filtrates. Evaporate the TFA solution under vacuum and triturate the residual product with dry ether at

0°C. Centrifugation of the crude rhodamine–PNA will give a pellet that can be purified by RP C_{18} HPLC using acetonitrile and 0.1% aqueous TFA buffer as solvents. HPLC will give two peaks corresponding to the two isomers of carboxytetramethyrhodamine. The calculated molecular weight is 5326.46 and M+1 = 5327.46

4. Notes

1. PNA monomers should be stored under dry, cold conditions. If the physical appearance of the monomers changes from a free-flowing powder form to aggregates, then the monomers should be dried *in vacuo* overnight before use.
2. Dry DMF is required in the synthesis of PNAs to dissolve the monomers and the activating reagent (HATU) under anhydrous conditions. The presence of moisture interferes with the purity and yield of the final products especially in the case of long PNAs (18-mers and longer). Dry DMF should be stored under nitrogen over dry 4 Å molecular sieves.
3. All resins should be stored under dry, cool conditions until their use.
4. Crown ether complexes with amines, N^{α}-amino acids, peptides of varying size, and side chain amino group of Lys and Arg have been prepared ***(15–23)***. The examples given here are only representative.
5. Evaporation of chloroform solutions is best accomplished by placing the solution in a round-bottom flask and use of a rotary evaporator.
6. **Recrystallization should be done in a fume hood away from sources of ignition, as both methanol and ethyl acetate are highly flammable.** Recovery of the crystals is most easily accomplished by filtration through a sintered glass funnel.
7. Absolute ethanol (100%, 200 proof) is the best choice.
8. Noncovalent protection of N^{α}-amino acids and the side chain amino group of Lys or Arg residues with crown ethers has most successfully been applied in the synthesis of peptides by the fragment condensation approach. This is illustrated here by the synthesis of NH_2-Phe-Gly-Gly-Pro-Asp-Leu-Tyr-OH. Single amino acid condensation in linear peptide synthesis often leads to undesirable oligomerization resulting from ineffective protection.
9. The optimal protocol requires the use of DCM as solvent for all the coupling reactions involving the crown ether complexes. The crown ether complexes are unstable in polar solvents such as DMF or DMSO. Consequently, use of DMF or DMSO as solvent in coupling reactions involving the crown ether complexes results in extensive oligomerization and other side product formation.
10. The efficiency of coupling to the crown ether complex is dependent on the nature of the amino acid in the N-terminus of the resin bound peptide. Competition for the crown ether molecule by primary amino groups compromises efficiency of coupling. Thus, the best results are obtained when the N-terminus amino acid is proline or other secondary amino acids.
11. Peptides larger than diglycine may require extension of coupling reaction time to 24 h.

12. In the case of PNAs containing consecutive identical bases, double coupling after the incorporation of the second base is necessary; otherwise a truncated product will be present.
13. Purine-rich PNA sequences require double coupling to improve the purity and yield of the final compound.
14. For analysis of PNA and PNA conjugates, an analytical HPLC equipped with a C18 300 Å reverse-phase column at a flow rate of 1.0 mL/min is recommended.
15. **Caution:** TFA is a highly corrosive irritant. Wearing proper protection for the hands and eyes is required. All operations involving TFA solutions should be performed in a well ventilated hood. Caution should also be exercised in making the TFA–*m*-cresol (4:1) solution for cleavage of the final product.

References

1. Merrifield, R. B. (1963) Solid-phase peptide synthesis I. *J. Am. Chem. Soc.* **85,** 2149–2154.
2. Al-Obeidi, F. A., Hruby, V. J., and Sawyer, T. K. (1998) Peptide and peptidomimetic libraries. *Mol. Biotech.* **9,** 205–223.
3. Dolle, R. E. (2000) Comprehensive survey of combinatorial library synthesis: 1999. *J. Combi.Chem.* **2,** 384–433.
4. Franzen, R. G. (2000) Recent advances in the preparation of heterocycles on solid support: a review of the literature. *J. Comb. Chem.* **2,** 195–214.
5. Hall, D. G., Manku, S., and Wang, F. (2001) Solution-and solid-phase strategies for the design, synthesis, and screening of libraries based on natural product templates: a comprehensive survey. *J. Comb. Chem.* **3,** 125–150.
6. Schultz, P. G. and Xiang, X.-D. (1998) Combinatorial approaches to materials science. *Curr. Opin. Solid State Mater. Sci.* **3,** 153–158.
7. Gennari, F., Seneci, P., and Miertus, S. (2000) Application of combinatorial technologies for catalyst design and development. *Catal. Rev. Sci. Eng.* **42,** 385–402.
8. Hudson, D. (1999) Matrix assisted synthetic transformations: a mosaic of diverse contributions. I. The pattern emerges. *J. Comb. Chem.* **1,** 333–360.
9. Hudson, D. (1999) Matrix assisted synthetic transformations: a mosaic of diverse contributions. II. The pattern emerges. *J. Comb. Chem.* **1,** 404–457.
10. Eggenweiler, H.-M. (1998) Linkers for solid-phase synthesis of small molecules: coupling and cleavage techniques. *Drug Discov. Today* **3,** 552–560.
11. Dueholm, K. L., Egholm, M., Behrens, C., Christensen, L., Hansen, H. F., Vulpius, T., et al. (1994) Synthesis of peptide nucleic acid monomers containing the four natural nucleobases: thymine, cytosine, adenine and guanine and their oligomerization. *J. Org. Chem.* **59,** 5767–5773.
12. Thomson, S. A., Josey, J. A., Cadilla, R., Gaul, M., Hassman, C. F., Luzzio, M. J., et al. (1995) Fmoc mediated synthesis of peptide nucleic acids. *Tetrahedron* **51,** 6179–6194.
13. Hyde, C. B., Welham, K. J., and Mascagni, P. (1989) The use of crown ethers in peptide chemistry. Part 2. Syntheses of dipeptide complexes with cyclic polyether

18-crown-6 and their derivatization with DMSO. *J. Chem. Soc., Perkin Trans. 2* **12,** 2011–2015.
14. Hyde, C. B. and Mascagni, P. (1990) The use of crown ethers in peptide chemistry. Part 3. Synthesis of an enkephalin derivative using 18-crown-6 as a non-covalent amino protecting group. *Tetrahedr. Lett.* **31,** 399–402.
15. Botti, P., Lucietto, P., Pinori, M., and Mascagni, P. (1993) The use of crown-ethers as non-covalent protecting groups for the synthesis of peptides, in *Innovation Perspectives on Solid Phase Synthesis, Collected Papers, 3rd International Symposium* (1994), meeting date 1993, pp. 459–462.
16. Botti, P., Ball, H. L., Rizzi, E., Lucietto, P., Pinori, M., and Mascagni, P. (1995) The use of crown ethers in peptide chemistry. IV. Solid phase synthesis of peptides using peptide fragments Na protected with 18-crown-6. *Tetrahedr. Lett.* **51,** 5447–5458.
17. Botti, P., Ball, H. L., Lucietto, P., Pinori, M., Rizzi, E., and Mascagni, P. (1996) The use of crown ethers in peptide chemistry. V. Solid-phase synthesis of peptides by the fragment condensation approach using crown ethers as non-covalent protecting groups. *J. Pept. Sci.* **2,** 371–380.
18. Barrett, A. G. M. and Lana, J. C. A. (1978) Selective acylation of amines using 18-crown-6. *J. Chem. Soc. Chem. Commun.* 471–472.
19. Ha, Y. L. and Chakraborty, A. K. (1992) Nature of the interactions of 18-crown-6 with ammonium cations: a computational study. *J. Phys. Chem.* **96,** 6410–6417.
20. Liou, C. C. and Brodbelt, J. S. (1992) Comparison of gas-phase proton and ammonium ion affinities of crown ethers and related acyclic analogs. *J. Am. Chem. Soc.* **114,** 6761–6764.
21. Hiraoka, M. (1982) *Crown Compounds: Their Characteristics and Applications.* Elsevier, Amestrdam.
22. Mascagni, P., Hyde, C. B., Charalambous, M. A., and Welham, K. J. (1987) The use of crown ethers in peptide chemistry. Part 1. Syntheses of amino acid complexes with the cyclic polyether 18-crown-6 and their oligomerisation in dicylohexylcarbodi-imide-containing solutions. *J.Chem. Soc. Perkin Trans.* **2,** 323–327.
23. Lansdorp, P. M., Verwoerd, N. P., van de Rijke, F. M., Dragowska, V., Little, M. T., Dirks, R. W., et al. (1996) Heterogeneity in telomere length of human chromosomes. *Hum. Mol. Genet.* **5,** 685–691.
24. Hamilton, S. E., Simmons, C. G., Kathiriya, I. S., and Corey, D. R. (1999) Cellular delivery of peptide nucleic acids and inhibition of human telomerase. *Chem. Biol.* **6,** 343–351.
25. Mayfield, L. D. and Corey, D. R. (1999) Enhancing solid phase synthesis by a noncovalent protection strategy-efficient coupling of rhodamine to resin-bound peptide nucleic acids. *Bioorg. Med. Chem. Lett.* **9,** 1419–1422.
26. Braasch, D. A. and Corey, D. R. (2001) Synthesis, analysis, purification, and intracellular delivery of peptide nucleic acids. *Methods* **23,** 97–107.
27. Mayfield, L. D. and Corey, D. R. (1999) Automated synthesis of peptide nucleic acids and peptide nucleic acid-peptide conjugate. *Analyt. Biochem.* **268,** 401–404.

2

Quality Control of Solid-Phase Synthesis by Mass Spectrometry

Jean-Louis Aubagnac, Robert Combarieu, Christine Enjalbal, and Jean Martinez

1. Introduction

Combinatorial chemistry *(1–7)* has drastically modified the drug discovery process by allowing the rapid simultaneous preparation of numerous organic molecules to feed bioassays. Most of the time, syntheses are carried out using solid-phase methodology *(8)*. The target compounds are built on an insoluble support (resins, plastic pins, etc). Reactions are driven to completion by the use of excess reagents. Purification is performed by extensive washing of the support. Finally, the molecules are released in solution upon appropriate chemical treatments.

Such a procedure is well established in the case of peptides, but solid-phase organic chemistry (SPOC) is more difficult. Optimization of the chemistry is required prior to library generation most of the time. Compound identification is complicated by the insolubility of the support. Release of the anchored structure in solution followed by standard spectroscopic analyses may impart delay and/or affect product integrity *(9)*. A direct monitoring of supported organic reactions is thus preferable to the "cleave and analyze" methodology. Nevertheless, it presents several constraints. A common resin bead loaded at 0.8 mmol/g commonly produces nanomole quantities of the desired compound, and only 1% of the molecules are located at the outer surface of the bead *(10)*. Very few materials, covalently bound to the insoluble support, are thus available for the analysis, which should ideally be nondestructive.

The relevance of mass spectrometry in the rehearsal phase of a combinatorial program is demonstrated through the control of various peptide syntheses. Fourier transform infra red (FTIR) *(11)* and cross polarization-magic angle

From: *Methods in Molecular Biology, Combinatorial Library Methods and Protocols*
Edited by: L. B. English © Humana Press Inc., Totowa, NJ

spinning nuclear magnetic resonnance (CP-MAS-NMR) spectroscopies are also suitable techniques *(12)*, but they lack the specificity or the sensitivity achievable by mass spectrometry.

Solid samples can be analyzed by mass spectrometry with techniques providing ionization by desorption *(13)* such as MALDI (matrix assisted laser desorption ionization) *(14)* and S-SIMS (static-secondary ion mass spectrometry) *(15)*. Ions are produced by energy deposition on the sample surface. The analysis can be performed at the bead level. Most of all, chemical images can be produced to localize specific compounds on the studied surfaces.

S-SIMS was found to be superior to MALDI for following supported organic synthesis for many reasons. First, cocrystallization of the solid sample with a matrix is required for MALDI experiments, which is not the case in S-SIMS (no sample conditioning). Second, libraries of organic molecules contain mostly low-molecular-weight compounds, which are not suitable for MALDI analysis owing to possible interference with the matrix ions. Finally, a specific photolabile linkage between the support and the built molecules is necessary to release the desired molecular ions in the gas phase upon laser irradiation. Standard resins allowing linkage of the compounds through an ester or an amide bond are directly amenable to S-SIMS analysis.

Characteristic ions of peptide chains (*see* **Note 1**) have been obtained by S-SIMS whatever the nature of the polymeric support *(16–18)*. N-Boc–protected peptides were synthesized on polystyrene resins *(16)*. Fmoc-protected peptides anchored to polyamide resins *(17)* were also studied, and a wide range of dipeptides were loaded on plastic pins *(18)*. All protecting groups (Boc, Fmoc, tBu, Z, Bn, Pht) gave characteristic ions in the positive mode, except Boc and tBu, which were not differentiated (*see* **Note 2**). The amino acids were evidenced by their corresponding immonium ions in the positive mode. These informative product ions were more abundant than ions related to the polymer, which require at least the rupture of two bonds *(19)*. Peptide synthesis was thus easily followed step-by-step. Coupling reactions were monitored by detection of the incoming residue immonium ion and of the N-protecting group ion. The deprotection reaction was evidenced by the absence of the latter ion. Nevertheless, the identification of a peptide at any stage of the preparation required that the whole peptide sequence, and not fragments, was released in the gas phase. In other words, orthogonality between the peptide-resin linkage and the internal peptide bonds was compulsory. The ester linkage was found suitable since the peptide carboxylate ion was identified in the negative mode. This bond was thus termed "SIMS-cleavable." The amide linkage was broken simultaneously with the internal peptide amide bond and so was not adequate for such studies (*see* **Note 3**).

The recourse to a "SIMS cleavable" bond allowed direct identification of support-bound peptides. Several results have illustrated this concept. As an

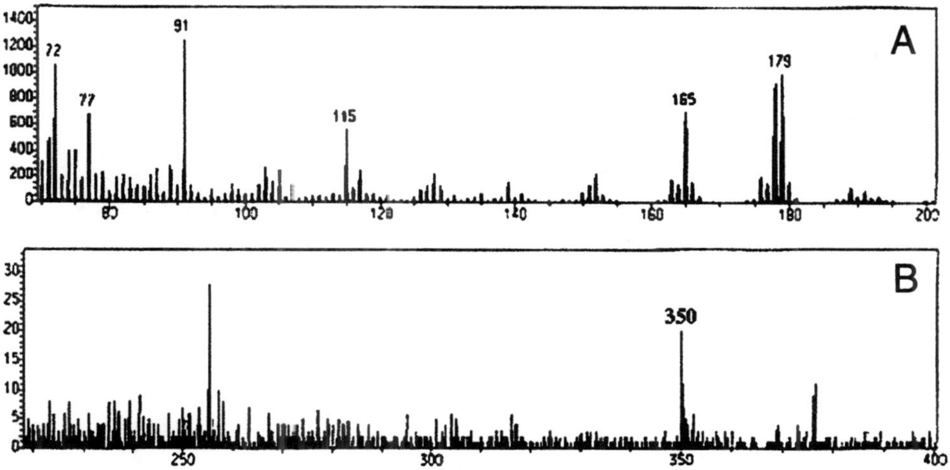

Fig. 1. **(A)** Positive S-SIMS spectrum of Fmoc-Met(O)$_2$-Ala-Val anchored to Wang resin: immonium ion of valine at m/z 72, Fmoc protection at m/z 165/178/179, polystyrene at m/z 77/91/115; **(B)** Negative S-SIMS spectrum of Fmoc-Met(O$_2$)-Val-Ala anchored to Wang resin: carboxylate ion H-Met(O$_2$)-Val-Ala-O$^-$ at m/z 350.

example, a tripeptide bearing an oxidized methionine, Fmoc-Met(O$_2$)-Ala-Val anchored to Wang resin, was subjected to S-SIMS bombardment and the spectra were recorded in both positive and negative modes (**Fig. 1**). Some immonium ions were present in the positive spectrum as expected (valine at *m/z* 72), but there was no information about the methionine residue. The negative spectrum provided the carboxylate ion of the whole peptide sequence (*m/z* 350), which showed, without any ambiguity, that methionine was completely oxidized.

The S-SIMS technique was found specific through the use of a S-SIMS cleavable bond. The technique was sensitive because fentomoles of growing peptides were analyzed in each experiment, and it was nondestructive ***(20)***. Indeed, only 1% of the molecules were located at the surface, and small areas of 20 × 20 μm^2 were selected and bombarded to generate a spectrum. So, the bead can be reused after the analysis.

Any organic molecule is suitable for S-SIMS analysis provided that stable ions could be produced. The domain of SPOC can now be explored. Different linkers are currently investigated to determine the specific lability of the molecule-support bond under S-SIMS bombardment whatever the compound and the type of insoluble support.

Imaging studies were also performed to identify mixtures of peptides in a single analysis in the search of a high-throughput process adapted to combina-

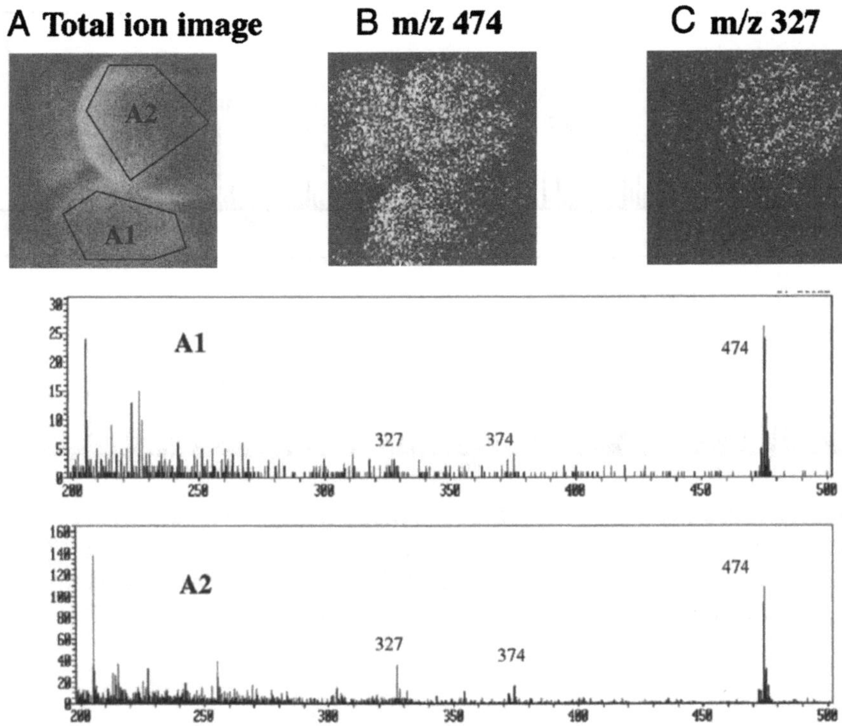

Fig. 2. **(A)** Total ion image showing two selected areas (*A1* and *A2*) each corresponding to one bead. The negative S-SIMS spectra generated from these two surfaces are given underneath. **(B)** Negative S-SIMS image of Boc-Pro-Phe-Leu (carboxylate ion at *m/z* 474); **(C)** Negative S-SIMS image of the deleted sequence Boc-Pro-Leu (carboxylate ion at *m/z* 327).

torial library profiling ***(21)***. Two types of mixtures can be envisaged. Beads, which were each loaded by the same molecules, were pooled or the beads could themselves bear different components (starting material, byproducts). For instance, the unwanted intramolecular cyclization of glutamic acid into pyroglutamic acid was evidenced by S-SIMS down to a level of only 15% of side-reaction ***(22)***. Incomplete coupling leading to truncated chains was also detected ***(23)***, and clear images were produced with only 9% of deleted sequences as displayed in **Fig. 2**.

2. Materials

2.1. Solid-Phase Peptide Synthesis

2.1.1. Synthesis of Boc-Protected Peptides

1. Carry out peptide syntheses on hydroxymethylpolystyrene resin loaded at 0.93 or 2.8 mmol/g (Novabiochem, Meudon, France).

2. L-configuration Boc-protected amino acids available from Senn Chemicals (Gentilly, France) and Propeptide (Vert le Petit, France).
3. Load first Boc-protected amino acid onto the resin according to the symmetrical anhydride procedure (dissolve 10 Eq of the residue in a minimum of dichloromethane).
4. Cool this solution in an ice-water bath and add 5 Eq of diisopropylcarbodiimide.
5. Stir the solution for 30 min at 4°C, filter, and concentrate under vacuum.
6. Dissolve the resulting symmetrical anhydride in dimethylformamide (DMF) and add to the resin with 0.1 Eq of dimethylaminopyridine.
7. Release the Boc protection by treatment with trifluoroacetic acid in dichloromethane.
8. Couple the second residue by 2 Eq of (benzotriazol-1-yloxy)tris(dimethylamino) phosphonium hexafluorophosphate (BOP) and diisopropylethylamine in dimethylformamide for 2 h.

2.1.2. Synthesis of Fmoc-Protected Peptides

1. Fmoc-protected amino acids available from Senn Chemicals (Gentilly, France).
2. 4-Methylbenz-hydrylamine (*MBHA*) *resin*: Carry out peptide syntheses on MBHA resin loaded at 0.8 mmol/g (Novabiochem, Meudon, France). Couple the amino acids by two equivalents of (BOP) and diisopropylethylamine in dimethylformamide for 2 h. Remove Fmoc protection with two treatments (3 and 15 min) of the resin with a solution of piperidine in DMF (20%, v/v).
3. *Wang resin*: Anchor the first amino acid to the resin (0.93 mmol/g, Novabiochem, Meudon, France) according to the symmetrical anhydride method. (The standard above-mentioned procedure was applied to build the sequence.)
4. *Chlorotrityl resin*: React the first amino acid overnight with the resin (1.5 mmol/g, Senn Chemicals, Gentilly, France) in the presence of N,N-diisopropylethylamine (DIEA). (The standard above-mentioned procedure was applied to build the sequence.)

2.1.3. Peptide Characterization

1. Check all syntheses prior to S-SIMS experiments by treating a few resin beads with hydrofluoric acid (HF) to release the built sequences in solution.
2. Identify the peptides with high performance liquid chromatography (HPLC) on an Alliance 2690 from Waters (Milford, MA) and electrospray mass spectrometry (ESI-MS) on a Platform II from Micromass (Manchester, UK).

2.2. Mass Spectrometry Instrumentation

1. Perform S-SIMS measurements on a TRIFT I spectrometer from the PHI-Evans Company (Eden Prairie, MN) equipped with a time-of-flight (TOF) analyzer.
2. Record spectra using a pulse (1 ns, 12 kHz) liquid metal source (^{69}Ga, 15 keV) operating in the bunched mode to provide good mass resolution ($m/\Delta m = 2000$ measured at *m/z* 43).
3. Perform charge compensation for all samples using a pulsing electron flood ($E_k = 20$ eV) at a rate of one electron pulse per five ion pulses (*see* **Note 1**).
4. Analyze surfaces in squares of 20×20 μm^2 to produce a S-SIMS spectrum.
5. Acquire all positive and negative spectra within 1–10 min with a fluence of less than 10^{12} ions/cm^2 ensuring static conditions on the sample.

6. For imaging studies, raster the primary ion beam on 400 × 400 μm^2 during 30 min to generate a complete mass spectrum at each pixel, and record a chemical image.
7. Use the "scatter" raster type, which is the one designed to be used for insulating samples: each pixel point is located as far from the previous and next pixel so as to spread the primary beam charge homogeneously.
8. Obtain mass spectra in an image from different selected areas by using simple drawing tools.

3. Methods

3.1 Sample Conditioning

1. At the end of the synthesis wash the resin beads with dichloromethane, ethanol, water, ethanol, and dichloromethane. Repeat this procedure three times.
2. Dry the resin beads overnight in a dessicator.
3. Fix an adhesive aluminum tape on a nonmagnetic stainless grid and place it in the cavity of the TOF-S-SIMS sample holder (the metallic grid prevents large variations in the extraction field over a large area insulator; it is possible, therefore, to move from one grid "window" to any of the other "windows" without any concern for retuning).
4. Sprinkle a few beads on the adhesive aluminum tape. (Do not touch the beads but manipulate them with tweezers.) The resin in excess is removed by an inert gas stream, and the remaining beads are well attached to the tape.
5. Insert the holder in the load lock of the mass spectrometer and pump it down until the required vacuum is reached.
6. Visualize the resin beads by a camera and select an area that contains well-defined beads of spherical appearance that are all roughly in the same plane. Record mass spectrometric data from this area.

3.2. Acquisition of a S-SIMS Spectrum

1. Choose one bead in the selected area, and define a surface of 20 × 20 μm^2 on the bead surface.
2. Trigger the primary bombardment. Examine the emitted secondary ions from the selected surface to modify the mass spectrometer tuning if required.
3. Start the acquisition. It should last 5 min.

3.3. Acquisition of a S-SIMS Image

1. Choose a surface in the selected area of 400 × 400 μm^2 containing a few beads.
2. Trigger the primary bombardment. Examine the emitted secondary ions from the selected surface to modify the mass spectrometer tuning if required.
3. Start the acquisition. It should last 30 min.
4. Generate the chemical images from the total ions (total image) or from various selected ions.
5. From any recorded image, select an area of interest in the bombarded surface (for instance one specific bead) and the corresponding S-SIMS spectrum will be displayed.

4. Notes

1. Owing to large charge effects on such insulating materials, charge compensation is required for all samples.
2. We have observed many similarities between the two desorption techniques: fast atom bombardment (FAB) and S-SIMS. The recorded ions in both positive and negative modes in S-SIMS could be deduced from the well-documented behavior of molecules in FAB. The amino acids that exhibited immonium ions were the same as the ones reported in the literature in FAB experiments ***(24)***. Fragmentations leading to ions characterizing the protecting groups were also identical ***(25,26)***.
3. The studied protecting groups and the corresponding recorded ions were as follows: Boc and tBu at *m/z* 57 ($C_4H_9^+$), Fmoc at *m/z* 165 ($C_{13}H_9^+$, $C_{13}H_9^-$), and *m/z* 179 ($C_{14}H_{13}^+$), Z at *m/z* 91 ($C_7H_7^+$), and Pht at *m/z* 160 as shown below.

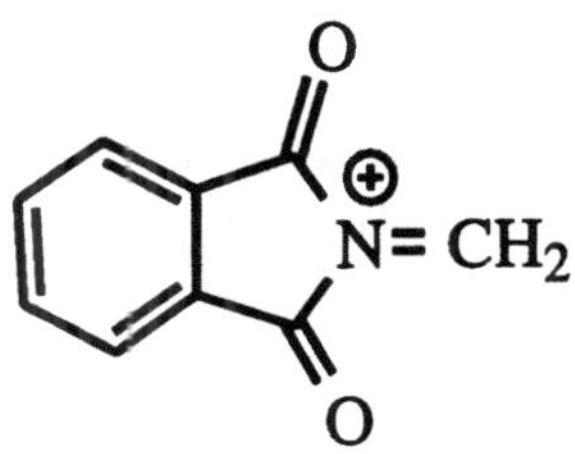

References

1. Czarnik, A. W. and Dewitt, S. H. (1997) *A practical guide to combinatorial chemistry*. American Chemical Society, Washington, DC.
2. Wilson, S. R. and Czarnik, A. W. (1997) *Combinatorial Chemistry—Synthesis and Application.* Wiley, New York, NY.
3. Bunin, B. A. (1998) *The Combinatorial Index.* Academic Press, London, UK.
4. Terrett, N. K. (1998) *Combinatorial Chemistry.* Oxford University Press, Oxford, UK.
5. Gordon, E. M. and Kervin, J. F. (1998) *Combinatorial Chemistry and Molecular Diversity in Drug Discovery.* Wiley, New York, NY.
6. Obrecht, D. and Villalgordo, J. M. (1998) *Solid-Supported Combinatorial and Parallel Synthesis of Small Molecular Weight Compound Libraries.* Tetrahedron Organic Chemistry Series, Volume 17, Pergamon, Elsevier, Oxford, UK.
7. Jung, G. (1999) *Combinatorial chemistry—Synthesis, Analysis, Screening.* Wiley-VCH, Weiheim, Germany.
8. Dolle, R. (2000) Comprehensive survey of combinatorial library synthesis: 1999. *J. Comb. Chem.* **2,** 383–433.
9. Metzger, J. W., Kempter, C., Weismuller, K.-H., and Jung, G. (1994) Electrospray MS and tandem MS of synthetic multicomponent peptide mixtures: determination of composition and purity. *Anal. Chem.* **219,** 261–277.
10. Yan, B., Fell, J. B., and Kumaravel, G. (1996) Progression of organic reactions on resin supports monitored by single bead FTIR microspectroscopy. *J. Org. Chem.* **61,** 7467–7472.
11. Yan, B. (1998) Monitoring the progress and the yield of solid-phase organic reactions directly on resin supports. *Acc. Chem. Res.* **31,** 621–630.

12. Shapiro, M. J. and Gounarides, J. S. (1999) NMR methods utilized in combinatorial chemistry research. *Prog. Nucl. Magn. Res. Spectros.* **35,** 153–200.
13. Busch, K. L. (1995) Desorption ionization mass spectrometry. *J. Mass Spectrom.* **30,** 233–240.
14. Karas, M., Bachmann, D., Bahr, U., and Hillenkamp, F. (1987) Matrix-assisted ultraviolet laser desorption of non-volatile compounds. *Int. J. Mass Spectrom. Ion Proc.* **78,** 53–68.
15. Benninghoven, A., Rudenauer, F. G., and Werner, H. W. (1987) *SIMS: Basic concepts, Instrumental Aspects, Applications and Trends.* Wiley, New York, NY.
16. Drouot, C., Enjalbal, C., Fulcrand, P., et al. (1996) Step-by-step control by time-of-flight secondary ion mass spectrometry of a peptide synthesis carried out on polymer beads. *Rapid Commun. Mass Spectrom.* **10,** 1509–1511.
17. Drouot, C., Enjalbal, C., Fulcrand, P., et al. (1997) Tof-SIMS analysis of polymer bound Fmoc-protected peptides. *Tetrahedron Lett.* **38,** 2455–2458.
18. Aubagnac, J.-L., Enjalbal, C., Subra, G., et al. (1998) Application of time-of-flight Secondary ion mass spectrometry to *in situ* monitoring of solid-phase peptide synthesis on the Multipin™ system. *J. Mass Spectrom.* **33,** 1094–1103.
19. Bertrand, P. and Weng, L.-T. (1996) Time-of-flight secondary ion mass spectrometry. *Mikrochim. Acta.* **13,** 167–182.
20. Enjalbal, C., Subra, G., Combarieu, R., Martinez, J., and Aubagnac, J.-L. (2000) Use of time of flight static-secondary ion mass spectrometry in peptide synthesis on solid support. *Rec. Res. Dev. Organic Chem.* **4,** 29–52.
21. Aubagnac, J.-L., Enjalbal, C., Drouot, C., Combarieu, R., and Martinez, J. (1999) Imaging time-of-flight secondary ion mass spectrometry of solid-phase peptide synthesis. *J. Mass Spectrom.* **34,** 749–754.
22. Enjalbal, C., Maux, D., Subra, G., Martinez, J., Combarieu, R., and Aubagnac, J.-L. (1999) Monitoring and quantification on solid support of a by-product formation during peptide synthesis by Tof-SIMS. *Tetrahedron Lett.* **40,** 6217–6220.
23. Enjalbal, C., Maux, D., Combarieu, R., Martinez, J., and Aubagnac, J-L. (2000) Mass spectrometry and combinatorial chemistry: New approaches for direct support-bound compound identification. *Combinatorial Chem. High Throughput Screening* **4,** 363–373.
24. Falick, A. M., Hines, W. M., Medzihradsky, K. F., Baldwin, M. A., and Gibson, B. W. (1993) Low-mass ions produced from peptides by high energy collision-induced dissociation in tandem mass spectrometry. *J. Am. Soc. Mass Spectrom.* **4,** 882–893.
25. Garner, G. V., Gordon, D. B., Tetler, L. W., and Sedgwick, R. D. (1983) FAB MS of Boc protected amino acids. *Org. Mass Spectrom.* **18,** 486–488.
26. Grandas, A., Pedroso, E., Figueras, A., Rivera, J., and Giralt, E. (1988) Fast atom bombardment mass spectrometry of protected peptide segments. *Biomed. Environm. Mass Spectrom.* **15,** 681–684.

3

Preparation of Encoded Combinatorial Libraries for Drug Discovery

Tao Guo and Doug W. Hobbs

1. Introduction

The revolution in genomics and proteomics is projected to expand the number of potential therapeutic targets to between 5,000 and 10,000 from the approximately 500 targets that have historically been used by the pharmaceutical industry in the development of drugs *(1,2)*. The research and development of a safe and effective drug is a slow and expensive process, which is currently estimated to take an average of 12 years and to have a risk adjusted cost of $500 million per drug *(3)*. The pharmaceutical industry is under intense pressure to bring novel drugs to market quickly and cost-effectively. Combinatorial chemistry has emerged during the past decade as a powerful tool to help accelerate the drug discovery process *(4–7)*. Combinatorial chemistry refers to methods for the high-throughput synthesis of a significant number (10^2 to $>10^6$) of compounds *(8)*. Among the various methods developed *(9–20)*, the solid-phase split-pool synthesis *(21–23)* is perhaps the most efficient approach for the rapid synthesis of a large number of compounds. In this approach, a library that usually contains >10,000 members can be constructed very rapidly from a small number of chemical building blocks. **Figure 1** illustrates the split-pool synthesis with a two step reaction A + B that uses three building blocks in step 1 (A_1, A_2, A_3) and three building blocks in step 2 (B_1, B_2, B_3). Nine products can be generated using only six reactions.

In a split-pool library, each resin bead contains a single compound. Widespread adoption of this technique has been hampered by the necessity of determining which structure is on which bead. A number of chemical and nonchemical encoding methods have been developed to help the structural determination in these libraries *(24–36)*. One chemical encoding method that was first invented

From: *Methods in Molecular Biology, Combinatorial Library Methods and Protocols*
Edited by L. B. English © Humana Press Inc., Totowa, NJ

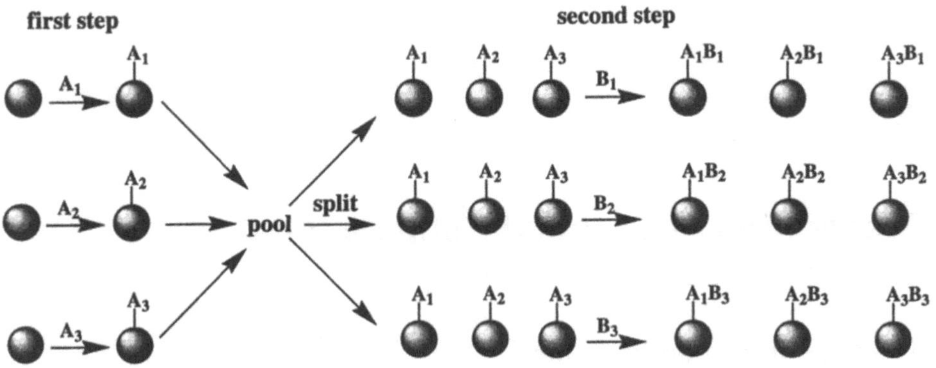

Fig. 1. The split-pool synthesis method.

by Still and co-workers at Columbia together with Wigler and co-workers at Cold Spring Harbor and later refined at Pharmacopeia uses a binary encoding protocol employing electrophoric molecular tags (ECLiPS™ technology) *(30–32)*. In this protocol, incorporation of each set of synthons is accompanied by the attachment of a unique binary set of electrophoric tags to the solid support during the library construction. **Figure 2** illustrates the synthesis of such an encoded library. The library synthesis is carried out by initial incorporation of the first set of synthons to the resin via an appropriate linker, followed by the attachment of tag/linker construct directly to the resin via carbene insertion. The resin is then pooled and split or directly divided *(37)* into portions for the incorporation of the second sets of synthons and binary tags. This process is repeated until the library synthesis is complete. The result of these operations is a collection of beads wherein the synthetic history of each bead is recorded with a unique binary code of tagging molecules. An orthogonal linkage strategy is used in the library synthesis to enable the release of compound independent from the tag molecules. The compound can then be evaluated in solution in any standard assay, while its identity can be determined separately by electron capture gas chromatography (EC/GC) analysis of the detached tags (**Fig. 3**).

The design of an encoded combinatorial library begins with defining the chemistry and evaluating the proposed structures with respect to the goal for the library (e.g., discovery or optimization). After an initial set of synthons are chosen, the library is enumerated *in silico* to produce a first-generation virtual library. A variety of calculations are performed on the virtual library to determine its overall drug-likeness and physical property profile *(38,39)*. Solid-phase reaction optimization and synthon paneling are simultaneously performed to determine the scope of the chemistry. As the optimal solid-phase reaction conditions are being established, the virtual library is refined to satisfy diver-

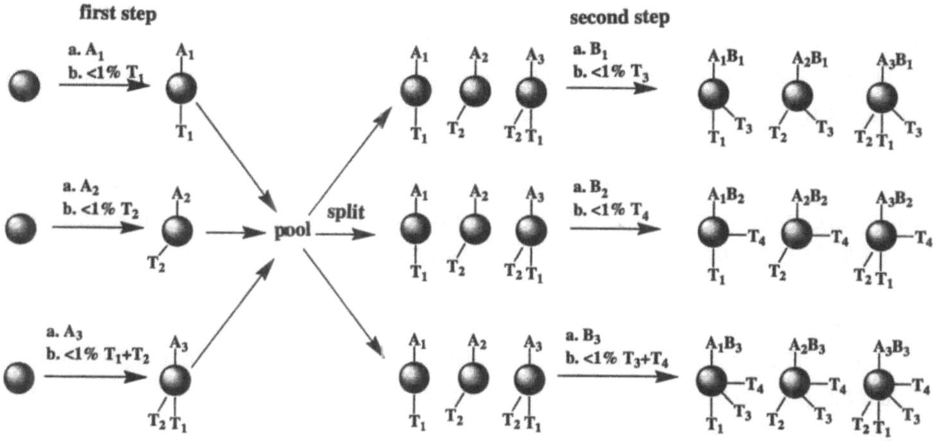

Fig. 2. The split-pool synthesis of an encoded combinatorial library.

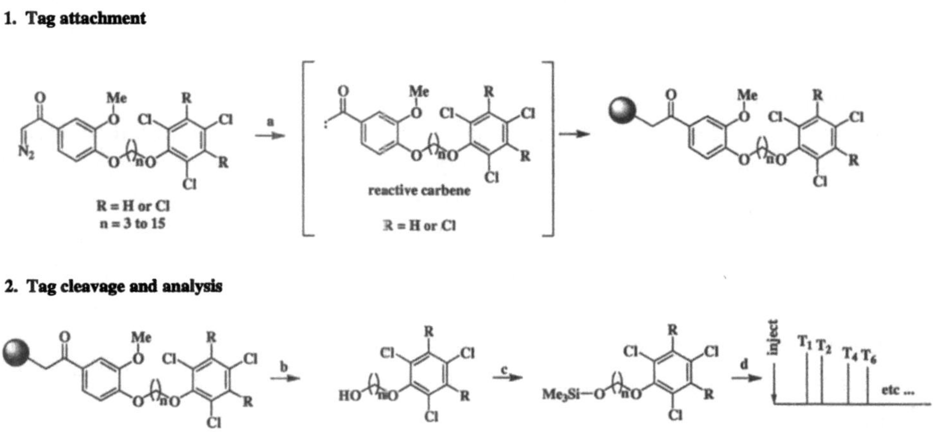

Fig. 3. Methods for tag attachment, detachment, and analysis. Reagents and conditions: (a) Resin, $[(CF_3CO_2)_2Rh]_2$, DCM, 25°C, 16 h; (b) $(NH_4)_2Ce(NO_3)_6$, hexane/CH_3CN/H_2O, 35°C, 5 h; (c) *N,O-bis*(trimethylsilyl)acetamide, hexane, 25°C, 10 min; (d) electron capture gas chromatography (EC/GC).

sity, physical property, and overall drug suitability criteria. A number of library quality control (QC) compounds are prepared prior to the library synthesis and are rigorously analyzed by mass spectrometry (MS) and quantitative high pressure liquid chromatography (HPLC) methods. The data from these QC compounds are used to estimate the optimal cleavage conditions, yield, and purity of the completed library. After library synthesis is complete, the quality of the library can be assessed by performing liquid chromatography mass spectrometry

(LC/MS) in conjunction with tag decode analysis on a statistical sampling of products from the library *(40)*. A typical encoded combinatorial library contains 10,000 to 100,000 compounds. Depending on the complexity of the chemistry and loading capacity of the beads, each compound is generally represented on 30–300 beads and each bead usually contains 200–60,000 picomoles of a single compound. High-throughput screening assays are usually carried out by first surveying one library equivalent of compounds using 5–30 compounds per well to identify the most active sublibrary followed by screening two or three library equivalents of the most active sublibrary at the single compound per well level.

Over 150 libraries totaling over 6 million compounds have been prepared at Pharmacopeia using the ECLiPS™ technology. Each library was based on one or multiple scaffolds. This large collection of diverse small-molecule compounds has proved to be a rich resource for drug discovery *(32,41–50)*. Three Pharmacopeia encoded combinatorial libraries, designated A, B, and C, are described here in detail to illustrate design, synthesis, screening, and structure activity relationship (SAR) analysis of encoded combinatorial libraries for drug discovery. Library A will illustrate the design considerations, library B the synthesis and screening procedures, and library C the SAR data analysis.

Library A is a discovery library aimed at identifying drug-like small molecule leads for G-protein coupled receptor (GPCR) targets *(50)*. Optimal diversity, good oral absorption properties, and solid-phase synthetic feasibility were all considered during the design phase *(8,38,39)*. Many cycles of design and property analysis were carried out *in silico* to arrive at the final version of the virtual library (**Fig. 4**). An actual LidDraw screen-shot of the final version of virtual library A is shown in **Fig. 5**, and the properties of the final version of the virtual library are depicted in **Fig. 6**.

Library B was designed and synthesized as an enzyme targeted library to identify inhibitors and SAR for aspartyl protease plasmepsin II, a key enzyme in the life cycle of the malarial parasite *Plasmodium falciparum* *(46,51)*. The encoded solid-phase synthesis of this library is illustrated in **Fig. 7**. The library was constructed in 4 combinatorial steps using 7 primary amines in the first step, 3 Boc-statines (known transition state mimetic for aspartyl proteases) in the second step, 31 Fmoc-amino acids in the third step, and 20 acylating agents in the fourth step, yielding an overall 13,020 final compounds. The four sets of synthons (R_A, R_B, R_C, and R_D) used in the library synthesis are listed in **Fig. 8**. To encode the library 10 molecular tags were employed (**Fig. 9**): 3 tags were used for the 7 R_A synthons, 2 tags for the 3 R_B synthons, and 5 tags for the 31 R_C synthons. The 20 R_D synthons in the fourth step were not encoded, but instead were stored in individual vials as sublibraries after the synthesis was complete. Screening of this library against plasmepsin II resulted in the discovery of potent and selective inhibitors as well as novel SAR *(46)*.

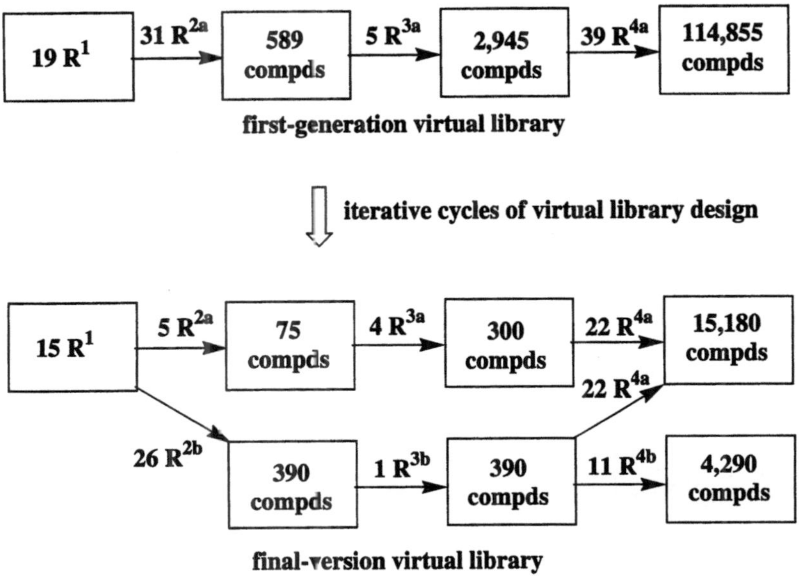

Fig. 4. Schematic illustration of virtual library design for library A.

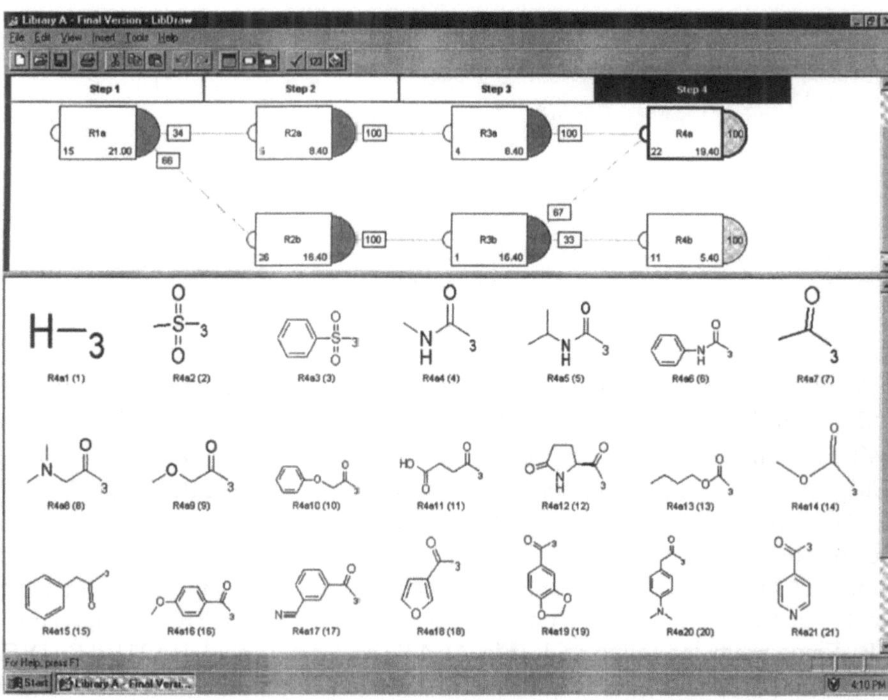

Fig. 5. A LibDraw program screen-shot of the final version of virtual library A.

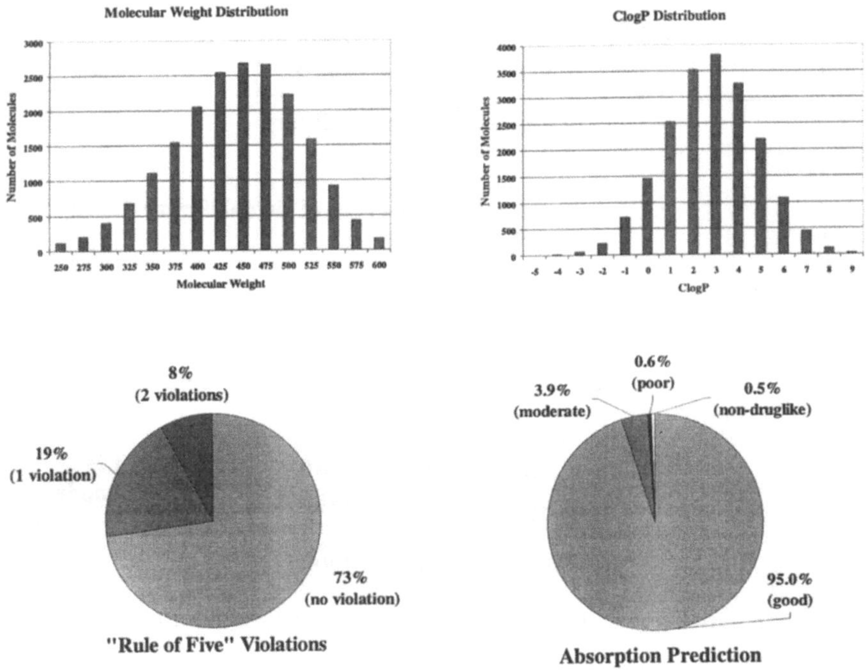

Fig. 6. Property analysis of the final version of virtual library A.

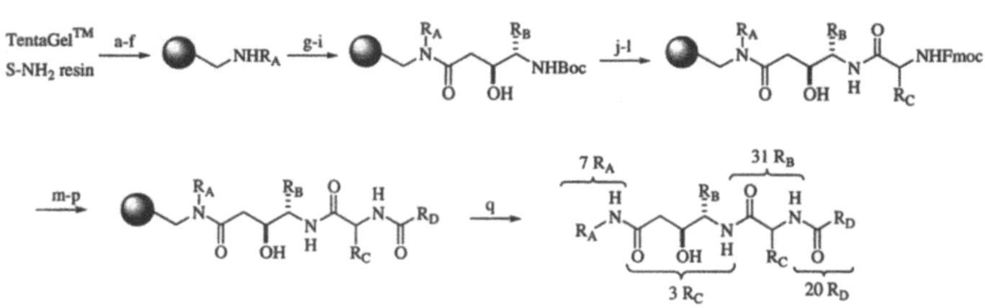

Fig. 7. Synthesis of the statine library B. Reagents and conditions: (a) TentaGel™ S-NH_2 resin (0.3 mmol/g) distributed into seven reaction vessels; (b) 3 Eq each Boc-Lys(Boc)-OH, HOBt, 5 Eq DIC, DCM; (c) encode using three tags; (d) 50% TFA/DCM, 1 h; (e) 5 Eq each 4-bromomethyl-3-nitrobenzoic acid, HOBt, 8 Eq DIC, DCM, 3 h; (f) one of seven R_A amines (**Fig. 4**): 10 Eq amine, THF, 8 h; (g) pool and split into three reaction vessels; (h) one of three R_B Boc-protected statines (**Fig. 4**): 4 Eq each statine, HATU, 8 Eq DIEA, DMF, 3 h; (i) encoded using two tags; (j) pool and split into 31 reaction vessels and encode using 5 tags; (k) 50% TFA/DCM, 1 h; (l) one of 31 R_C Fmoc-protected amino acids (**Fig. 4**): 4 Eq each amino acid, HATU, 8 Eq DIEA, DMF, 6 h; (m) pool; (n) 30% piperidine/DMF, 1 h; (o) split into 20 reaction vessels; (p) one of 20 R_D acylation agents (**Fig. 4**): 4 Eq each of R_DCO_2H, HATU, 8 Eq DIEA, 6 h; (q) hν (365 nm), MeOH, 50°C, 2.5 h.

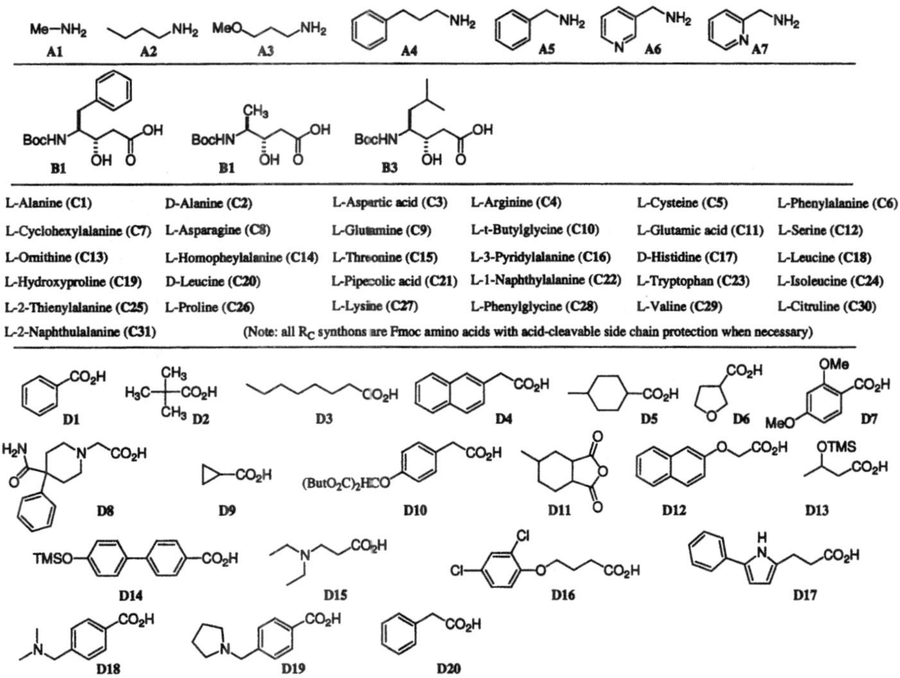

Fig. 8. Synthons for the statine library B.

Library C was designed and synthesized as an optimization library for a GPCR target in order to find small molecule agonists *(52)*. Screening of this library resulted in the discovery of potent and selective compounds as well as novel SAR for the target. **Figure 10** shows the generic structure of this library along with a 3D plot of the SAR found in one sublibrary.

2. Materials

2.1. Library Design

1. LibDraw (library drawing program, Pharmacopeia, Inc., Princeton, NJ).
2. LibProp (library enumeration and property calculation program, Pharmacopeia, Inc., Princeton, NJ).
3. Excel (Microsoft Corp., Seattle, WA).

2.2. Library Synthesis

1. Apparatus: glass shaking vessels (small: 20 mL, medium: 100 mL, large: 200 mL, Pharmacopeia, Inc., Princeton, NJ), Burrell wrist action shaker (Fisher Scientific, Pittsburgh, PA).
2. Resin: TentaGel™ S-NH_2 resin, 0.29 mmol/g, 180–220 µm (Rapp Polymere GmbH, Tübingen, Germany).

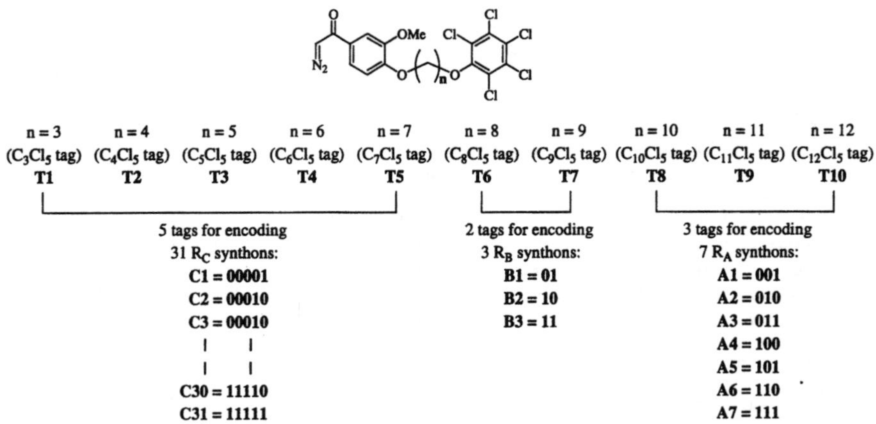

Fig. 9. Tags and tagging strategy for the statine library B.

3. Chemical building blocks: Boc-Lys(Boc)-OH, 4-bromomethyl-3-nitrobenzoic acid, R_A amines A1–A7, Boc-statines B1–B3, Fmoc-amino acids C1–C31, and acylating agents D1–D20 (**Fig. 8**).
4. Molecular tags: diazoketone tags T1–T10 (**Fig. 9**).
5. Chemical reagents: 1,3-diisopropylcarbodiimide (DIC), 1-hydroxybenzotriazole (HOBt), *O*-(7-azabenzotriazol-1-yl)-*N,N,N'N'*-tetramethyluronium hexafluorophosphate (HATU), *N,N*-diisopropylethylamine (DIEA), triethylamine (Et_3N), rhodium(II) trifluoroacetate dimer ($[(CF_3CO_2)_2Rh]_2$), trifluoroacetic acid (TFA), piperidine.
6. Solvents: acetonitrile (CH_3CN), *N,N*-dimethylformamide (DMF), dichloromethane (DCM), methanol (MeOH), ethanol (EtOH), ethyl acetate (EtOAc), water (H_2O).
7. Solution for removing acid-labile protecting groups: TFA/phenol/thiophenol/ethanedithiol/water (82:5:5:3:5).
8. Ninhydrin test reagents: *(1)* phenol/EtOH (7:3), *(2)* 0.2 m*M* potassium cyanide (KCN) in pyridine, *(3)* 0.28 *M* ninhydrin in EtOH.

2.3. Library Screening

1. Apparatus: UV light chamber (Pharmacopeia, Inc., Princeton, NJ), Genevac (Genevac, Ltd., Ipswich, UK), Tecan SLT FluoStar fluorescence plate reader (Tecan U.S., Research Triangle Park, NC), Sonicator, 96-well filter-bottom plates, 96-well assay plates.
2. Plasmapsin II (from Dr. Daniel E. Goldberg, Howard Hughes Medical Institute, Washington University School of Medicine, St. Louis, MO).
3. 4-(4-Dimethylaminophenylazo)benzoyl (DABCYL)-γ-aminobutyric acid-Glu-Arg-Met-Phe-Leu-Ser-Phe-Pro-EDANS (AnaSpec, Inc., San Jose, CA).
4. Bovine serum albumin (BSA).
5. Sodium acetate; Tween 20; glycerol; 1 *M* Tris-HCl (pH 8.5); dimethyl sulfoxide (DMSO); MeOH.

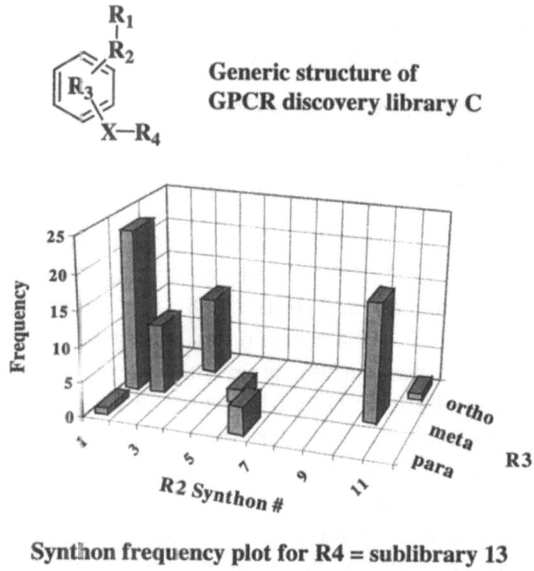

Fig. 10. Combinatorial SAR: R2 synthon selection as a function of R3 from screening library C.

2.4. Compound Decoding

1. Apparatus: Hewlett Packard 5890/ECD gas chromatography (GC) system, DB-1 GC column: 15 m × 0.25 mm id, 0.25 μm film, GC vials (all from Agilent Technologies, Inc., Piscataway, NJ).
2. 96-well filter-bottom plates, 96-well assay plates.
3. 0.3 *M* ceric ammonium nitrate $[(NH_4)_2Ce(NO_3)_6]$ solution in H_2O.
4. Octane.
5. *N,O-bis*(trimethylsilyl)-acetamide.

2.5. SAR Data Analysis

1. Excel (Microsoft Corp., Seattle, WA).

3. Methods

3.1. Library Design

1. Create virtual libraries using LibDraw (**Fig. 5**), a program developed internally at Pharmacopeia (*see* **Note 1**).
2. Calculate library properties using LibProp, another internally developed software program at Pharmacopeia (*see* **Note 2**).
3. Display library properties as bar graphs and/or pie charts using Microsoft Excel for visual inspection. Refine the virtual libraries until an acceptable property distribution is achieved (**Fig. 6**, *see* **Note 3**).

4. Continue the *in silico* analysis process until an optimal balance between diversity and drug-likeness is achieved (*see* **Note 4**).

3.2. Library Synthesis and Encoding

3.2.1. Resin Double-Loading, Attachment of Photolabile Linker, Encoding and Incorporation of R_A Amine Synthons

1. Suspend TentaGel™ S-NH_2 resin (180–220 mm, 0.29 mmol/g, 10 g, 2.9 mmol) in 150 mL of 9:1 (v/v) DMF/DCM in a large shaking vessel. Add Boc-Lys(Boc)-OH (5.12 g, 8.7 mmol, 3 eq) and HOBt (1.18 g, 8.7 mmol, 3 Eq) followed by the addition of DIC (2.73 mL, 17.4 mmol, 6 Eq). Shake the mixture for 16 h at 25°C. Drain the mixture and then wash the resin with 150 mL each of DMF (3×), MeOH (3×), and DCM (3×). Perform Nihydrin test for an aliquot of the resin; a negative result indicates complete coupling. Dry the resin *in vacuo* and then divide it into seven equal portions (0.83 mmol, double loading, *see* **Note 5**). Place each portion into seven medium shaking vessels.
2. According to the tagging scheme for the seven first-step synthons (**Fig. 9**), treat the resin in the seven vessels with one or more of the T8–T10 tags (*see* **Note 6**). For example, suspend the resin in vessel 1 (for synthon A1, 2.14 g, 0.83 mmol double-loading) in 50 mL of EtOAc and add a solution of T10 ($C_{12}Cl_5$ tag, 0.16 g, 7.5% of resin mass) in DCM (1.3 mL). Agitate the mixture for 2 h, then add 2.6 mL of a 0.2 mg/mL solution of $[(CF_3CO_2)_2Rh]_2$ in DCM and agitate the mixture at 25°C for 16 h. Drain the mixture and then wash the resin with 50 mL each of DCM (4×), MeOH (2×), and DCM (4×).
3. After all tagging reactions are complete, suspend the resin in 50% TFA/DCM and shake for 1 h. Drain the mixture and wash the resin with 50 mL each of DCM (3×), MeOH (3×), 20% Et_3N/MeOH (1×), MeOH (3×), DMF (3×), and DCM (3×).
4. Resuspend the resin in DCM (25 mL) and then add a preincubated (45 min) solution of 4-bromomethyl-3-nitrobenzoic acid (0.83 g, 3.21 mmol, 3.9 Eq), HOBt (0.43 g, 3.21 mmol, 3.9 Eq), and DIC (1.0 mL, 6.42 mmol, 7.8 Eq) in 25 mL of DCM. Shake the mixture at 25°C for 3 h. Drain the mixture and wash the resin with DCM (3× 50 mL). Perform this operation in tandem for each of the seven vessels of tagged resin (*see* **Note 7**).
5. Add 10.7 mmol (12.9 Eq) of a primary amine (*see* **Fig. 4** for the list of seven R_A amines) to a suspension of the 2-nitrobenzylbromide resin (0.83 mmol) in 50 mL of THF in a medium shaking flask and shake the mixture at 25°C for 16 h. Drain the mixture and then wash the resin with 50 mL each of DMF (3×), MeOH (3×), 10% TFA/MeOH (1×), MeOH (3×), DMF (3×), and DCM (3×).

3.2.2. Incorporation and Encoding of R_B Boc-Statine Synthons

1. Combine and mix the secondary amine resin and then divide the resin into three batches. Suspend each batch of the resin (1.9 mmol), independently, in 50 mL of DMF in a medium shaking vessel. Treat the resin with one of the three Boc-

protected statine R_B synthons (**Fig. 4**, 4.78 mmol, 2.5 Eq), DIEA (1.66 mL, 9.56 mmol, 5.0 Eq), and then HATU (1.82 g, 4.78 mmol, 2.5 mmol). Shake the mixture at 25°C for 6 h. Drain the mixture and wash the resin with 50 mL each of DMF (3×), MeOH (3×), DMF (3×), and DCM (3×).

2. According to the tagging scheme for the three second-step synthons (**Fig. 9**), treat the resin in the three vessels with one or two of the T6–T7 tags. For example, suspend the resin in vessel 1 (for synthon B1, approx 3.7 g, 1.9 mmol) in 85 mL of EtOAc and add a solution of T7 (C_9Cl_5 tag, 0.30 g, 8% of resin mass) in 2.5 mL of DCM. Agitate the mixture for 2 h and then add 4.7 mL of a 1.5 mg/mL solution of $[(CF_3CO_2)_2Rh]_2$ in DCM. Shake the mixture at 25°C for 16 h. Drain the mixture and wash the resin with 90 mL each of DCM (4×), MeOH (2×), and DCM (4×).

3.2.3. Encoding and Incorporation of R_C Fmoc-Amino Acid Synthons

1. Pool the resin from the second step as a suspension in DCM (200 mL) and mix it into homogeneity. After draining the solvent, dry the resin *in vacuo*. Split a portion (5.58 g, 3.4 mmol, *see* **Note 8**) of the resin equally into 31 small reaction vessels, each containing 0.18 g (0.11 mmol) of the resin.
2. According to the tagging scheme for the 31 third-step synthons (**Fig. 9**), treat the resin in the 31 vessels with one or more of the T1–T5 tags. For example, suspend the resin in vessel 1 (for synthon C1, 0.18 g, 0.11 mmol) in 5 mL of EtOAc and add a solution of T5 (C_7Cl_5 tag, 10 mg, 5.5% of resin mass) in 100 μL of DCM. Agitate the mixture for 2 h and then add 220 μL of a 1.5 mg/mL solution of $[(CF_3CO_2)_2Rh]_2$ in DCM. Shake the mixture at 25°C for 16 h. Drain the mixture and wash the resin with 5 mL each of DCM (4×), MeOH (2×), and DCM (4×).
3. Treat the resin in each of the 31 vessels with a unique R_C Fmoc-amino acid synthon. For example, add a solution of Fmoc-L-alanine (50 mg, 0.16 mmol, 1.5 Eq) and HATU (61 mg, 0.16 mmol, 1.5 Eq) in 8 mL of DMF to the resin in vessel 1 (0.18 g, 0.11 mmol). Agitate the suspension at 25°C for 10 min and then add DIEA (56 μL, 0.32 mmol, 3 Eq). Shake the mixture at 25°C. Monitor the coupling reaction in the vessel using ninhydrin test to determine the level of the amine functionality remaining. Upon completion of the coupling reaction (2 h, negative ninhydrin test), drain the mixture and wash the resin with 10 mL each of DMF (3×), MeOH (3×), and DCM (3×). Perform this procedure in tandem for each of the R_C Fmoc-amino acid synthons listed in **Fig. 8**.

3.2.4. Fmoc-Deprotection and Incorporation of R_D Acylation Agents to Give 20 Sublibraries

1. Combine the resin from step 3 into a large shaking vessel. Add a solution of 30% piperidine in DMF (100 mL) and shake the suspension at 25°C for 1 h. Drain the mixture and wash the resin with 100 mL each of DMF (2×), DCM (2×), MeOH (3×), and DCM (5×).
2. Dry the resin *in vacuo* and then split equally into 20 small shaking vessels, providing 0.28 g (0.18 mmol) of resin in each vessel. Treat the resin in each vessel

with one of the 20 R_D acylation reagents. For example, add a solution of benzoic acid (37 mg, 0.3 mmol, 1.7 Eq), HATU (137 mg, 0.36 mmol, 2 Eq), and DIEA (153 μL, 0.88 mmol, 4.9 Eq) in DMF (7 mL) to the resin in vessel 1. Shake the mixture at 25°C for 1 h to give a negative Ninhydrin test. Drain the mixture and wash the resin with 10 mL each of DMF (2×), MeOH (3×), and DCM (5×).

3. Shake the resin with a 10 mL solution of TFA/phenol/thiophenol/ ethanedithiol/ water (82:5:5:3:5) at 25°C for 1.5 h to remove all of the protecting groups on R_C amino acid side chains and on the R_D acylating agents. Drain the mixture and wash the resin with 10 mL each of 50% TFA/water (2×), DMF (2×), MeOH (4×), DMF (2×), and DCM (5×).
4. Dry the resin *in vacuo* and store the resin bound compounds as sublibrary 1. Perform the coupling procedure for the resin in all the reaction vessels except vessel 11 using one of the R_D carboxylic acids listed in **Fig. 8**. For vessel 11, treat the resin with D11 anhydride synthon (0.36 mmol, 2 Eq) in 7 mL of DMF at 45°C for 8 h. Store each of the final resin batches separately as an individual sublibrary, thereby obviating the need for encoding.

3.3. Library Screening

3.3.1. Photolytic Cleavage of Products from Resin Beads

1. Array the resin beads from the sublibraries of Library B into 96-well filter-bottom plates (20 beads per well for initial survey screening, or a single bead per well for follow-up analysis) using an automated bead arraying apparatus.
2. Suspend the dried beads in each well in 150 μL of MeOH. Irradiate the mixture at 365 nm for 30 min at 50°C employing a custom UV light chamber and then incubate the mixture for an additional 2 h. Filter the mixture and collect the eluent into a 96-well assay plate. Dry the mixture in Genevac (0.1 Torr) for 2 h at 40°C to give the dried compounds.

3.3.2. Plasmepsin II Assay

1. Add 25 μL of the assay mixture that contains 50 m*M* sodium acetate (pH 5.0), 0.01% Tween 20, 12.5% glycerol, 1 mg/mL BSA, and 12 μ*M* plasmapsin II substrate DABCYL-γ-aminobutyric acid-Glu-Arg-Met-Phe-Leu-Ser-Phe-Pro-EDANS into each well of the 96-well microtiter plate containing dried compounds or empty control wells. Sonicate the plates to solubilize the compounds.
2. Initiate the enzymatic reaction with the addition of 25 μL of 8 n*M* plasmapsin II in an aqueous buffer that contains 50 m*M* sodium acetate (pH 5.0), 0.01% Tween 20, 1 mg/mL BSA, and 12.5% glycerol. Incubate the assay mixture at 25°C for 10 min and then quench the reaction by the addition of 25 μL of 1 *M* Tris-HCl (pH 8.5 and containing 50% DMSO). Record the EDANS fluorescence using a Tecan SLT FluoStar fluorescence plate reader equipped with a 350 nm excitation filter and a 510 nm emission filter.

3.4. Compound Decoding

1. Incubate each single bead in one well of a 96-well plate with 10 µL of a freshly prepared 0.3 *M* aqueous solution of $(NH_4)_2Ce(NO_3)_6$ and 50 µL of octane at 25°C for 1 h to cleave the tag molecules.
2. Transfer the octane extracts of the tag alcohols (35 µL) into GC vials and then add *N,O*-bis(trimethylsilyl)-acetamide (5 µL). Incubate the mixture at 25°C for at least 10 min to convert the tag alcohols to their corresponding trimethylsilyl ethers.
3. Inject the tag trimethylsilyl ethers (1 µL) into the HP5890/ECD system using a DB-1 column (15 m × 0.25 mm id, 0.25 µm film). Apply a temperature ramp of 200–325°C in 5 min and then maintain the temperature at 325°C for 10 min. Set the electron capture detector at 400°C and the auxiliary gas at 35 psi. One complete chromatogram run takes 15 min.
4. Analyze the EC/GC chromatogram of tag molecules to generate the compound structure.

3.5. SAR Data Analysis

1. After decoding, plot the frequency of synthons found in the decoded structures in 2D or 3D bar graphs using Microsoft Excel and analyze SAR (*see* **Note 9**).
2. Perform resynthesis of the active compounds in greater quantities to confirm activity through multi-point IC_{50}, or K_i determination.

4. Notes

1. LibDraw allows the variable chemical building blocks to be drawn as fragments, then connects the fragments to create the virtual products according to a specific recombination scheme. Other programs may be substituted, providing they allow convenient reorganization of the split-pool strategy as well as enumeration of library members.
2. LibProp was used to calculate various properties, such as molecular weight, logP, hydrogen bond donor and acceptor numbers, and predicted oral absorption. Other programs and properties may be substituted. The objective is to compare the property distribution of the virtual library with a set of "ideal" properties *(38,39)*.
3. This can be done by either modifying the choice of synthons for one or more steps or by altering the splitting strategy to avoid the combination of specific synthons. For example, library A (**Fig. 4**) was rearranged to prevent the most lipophilic synthons in step 1 from combining with the most lipophilic synthons in step 2. The second-generation virtual library created by this reorganization has a much better property distribution profile (**Fig. 6**).
4. A balance between diversity and drug-likeness needs to be reached. Generally, 75% of the compounds should be predicted to have good oral bioavailability.
5. Lysine is used to double the bead loading of the resin.
6. The tagging reaction can be performed using one or more tags at the same time.

7. Since photolabile linker is sensitive to light all the reactions need to be carried out in an unlighted hood.
8. The remaining portion of the resin was for preparing another library.
9. **Figure 10** is an example of a 3D plot showing the synthon preferences for library C. This type of multidimensional analysis allows the identification of interrelationships between variables in the library. In the case of library C, the majority of active compounds were found in the series where R3 represents a *meta* orientation on the aromatic core. Within the *meta* series, there is a preference for compounds where R2 = synthons 1, 2, and 10. The key finding from this chart, however, is that the SAR is strikingly different when R3 represents an *ortho* orientation. In the *ortho* series, R2 strongly prefers synthon 3, which is not observed at all in the *meta* series. Similarly, the *para* series also exhibits distinct SAR. Little activity was observed when R3 = *para*-substituted, except in combination with R2 = synthon 1 and 6. The combinatorial SAR revealed here seem to indicate that regional optimization as practiced by traditional medicinal chemistry may be an inappropriate strategy for certain biological targets.

Acknowledgment

Our colleagues at Pharmacopeia are thanked for their contributions to the materials discussed in this chapter.

References

1. Edwards, A. M., Arrowsmith, C. H., and Pallieres, B. D. (2000) Proteomics: new tools for a new era. Bridging the gap between genomics and drug discovery. *Mod. Drug Disc.* **Sept,** 55–60.
2. Drews, J. (2000) Drug discovery: a historical perspective. *Science* **287,** 1960–1964.
3. Lipper, R. A. (1999) E pluribus product. *Mod. Drug Disc.* **Jan/Feb,** 55–60.
4. Floyd, C. D., Leblanc, C., and Whittaker, M. (1999) Combinatorial chemistry as a tool for drug discovery, in *Progress in Medicinal Chemistry* (King, F. D. and Oxford, A. W., eds.), Elsevier Science, Amsterdam, Vol. 36, pp. 91–168.
5. Dolle, R. E. (2000) Comprehensive survey of combinatorial library synthesis: 1999. *J. Comb. Chem.* **2,** 383–433.
6. Dolle, R. E. and Nelson, K. H. J. (1999) Comprehensive survey of combinatorial library synthesis: 1998. *J. Comb. Chem.* **1,** 235–282.
7. Dolle, R. E. (1999) Comprehensive survey of chemical libraries yielding enzyme inhibitors, receptor agonists and antagonists, and other biologically active agents: 1992 through 1997. *Annu. Rep. Comb. Chem. Mol. Diversity* **2,** 93–127.
8. Fecik, R. A., Frank, K. E., Gentry, E. J., Menon, S. R., Mitscher, L. A., and Telikepalli, H. (1998) The search for orally active medications through combinatorial chemistry. *Med. Res. Rev.* **18,** 149–185.
9. Geysen, H. M., Meloen, R. H., and Barteling, S. J. (1984) Use of peptide synthesis to probe viral antigens for epitopes to a resolution of a single amino acid. *Proc. Natl. Acad. Sci. USA* **81,** 3998–4002.
10. Franzen, R. G. (2000) Recent advances in the preparation of heterocycles on solid support: a review of the literature. *J. Comb. Chem.* **2,** 195–214.

11. Suto, M. J. (1999) Developments in solution-phase combinatorial chemistry. *Curr. Opin. Drug Disc. Develop.* **2,** 377–384.
12. Parlow, J. J., Devraj, R. V., and South, M. S. (1999) Solution-phase chemical library synthesis using polymer-assisted purification techniques. *Curr. Opin. Chem. Biol.* **3,** 320–336.
13. Hermkens, P., Ottenheijm, H., and Rees, D. (1996) Solid-phase organic reactions: a review of recent literature. *Tetrahedron* **52,** 4527–4554.
14. Hermkens, P., Ottenheijm, H., and Rees, D. (1997) Solid-phase organic reactions II: a review of the literature Nov 95–Nov 96. *Tetrahedron* **53,** 5643–5678.
15. Booth, S., Hermkens, P. H. H., Ottenheijm, H. C. J., and Rees, D. C. (1998) Solid-phase organic reactions III: a review of the literature Nov 96–Dec 97. *Tetrahedron* **54,** 15,385–15,443.
16. Früchtel, J. S. and Jung, G. (1996) Organic chemistry on solid supports. *Angew. Chem. Int. Ed. Engl.* **35,** 17–41.
17. Osborn, H. M. I. and Khan, T. H. (1999) Recent developments in polymer supported synthesis of oligosaccharides and glycopeptides. *Tetrahedron* **55,** 1807–1850.
18. Lorsbach, B. A. and Kurth, M. J. (1999) Carbon-carbon bond forming solid-phase reactions. *Chem. Rev.* **99,** 1549–1581.
19. Gordon, K. and Balasubramanian, S. (1999) Recent advances in solid-phase chemical methodologies. *Curr. Opin. Drug Disc. Develop.* **2,** 342–349.
20. Schreiber, S. L. (2000) Target-oriented and diversity-oriented organic synthesis in drug discovery. *Science* **287,** 1964–1969.
21. Furka, A. (1995) History of combinatorial chemistry. *Drug Dev. Res.* **36,** 1–12.
22. Lam, K. S., Salmon, S. E., Hersh, E. M., Hruby, V. J., Kazmierski, W. M., and Knapp, R. J. (1991) A new type of synthetic peptide library for identifying ligand-binding activity. *Nature* **354,** 82–84.
23. Houghten, R. A., Pinilla, C., Blondelle, S. E., Appel, J. R., Dooley, C. T., and Cuervo, J. H. (1991) Generation and use of synthetic peptide combinatorial libraries for basic research and drug discovery. *Nature* **354,** 84–86.
24. Brenner, S. and Lerner, R. A. (1992) Encoded combinatorial chemistry. *Proc. Natl. Acad. Sci. USA* **89,** 5381–5383.
25. Needels, M. C., Jones, D. G., Tate, E. H., Heinkel, G. L., Kochersperger, L. M., Dower, W. J., Barrett, R. W., and Gallop, M. A. (1993) Generation and screening of an oligonucleotide-encoded synthetic peptide library. *Proc. Natl. Acad. Sci. USA* **90,** 10,700–10,704.
26. Nielsen, J., Brenner, S., and Janda, K. D. (1993) Synthetic methods for the implementation of encoded combinatorial chemistry. *J. Am. Chem. Soc.* **115,** 9812–9813.
27. Kerr, J. M., Banville, S. C., and Zuckermann, R. N. (1993) Encoded combinatorial peptide libraries containing non-natural amino acids. *J. Am. Chem. Soc.* **115,** 2529–2531.
28. Nikolaiev, V., Stierandova, A., Krchnak, V., Seligmann, B., Lam, K. S., Salmon, S. E., and Lebl, M. (1993) Peptide-encoding for structure determination of nonsequenceable polymers within libraries synthesized and tested on solid-phase supports. *Pept. Res.* **6,** 161–170.

29. Krchnak, V., Wichsel, A., Cabel, D., and Lebl, M. (1995) Linear presentation of variable side-chain spacing in a highly diverse combinatorial library. *Pept. Res.* **8,** 198–205.
30. Ohlmeyer, M. H. J., Swanson, R. N., Dillard, L., Reader, J. C., Asouline, G., Kobayashi, R., Wigler, M., and Still, W. C. (1993) Complex synthetic chemical libraries indexed with molecular tags. *Proc. Natl. Acad. Sci. USA* **90,** 10,922–10,926.
31. Nestler, H. P., Bartlett, P. A., and Still, W. C. (1994) A general method for molecular tagging of encoded combinatorial chemistry libraries. *J. Org. Chem.* **59,** 4723–4724.
32. Burbaum, J. J., Ohlmeyer, M. H. J., Reader, J. C., Henderson, I., Dillard, L. W., Li, G., Randle, T. L., Sigal, N. H., Chelsky, D., and Baldwin, J. J. (1995) A paradigm for drug discovery employing encoded combinatorial libraries. *Proc. Natl. Acad. Sci. USA* **92,** 6027–6031.
33. Ni, Z.-J., Maclean, D., Holmes, C. P., Murphy, M. M., Ruhland, B., Jacobs, J. W., Gordon, E. M., and Gallop, M. A. (1996) Versatile approach to encoding combinatorial organic syntheses using chemically robust secondary amine tags. *J. Med. Chem.* **39,** 1601–1608.
34. Moran, E. J., Sarshar, S., Cargill, J. F., Shahbaz, M. M., Lio, A., Mjalli, A. M. M., and Armstrong, R. W. (1995) Radio frequency tag encoded combinatorial library method for the discovery of tripeptide-substituted cinnamic acid inhibitors of the protein tyrosine phosphatase PTP1B. *J. Am. Chem. Soc.* **117,** 10,787–10,788.
35. Nicolaou, K. C., Xiao, X.-Y., Parandoosh, Z., Senyei, A., and Nova, M. P. (1995) Radiofrequency encoded combinatorial chemistry. *Angew. Chem., Int. Ed. Engl.* **34,** 2289–2291.
36. Nicolaou, K. C., Pfefferkorn, J. A., Mitchell, H. J., Roecker, A. J., Barluenga, S., Cao, G. Q., Affleck, R. L., and Lillig, J. E. (2000) Natural product-like combinatorial libraries based on privileged structures. 2. Construction of a 10000-membered benzopyran library by directed split-and-pool chemistry using NanoKans and optical encoding. *J. Am. Chem. Soc.* **122,** 9954–9967.
37. Baldwin, J. J. and Horlbeck, E. (1997) Encoded libraries may be created using split-pool or direct divide synthesis. *US Patent 5,663,046.*
38. Lipinski, C. A., Lombardo, F., Dominy, B. W., and Feeney, P. J. (1997) Experimental and computational approaches to estimate solubility and permeability in drug discovery and development settings. *Adv. Drug Delivery Rev.* **23,** 3–25.
39. Egan, W. J., Merz, K. M., Jr., and Baldwin, J. J. (2000) Prediction of drug absorption using multivariate statistics. *J. Med. Chem.* **43,** 3867–3877.
40. Dolle, R. E., Guo, J., O'Brien, L., Jin, Y., Piznik, M., Bowman, K. J., Li, W., Egan, W. J., Cavallaro, C. L., Roughton, A. L., Zhao, Q., Reader, J. C., Orlowski, M., Jacob-Samuel, B., and Carroll C. D. (2000) A statistical-based approach to assessing the fidelity of combinatorial libraries encoded with electrophoric molecular tags. Development and application of tag decode-assisted single bead LC/MS analysis. *J. Comb. Chem.* **2,** 716–731.
41. Baldwin, J. J., Burbaum, J. J., Chelsky, D., Dillard, L. W., Henderson, I., Li, G., Ohlmeyer, M. H. J., Randle, T. L., and Reader, J. C. (1995) Combinatorial libraries encoded with electrophoric tags. *Eur. J. Med. Chem.* **30,** 349s–358s.

42. Baldwin, J. J. (1996) Design, synthesis and use of binary encoded synthetic chemical libraries. *Mol. Diversity* **2,** 81–88.
43. Chabala, J. C., Baldwin, J. J., Burbaum, J. J., Chelsky, D., Dilliard, L., Henderson, I., Li, G., Ohlmeyer, M. H. J., Randle, T. L., Reader, J. C., Rokosz, L., and Sigal, N. H. (1995) Binary encoded small-molecule libraries in drug discovery and optimization. *Persp. Drug Disc. Des.* **2,** 305–318.
44. Appell, K. C., Chung, T. D. Y., Ohlmeyer, M. J. H., Sigal, N. H., Baldwin, J. J., and Chelsky, D. (1996) Biological screening of a large combinatorial library. *J. Biomol. Screening* **1,** 27–31.
45. Appell, K. C., Chung, T. D. Y., Solly, K. J., and Chelsky, D. (1998) Biological characterization of neurokinin antagonists discovered through screening of a combinatorial library. *J. Biomol. Screening* **3,** 19–27.
46. Carroll, C. D., Patel, H., Johnson, T. O., Guo, T., Orlowski, M., He, Z.-M., Cavallaro, C. L., Guo, J., Oksman, A., Gluzman, I. Y., Connelly, J., Chelsky, D., Goldberg, D. E., and Dolle, R. E. (1998) Identification of potent inhibitors of plasmodium falciparum plasmepsin II from an encoded statine combinatorial library. *Bioorg. Med. Chem. Lett.* **8,** 2315–2320.
47. Horlick, R. A., Ohlmeyer, M. H., Stroke, I. L., Strohl, B., Pan, G., Schilling, A. E., Paradkar, V., Quintero, J. G., You, M., Riviello, C., Thorn, M. B., Damaj, B., Fitzpatrick, V. D., Dolle, R. E., Webb, M. L., Baldwin, J. J., and Sigal, N. H. (1999) Small molecule antagonists of the bradykinin B1 receptor. *Immunopharmacology* **43,** 169–177.
48. McMillan, K., Adler, M., Auld, D. S., Baldwin, J. J., Blasko, E., Browne, L. J., Chelsky, D., Davey, D., Dolle, R. E., Eagen, K. A., Erickson, S., Feldman, R. I., Glaser, C. B., Mallari, C., Morrissey, M. M., Ohlmeyer, M. H. J., Pan, G., Parkinson, J. F., Phillips, G. B., Polokoff, M. A., Sigal, N. H., Vergona, R., Whitlow, M., Young, T. A., and Devlin, J. J. (2000) Allosteric inhibitors of inducible nitric oxide synthase dimerization discovered via combinatorial chemistry. *Proc. Natl. Acad. Sci. USA* **97,** 1506–1511.
49. Li, G. and Guo, T. (2000) ECLiPS™ technology for drug discovery, in *Frontiers of Biotechnology & Pharmaceuticals* (Zhao, K., Reiner, J., and Chen, S.-H., eds.), Science Press, New York, Vol. 1, pp. 150–163.
50. Hobbs, D. and Guo, T. (2001) Library design concepts and implementation strategies, in *Combinatorial Library Design and Evaluation* (Ghose, A. K., Viswanadhan and V. N., eds.), Marcel Dekker, New York, pp. 1–50.
51. Dolle, R. E., Guo, T., Johnson, T. O., Patel, H. K., Tao, S., and He, Z. M. (1999) Combinatorial hydroxy-amino acid amide libraries. *US Patent 5,972,719.*
52. Guo, T. (2000) Encoded combinatorial libraries in drug discovery. *219th ACS National Meeting, San Francisco, March 26–30, 2000,* ORGN-218.

4

Simple Tools for Manual Parallel Solid Phase Synthesis

Viktor Krchňák and Andrew Burritt

1. Introduction

An inherent feature of parallel solid phase synthesis is the need to handle a large number of reaction vessels at the same time. Consequently, in order to make demanding synthetic tasks manageable, two categories of synthesizers, manual and automated, have been designed and produced. The main feature of a manual synthesizer is the integration of reaction vessels and common steps during synthesis. Reaction vessels are combined into so-called reaction blocks that enable performing specific operations (e.g., washing resin beads, adding common reagents, incubation) in all integrated reaction vessels at the same time. An automated synthesizer offers full automation of the entire synthetic process. The reaction vessels can be controlled on an individual basis and independent protocols can be performed in different vessels. Semiautomatic instruments feature integration and automation of the most commonly occurring steps. Even though full automation brings numerous advantages, the throughput of "manual" laboratories does not need to suffer. Without any expensive automated devices, production may still reach a thousand compounds per day.

In order to be able to select the most suitable instrumentation for a solid phase combinatorial synthesis, one has to answer three basic questions:

1. What is the projected throughput of compounds? This can vary from a single chemical entity per week/month to several thousand compounds per day.
2. What is the quantity of each compound needed? Some research projects may require as much as 50 mg of HPLC purified material, others may be satisfied with one hundred picomoles of compound cleaved from a single bead.

From: *Methods in Molecular Biology, Combinatorial Library Methods and Protocols*
Edited by: L. B. English © Humana Press Inc., Totowa, NJ

Table 1
Characteristics of Various Synthetic Scenarios

Scenario	Number of compounds	Resin load	Quantity	Synthesizer
#1	4	3–30 g	0.5–5 g	La Marast
#2	6–48	50 mg–3 g	12.5–750 mg	Domino Block
#3	96	5–50 mg	1.25–12.5 mg	Don Cucna

3. What is the available budget for instrumentation? Prices of apparatus can range from several hundred dollars, for a basic manual synthesis system, to several million dollars for a fully integrated custom designed robotic synthesizer.

In this chapter, three inexpensive manual synthesizers for parallel solid phase synthesis that can accommodate three different synthetic scenarios are described. The scenarios differ in the number of compounds synthesized in parallel and in the amount of compound synthesized in one reaction vessel. All instruments are designed to be personal tools for chemists. Operation of the apparatus is very simple, and does not need any special training course, programming, etc. The chemist should have the apparatus available at any time and use them in the same way that one uses other laboratory tools, such as a rotary evaporator, magnetic stirrer, or TLC chamber.

2. Materials and Methods

To cover the differing requirements of various research projects, we have arbitrarily divided potential synthetic throughput into three categories. A theoretical quantity of each compound to be synthesized has been assigned to each of the scenarios, as there has to be a certain relationship between the number of compounds synthesized concurrently and the quantity of each compound required. **Table 1** shows representative characteristics of the three synthetic situations that we believe are the most often encountered requirements in research projects. "Resin load" refers to the amount of resin in one reaction vessel; "quantity" is the calculated amount of compound obtained with a resin substitution of 1 mmol/g. An average molecular weight of 500 Da, and a 50% yield have been assumed. Obviously, the synthetic throughput is not limited to the number of compounds exemplified in each scenario (e.g., 96 compounds for scenario # 3 in one batch). One synthetic batch may typically include more that 10 integrated reaction vessels (960 compounds) and synthetic batches can frequently be nested.

In the following paragraphs we describe synthesizers that can accomplish the three examples defined above. To meet the demand for simple and inexpensive tools that allow high-throughput parallel solid phase synthesis, a new concept for liquid exchange was developed. The most time consuming operation of solid phase synthesis is washing the resin beads, since resin in each reaction vessel has to be washed and conditioned numerous times before the next chemical transformation can be performed.

From the instrumental point of view, washing of resin beads involves three steps:

1. Draining the reaction vessel (separation of the liquid phase from the resin beads).
2. Addition of the washing solvent (solvent transfer from container into the reaction vessel).
3. Equilibration of resin with the incoming solvent (mixing the resin slurry).

The most common method for draining the reaction vessel (separation of the liquid phase) is filtration through a porous material (glass, Teflon, polypropylene) using a difference in relative pressure between the reaction vessel and the waste container. The reaction vessels generally either are connected to an evacuated waste container or are pressurized with air/nitrogen. The most common problem with this technique is the clogging of the filter. Even partial clogging causes the air to pass through the reaction vessel with the lowest resistance, i.e., through the resin/filter of an already drained reaction vessel. As a result of this, the partially clogged reaction vessel drains either very slowly or not at all.

In order to make washing resin beads very simple and to solve some of the inherent problems connected with the liquid transfer, we have developed and used two new concepts in our synthesizers. The liquid exchange in the La Marast and Domino Blocks is based on the evacuated reaction vessel concept *(1)*. The Don Cucna synthesizer uses a suction principle for draining reaction vessels and a standard polypropylene 96-well plate as an integrated disposable reaction vessel *(2)*. All synthesizers are commercially available (Torviq, Tucson, AZ, www.torviq.com).

2.1. La Marast Synthesizer

2.1.1. Description of the Synthesizer

The principle of liquid transfer in the La Marast reaction vessel is very simple. In order to drain the reaction vessel, the single outlet of the vessel is connected to an evacuated waste container via a selection valve (**Fig. 1**). The air (or inert gas, if required) present in the reaction vessel starts to expand and pushes solvent from the reaction vessel into the evacuated waste container. At the same time, the reaction vessel is evacuated. The selection valve is then switched and the evacuated reaction vessel is connected to a solvent reservoir. The liquid in the reservoir (which was under atmospheric pressure) flows into the reaction vessel. Thus, the only operation necessary to wash the resin beads in the reaction vessel is turning a handle of a valve.

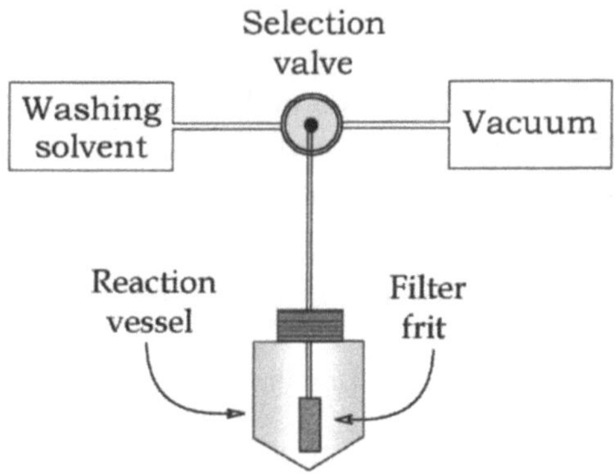

Fig. 1. Principle of liquid exchange in the La Marast reaction vessel.

Conical centrifuge tubes (120 mL, 210 mL, and 450 mL) serve as reaction vessels (**Fig. 2**). A female Luer bulkhead fitting, with external thread, is attached to the cap and Teflon tubing is used to connect the fitting with a porous polypropylene filter frit. The reaction vessels are placed into custom-made holders. The footprint of a holder matches that of a standard 96-well plate. This allows the use of commercially available titer plate shakers rather than special custom-built shakers.

In order to handle four reaction vessels in parallel and to have a choice from four different washing solvents, the La Marast synthesizer is constructed with three four-port distribution Teflon valves (**Fig. 3**). Four reaction vessels are connected, via Teflon tubing, with the four ports of the V1 valve. The common port of the V1 valve is connected to the common port of the four-port V2 valve. The port on the right-hand side is connected to the evacuated waste container. An empty syringe is attached to the upper port of the V2 valve. The lower port is left empty. The left-hand side port is connected to the common port of the V3 valve. The four ports of the V3 valve are connected to four reservoirs with solvents. All connections and wetted parts of valves are made of Teflon. The disposable reaction vessels are made of polypropylene. If dictated by a chemical protocol, Teflon reaction vessels could be used.

2.1.2. Description of Operation

The V1 valve is connected to the first reaction vessel. At this point, the valve V2 is connected to the reagent port. The V2 valve is moved to the position open to the waste container and the vessel is emptied/evacuated. The solvent is selected using the V3 valve, the V2 valve is connected to the V3 valve, and solvent starts to fill the vessel. When a sufficient volume of solvent is introduced into the vessel, the valve V2 is connected to the air/gas port and the pressure in the reaction vessel is equili-

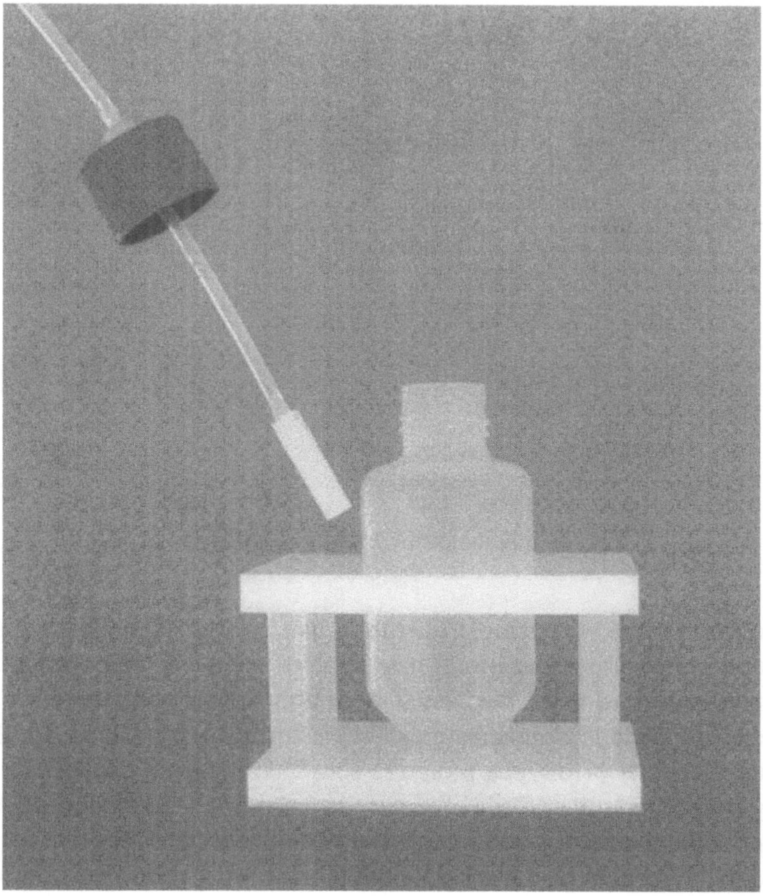

Fig. 2. The La Marast reaction vessel.

brated with the atmosphere. A balloon filled with an inert gas is connected to the air/gas port if there is a need to wash the resin under an inert atmosphere.

After washing and/or conditioning the resin, the solution of the reagent(s) is added from the reagent port using the reagent syringe. The vessel is evacuated, the solution is transferred into the reagent syringe, and the V2 valve is connected to the reagent syringe. Alternatively, the reaction vessel could be opened and the reagent manually introduced into the vessel. The reaction vessels are typically left shaking during the reaction time on the plate shaker. Alternatively, they may be disconnected from the V1 valve, the outlet closed by a male Luer plug, and the vessels shaken on another shaker. The next set of four vessels could then be connected to the synthesizer for resin washing/ conditioning.

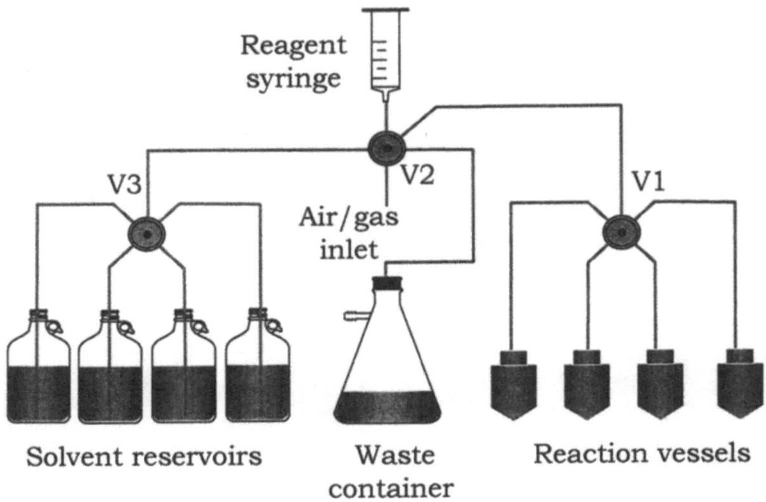

Fig. 3. Scheme of the La Marast synthesizer.

In addition to its very simple operation, the benefit of the La Marast (and also Domino Block) synthesizer is its complete enclosure. The contents of the reaction vessels need never be opened to the atmosphere, and, therefore, chemistry requiring an inert atmosphere can be performed. Dry solvent for the conditioning of resins or oxygen/moisture-sensitive reagents can be introduced directly from a septum-sealed container via a septum needle connected to any "Solvent" line. Alternatively, an enclosed syringe with the sensitive solution is coupled to the reagent port and the solution is introduced into reaction vessels.

2.2. Domino Block Synthesizer

2.2.1. Description of the Synthesizer

A "Domino Block" is a liquid distribution manifold that is made of Teflon and has a footprint of a 96-well plate (**Fig. 4**). The Domino Block has two functions: (i) to clamp (hold) reaction vessels and (ii) to connect all reaction vessels to one common port. This common port is used to introduce and remove solvent. Plastic polypropylene syringes, equipped with a porous disk (porosity 60 μm) *(3)* at the bottom of the syringe barrel, are used as reaction vessels. The syringes, charged with resin, are attached via a male Luer fitting to the female Luer-lock fittings of the Domino Block. Domino Blocks for 6 (2 rows of 3 syringes), 12 (4 rows of 3 syringes), and 24 (6 rows of 4 syringes) reaction vessels are available (**Fig. 5**). The 6-block accommodates syringe sizes up to 50 mL; the 12-block, syringes sizes up to 20 mL and the 24-block is useful for 3- and 5-mL syringes.

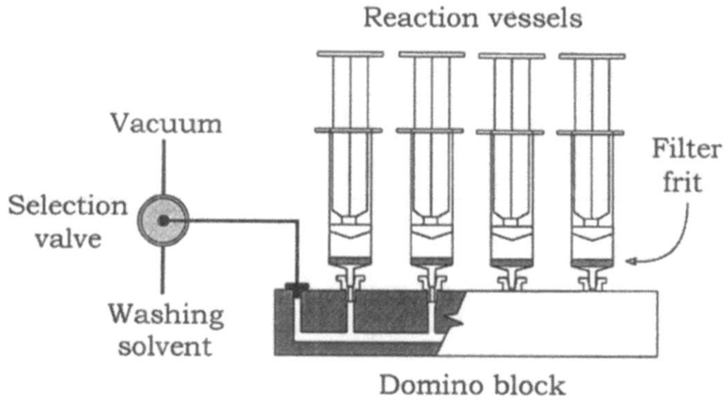

Fig. 4. Schematic drawing of a Domino Block.

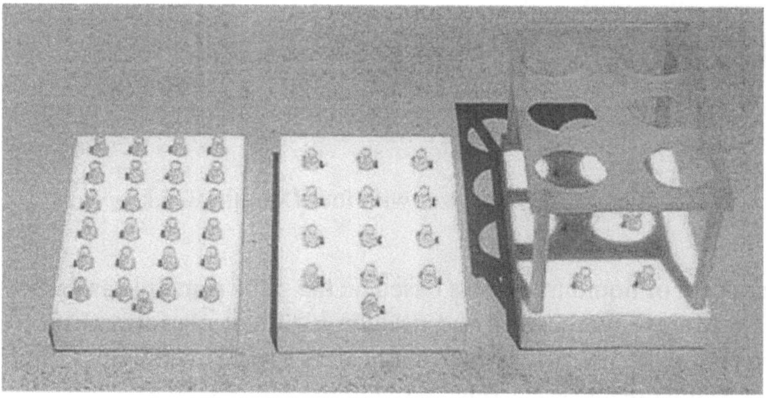

Fig. 5. Domino Blocks for 24, 12, and 6 reaction vessels. The 6-reaction vessel block is shown with a syringe holder.

Table 2 illustrates the five different syringe sizes that are practical for use with the Domino Block synthesizer and the typical maximum resin load for each syringe. Polypropylene syringes can be used at elevated temperatures (up to 90°C), sonicated, or even used in a microwave irradiation apparatus. However, polypropylene syringes do not tolerate prolonged (overnight) exposure to certain solvents, including methylene chloride, tetrahydrofuran, and dioxane. The syringe barrel tends to swell and the plunger then does not provide a leak-proof seal. In this case, the resin can be transferred to a new syringe or the plunger can be moved to a position that was previously not in contact with the solvent (enlarging the working volume).

Table 2
Typical Resin Load Per Syringe

Syringe volume	3 mL	5 mL	10 mL	20 mL	50 mL
Quantity of resin	100 mg	300 mg	500 mg	1 g	3 g

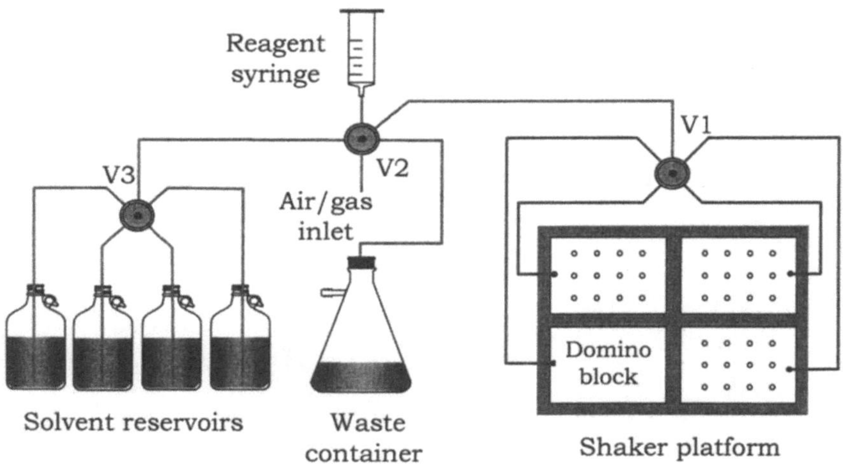

Fig. 6. A synthesizer with four Domino Blocks.

The concept of liquid transfer is based on the same principle as that described for the La Marast synthesizer and uses an identical platform (**Fig. 6**). Obviously, having the same platform, a synthesizer can be used with a combination of the La Marast reaction vessels and Domino Blocks.

2.2.2. Description of Operation

Domino Blocks, fitted with syringes containing resin and solvent, are placed on a titer plate shaker (e.g., Lab-Line, Dubuque, Iowa; www.barnsteadthermolyne.com) and connected to the appropriate port of the V1 valve (**Figs. 6** and **7**, *see* **Notes 1–3**). Domino Blocks are typically shaken during the entire washing cycle. The distribution valve, V2, is turned to the "Waste" port to connect the Domino Block and the reaction vessels to an evacuated waste container (*see* **Note 4**). After the syringes are emptied, which typically takes less than 10 s, the valve is turned to the "Air" port for about 1 s. A small amount of air is drawn into the syringes via this port, which is equipped with a syringe filter (the only reason for passing the air through the filter is to limit the airflow and to make the introduction of the correct amount of air easier). The amount of air depends on the number and size of the syringes on the Domino Block. As a rule of thumb,

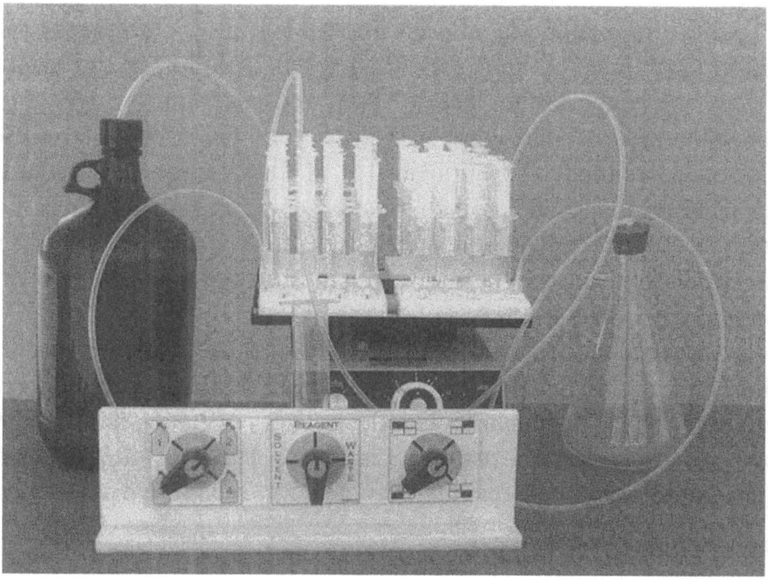

Fig. 7. The Domino Blocks synthesizer.

one third of the actual syringe volume filled with air secures fast and reliable operation. A balloon filled with an inert gas can be connected to the “Air” port if an inert atmosphere is required.

The valve V2 is then turned to the “Solvent” port to connect the Domino Block with the solvent container. The reaction vessels are filled with the solvent. An alternative way of introducing an air gap into syringes can be used. After the syringes are evacuated, the valve is quickly turned to the “Solvent” port. Before the syringes are completely filled with the solvent, the valve is turned to the “Reagent” port. Air enters the syringes via the “Reagent” port. It is important to always introduce enough air into the reaction vessels. The next evacuation cycle relies on expansion of the air present in reaction vessels. If required, instead of air, an inert gas can be introduced, e.g., from a balloon filled with nitrogen attached to the “Air” port.

During the addition and removal of liquid, the Domino Blocks are shaken to equilibrate the resin beads with the fresh solvent. While the solvent is exchanged in the reaction vessels of one Domino Block, the resin in the other three blocks is shaken with the washing solvent. It is necessary that the resin beads be exposed to the fresh solvent for at least 1 min to allow diffusion of soluble compounds out of the beads. (This time depends largely on the bead size and increases dramatically with the bead diameter, *see* **Notes 5–7**).

To introduce a common reagent (e.g., a solution of activated N-protected amino acid during peptide synthesis) into all syringes, the reagent syringe is attached to the "Reagent" port and filled with a solution of the reagent. The valve V2 is connected to the "Waste" port, and, after the syringes are evacuated, the valve is switched to the "Reagent " port. The syringes are filled with the solution of reagent. The uniformity of distribution among individual syringes is better than ±20%. Since most organic reactions on the solid phase use a high excess of reagents, the distribution error is acceptable. When accurate reagent distribution is required, syringes are charged with the appropriate solution manually on a one by one basis.

In order to receive different reagent solutions, syringes are disconnected from the Domino Block after the last wash and reagent solutions are drawn into individual syringes manually. Gentle, but very efficient, mixing during the reactions may be achieved using a Labquake Tube Rotator (Thermolyne, Dubuque, Iowa; www.barnsteadthermolyne.com, *see* **Note 8**).

2.3. Don Cucna Synthesizer

The customary way to separate resin beads from a liquid is filtration through a porous material, usually a glass or plastic frit. One problem associated with this type of filtration technique is the potential for clogging of the porous material. Although this problem may not be critical when a small number of reaction vessels are handled at the same time, increasing the number of vessels considerably enhances the risk of ruining a synthesis due to the clogging of a filter. The technique described below circumvents the problems associated with use of filters during solid phase synthesis in a 96-well format.

2.3.1. Description of the Synthesizer

A method has been developed that enables solid phase organic synthesis to be performed in microtiter wells not equipped with any kind of porous material at the bottom to facilitate the separation of solid resin beads from a solvent. The concept of washing resin beads in the Don Cucna synthesizer was developed by the need for a reliable and fast operational cycle applicable to a hundred reaction vessels at the same time. The simplest compact reaction block for solid phase synthesis is the 96-well plate. The suction (aspirating) principle of the Don Cucna synthesizer is based on the fact that in most solvents used in solid phase synthesis the resin beads settle to the bottom of the wells of the plates. The settling of the resin is relatively fast (tens of seconds). After the resin beads have settled, stainless-steel needles connected to an evacuated waste container are slowly immersed into the wells of a plate (**Fig. 8**). The needles remove the liquid from above the surface of the resin without disturbing the resin bed. For washing the resin beads in 96-well microtiter plates, two

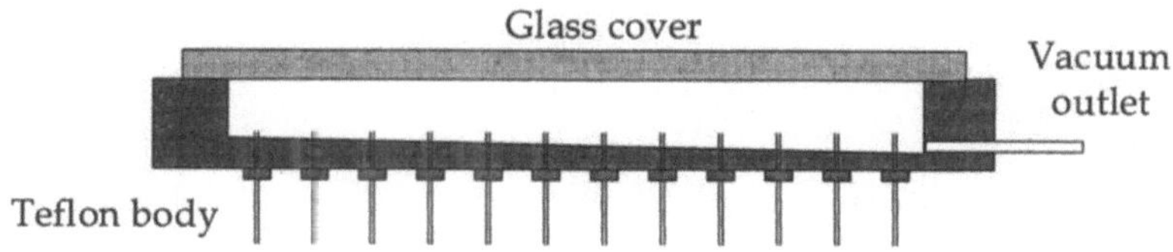

Fig. 8. Scheme of the aspirator.

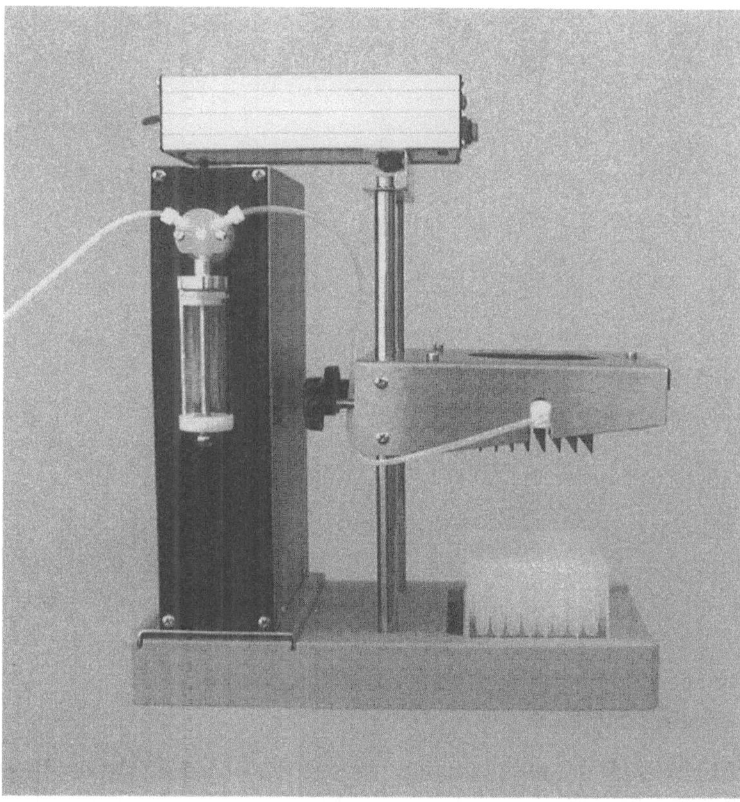

Fig. 9. The 96-well dispenser.

manually operated dedicated tools were designed: the dispenser for liquid delivery (**Fig. 9**) and the aspirator for removing the liquid (**Fig. 10**).

Each washing cycle is very fast and is limited by the time necessary to equilibrate the resin beads with the fresh washing solvent. The dispensing and aspiration of solvent takes only seconds. Typically a syringe pump delivers the solvent; however, the liquid can be delivered from a pressurized solvent container or manually by repeatable syringe. When a small

Fig. 10. The 96-well aspirator.

amount of resin (5–10 mg) is used, the stream of liquid mixes the resin. If a large amount of resin is used, external mixing (e.g., shaking the plate on a titer plate shaker) is required.

The disadvantages of the aspiration concept are: (i) less efficient washing, as some solvent is left among the settled resin beads, which typically requires more washes than when compared to a filtration washing protocol; (ii) solvent or solvent mixtures used for washing have to have a lower density than the resin beads. The considerable advantages are: (i) very fast and reliable washing; (ii) there is no frit clogging. Resin beads that do not completely settle may, in the worst case, destroy the synthesis in one reaction vessel (well), not the entire batch.

2.3.2. Description of Operation

The resin beads can be distributed into 96-well plates either in a dry state or as slurry. To distribute the same resin (or resin-bound intermediate) into all 96-wells of plate, one can use the FlexChem Resin Loader (Robin Scientific, Sunnyvale, CA, www.robsci.com) for dry resin. Different resin-bound intermediates can be distributed manually into individual rows or columns of the plate as slurry in a suitable solvent. The following is a method suitable for plating eight different resin-bound intermediates into eight rows of a microtiter plate.

In order to distribute 20 mg of resin beads into each well of one row, a slurry of the resin-bound intermediate (240 mg) in DMF (2.5 mL) is made in a trough (Matrix, Lowell, MA). The slurry is mixed using a 12-channel pipet, equipped with wide orifice tips, by rapidly pipeting the suspension in and out of the tips several times to make the suspension of resin homogeneous. Then 200 μL of the slurry is drawn into the large orifice tips and transferred into wells of the first row of a deep-well plate (Matrix, Lowell, MA). In order to maximize resin transfer, an additional 2.5 mL of DMF is added to the small slurry volume remaining in the trough and then transferred into the same deep well plate (400 μL total volume of resin slurry in each well). The above process is then repeated until the remaining seven resin-bound intermediates have been distributed into the appropriate rows of the 96-well microtiter plate.

The excess liquid in the plate is removed using the aspirator. To remove the solvent, the plate is placed on the platform of the aspirating device and the manifold is slowly lowered. The end point is adjusted such that the needles nearly touch the resin surface (0.5–1 mm distance). The manifold is then lifted to its upright position. If required, the resin is preconditioned (washed) with the solvent to be used in the subsequent chemical transformation. The plate is placed under the 96-well dispensing device and the solvent is delivered into the wells (25 mL per plate) using a syringe pump (Cavro, Sunnyvale, CA, www.cavro.com). The stream of solvent is strong enough to mix the resin beads with the incoming solvent. After the resin settles, usually taking less than 1 min, the solvent is aspirated and the washing cycle repeated.

The next step is the distribution of building blocks and reagents. Twelve stock solutions of building blocks are prepared and an aliquot (approx 100–500 μL) is distributed manually into wells using a 12-channel pipet. The plates are closed by 96-well cap mats (Matrix, Lowell, MA) and shaken on an orbital shaker. Alternatively, the plates are sealed by an automatic plate sealer (Abgene, Surrey, UK, www.abgene.com).

After the chemical transformation is finished, the plates are washed with solvents using the 96-channel aspirator and dispenser, and the resin is then air-dried. Cleavage of the target compounds from the solid support is usually

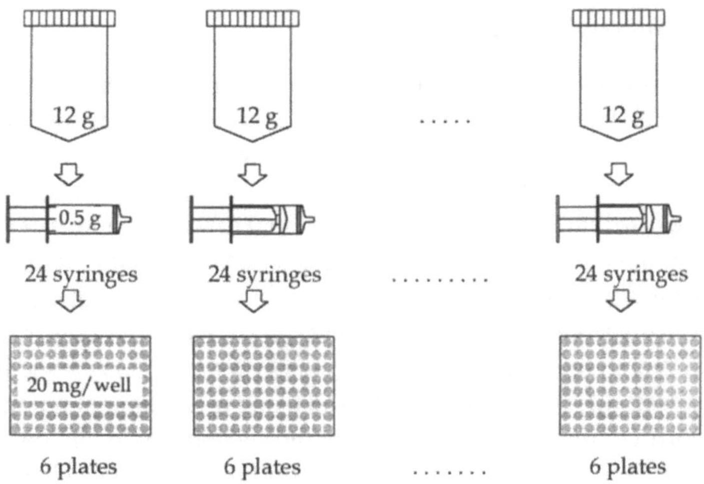

Fig. 11. Combinatorial solid phase synthesis using the split-split concept.

performed by a liquid cocktail, TFA being the most often used reagent (*see* also Chapter 5). The cleavage reagents are removed and the resulting target compounds extracted with a suitable solvent. Solutions of the extracted compounds are filtered using a filter plate (Orochem Technologies, Westmont, IL, www.orochem.com) and the resulting extracts are evaporated using a SpeedVac (Savant, Holbrook, NY, www.savec.com).

2.4. Split–Split Concept

Synthesis of combinatorial arrays of compounds can, in principle, be performed three different ways. The "Split and Mix " (split and pool/recombine) concept introduced by Furka ***(4–6)*** is the most efficient method for preparation of sizable libraries (tens of thousands of compounds). However, the technique requires tracking of the chemical history of the resin beads and limits the quantity of synthesized material to the loading per solid phase unit (one unit can be represented by one particle, bead, lantern, etc., or one container, T-bag, Kan, etc.). Parallel synthesis, on the other hand, can supply any quantity per compound, but it requires handling large numbers of reaction vessels at one time.

The hybrid approach, the "Split–Split" (also referred to as a "split only") concept ***(2,7–9)*** can be described as a parallel synthesis with a reduced number of reaction vessels (**Fig. 11**). Synthesis of sizable combinatorial libraries by the split–split method requires handling large numbers of reaction vessels only in the last combinatorial step. The synthesis starts with x reaction vessels, where x equals the number building blocks used in the first combinatorial steps. The next combinatorial step uses $x \times y$ reaction vessels; y is the number of building

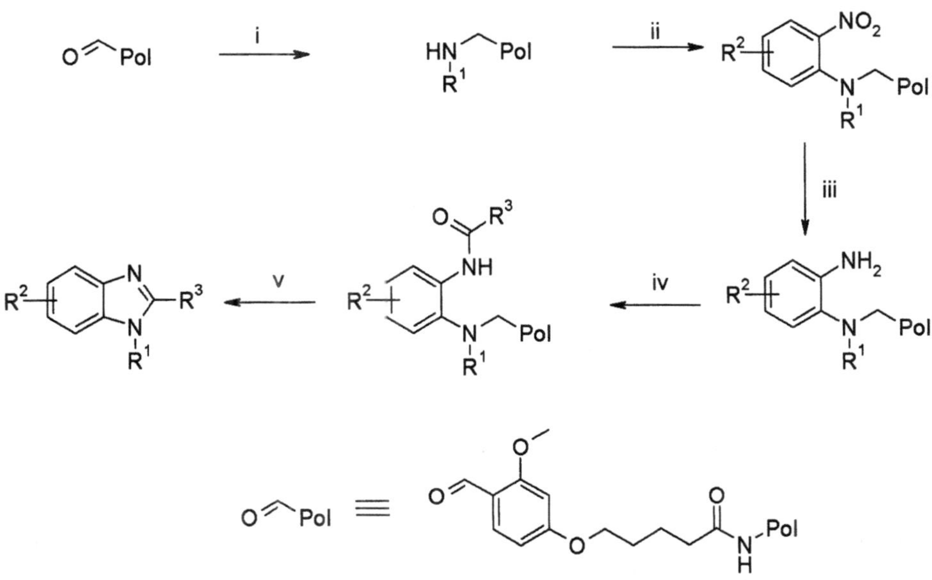

Scheme 1. A solid phase traceless synthesis of benzimidazoles. Reagents: (i) amine/$NaBH(AcO)_3$ in DMF/AcOH; (ii) *o*-fluoronitrobenzene, DMSO, rt, overnight; (iii) $SnCl_2 \cdot 2H_2O$ in NMP, rt, overnight; (iv) acid chloride/DIEA in DCM, rt, overnight; (v) AcOH, 80°C, overnight.

blocks in the second combinatorial step. The number of reaction vessels in the third combinatorial step is equal to the number of compounds synthesized in the library.

The principle of the split–split method is illustrated by an example of a three combinatorial step benzimidazole library that uses 24 amines, 12 *o*-fluoronitrobenzenes, and 48 acids, producing a library of 13,824 compounds (**Scheme 1**) *(10)*. The first step can be performed in 24 La Marast reaction vessels, each containing 12 g of resin. For the second combinatorial step, resin from each La Marast reaction vessel is split into 24 syringes and the Domino Block synthesizer is used to handle the syringes. The last combinatorial step is performed in wells of 96-well plates, with the resin from each syringe being split into two rows (24 wells) of a plate.

Plating resin beads from syringes into 96-well plates follows an algorithm that allows manual distribution of building blocks. It is impractical to distribute one resin bound intermediate into half of a plate for reaction with 48 acids. A more convenient way of plating beads with different resin-bound intermediates is illustrated on **Fig. 12**. Each row of a plate receives a different intermediate.

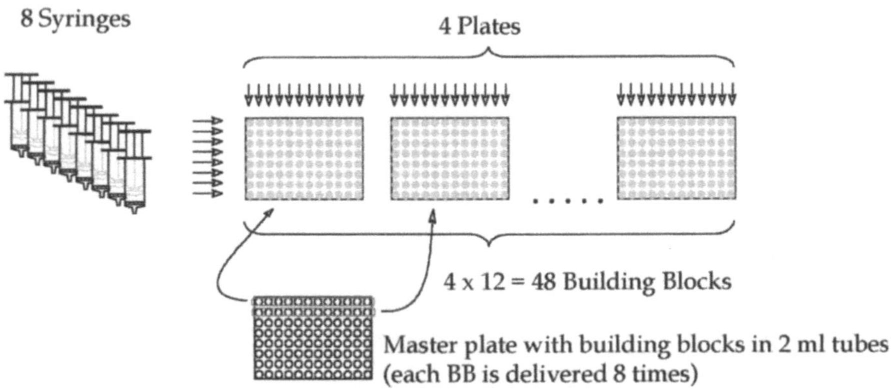

Fig. 12. Plating resin beads for manual distribution of building blocks

Consequently, the building blocks can be distributed, one per column, using a 12-channel pipet (*see* **Notes 9** and **10**).

3. Conclusion

Parallel solid phase organic synthesis has become an integral part of many laboratories involved in organic synthesis. Consequently, reliable, simple, and affordable tools are needed. There is no simple solution as to what instrumentation is best suited to perform organic solid phase synthesis. The instrumentation is a tool and the selection of the most appropriate tool should be dictated by the project needs. Three manual synthesizers are described in this chapter that are convenient for performing the three synthetic scenarios presented. For many laboratories and projects, the Domino Block synthesizers satisfy most requirements. La Marast synthesizers fulfill the need for larger research quantities and can be easily integrated with Domino Blocks. The Don Cucna apparatus allows expansion of the compound throughput to up to a thousand compounds per day in milligram quantities.

4. Notes

1. Combinatorial solid phase synthesis using Domino Blocks can be performed in a simple manner. One Domino Block integrates syringes that undergo the same chemical transformation in the first combinatorial step. After finishing the first step, syringes containing different resin-bound intermediates that receive the same reagents in the second combinatorial step are connected to one Domino Block.
2. The volume of reaction vessels is adjustable by moving the syringe plunger. Consequently, different amounts of resin can be placed into individual syringes on the same block. Different sizes of syringes can also be attached to the same block;

they will always be filled, as the volume of the evacuated reaction vessel determines the volume of liquid introduced.

3. It is not necessary to work with a fully loaded Domino Block. The remaining positions on a Domino Block can be closed using male Luer plugs.
4. The vacuum typically does not move the plunger and there is no need to fix the position of the plunger. However, the syringe must be wetted by solvent before use in order to restrict plunger movement.
5. Syringes can be left shaking on Domino Blocks during the course of a reaction, however, we close the syringes with syringe pressure caps for longer reaction times (overnight), or if the reaction requires an elevated temperature (capped syringes can be shaken in an incubator at up to 90°C, depending on the reaction solvent). Domino Blocks do not need to be engaged exclusively with a particular synthetic batch for the whole synthesis, nested synthetic batches may be handled.
6. Syringes without plungers can also be used on Domino Blocks. To wash the resin, the washing solvent is added from a squeeze bottle into all of the syringes and then the Domino Block is connected to the evacuated waste container. This technique is particularly useful when a reagent is added, or a gas liberated, during a reaction.
7. When the liquid from one reaction vessel is removed, a vacuum is created inside the particular reaction vessel. There is no airflow through the resin and the vacuum is fully engaged for filtering a liquid through a slower (partially clogged) filter.
8. The La Marast and Domino Block synthesizers share the same principle and platform. As a result of this, Domino Blocks can be combined with the La Marast reaction vessels on the same shaker, sharing the same valve system. Any port of the V1 valve can be connected either to a Domino Block or to a La Marast reaction vessel (compare **Figs. 3** and **6**).
9. One aspect worth mentioning is quality control (QC) during a library synthesis. Large combinatorial compound arrays synthesized in parallel fashion increase the amount of quality control required to analyze all of the intermediates produced during the synthesis. The split–split approach dramatically decreases the number of intermediates during the synthesis and therefore the purity of all intermediates can generally be determined.
10. The orthogonal distribution of resin-bound intermediates and building blocks for the last combinatorial step allows a simple way of increasing the confidence that the major peak, observed on HPLC analysis of each well, corresponds to the expected product. A line graph is constructed correlating the column identifier with the retention time observed for the major peak of each well in a particular row of a 96-well plate (**Fig. 13**). A similar graph is constructed for each of the remaining rows of the same plate. By comparing the trend observed across a given row with those of the other seven rows of the plate, a correlation or general pattern can, in most cases, be observed. If all compounds from each intermediate follow the same trend, they may be considered to be the expected products. This is not an

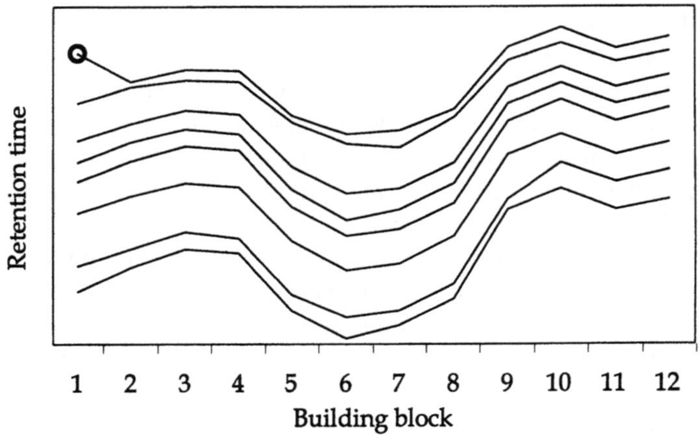

Fig. 13. Trend of retention times for eight resin-bound intermediates reacted with 12 building blocks. The compound marked by a circle is not the expected product. Its retention time does not follow the common trend.

absolute proof, however, the probability that a certain combination of building blocks form an unexpected product with the retention time that corresponds to the expected product and, at the same time, shows the expected molecular ion on MS analysis, is relatively low.

References

1. Krchňák, V. and Padera, V. (1998). The domino blocks: a simple solution for parallel solid phase organic synthesis. *Bioorg. Med. Chem. Lett.* **22,** 3261–3264.
2. Krchňák, V., Weichsel, A. S., Lebl, M., and Felder, S. (1997). Automated solid-phase organic synthesis in micro-plate wells. Synthesis of N-(alkoxy-acyl)amino alcohols. *Bioorg. Med. Chem. Lett.* **7,** 1013–1016.
3. Krchňák, V., Vagner, J., and Mach, O. (1989). Multiple continuous-flow solid-phase peptide synthesis. Synthesis of an HIV antigenic peptide and its omission analogues. *Int. J. Pept. Protein Res.* **33,** 209–213.
4. Furka, A., Sebestyen, F., Asgedom, M., and Dibó, G. (1988). More peptides by less labour. (Poster Presentation). *10th International Symposium on Medicinal Chemistry (Budapest),* p. 288.
5. Furka, A., Sebestyen, F., Asgedom, M., and Dibó, G. (1988). Cornucopia of peptides by synthesis. (Poster Presentation). *14th International Congress of Biochemistry (Prague),* **5,** p. 47.
6. Furka, A., Sebestyen, F., Asgedom, M., and Dibó, G. (1991). General method for rapid synthesis of multicomponent peptide mixtures. *Int. J. Pept. Protein Res.* **37,** 487–493.
7. Krchňák, V. (1998). Semi-automated high throughput combinatorial solid-phase organic synthesis. *Biotechnol. Bioeng. (Comb. Chem.)* **61,** 135–141.

8. Brooking, P., Doran, A., Grimsey, P., Hird, N. W., MacLachlan, W. S., and Vimal, M. (1999). Split-Split. A multiple synthesiser approach to efficient automated parallel synthesis. *Tetrahedron Lett.* **40,** 1405–1408.
9. Lebl, M., Krchňák, V., Ibrahim, G., et al. (1999). Solid-phase synthesis of large tetrahydroisoquinolinone arrays by two different approaches. *Synthesis* 1971–1978.
10. Smith, J. and Krchňák, V. (1999). A solid phase traceless synthesis of benzimidazoles with three combinatorial steps. *Tetrahedron Lett.* **40,** 7633–7636.

5

Cleavage of Compounds from Solid Phase by Gaseous Reagents

Viktor Krchňák

1. Introduction

The recent interest in organic solid phase synthesis, triggered by the advent of chemical combinatorial methods *(1–7)*, also accelerated methodology development. Simplification of chemical protocols, their robustness, and amenability to handling large arrays of compounds, prepared by combinatorial/parallel solid phase synthesis, is one area that witnessed numerous novel contributions. This chapter describes an apparatus and method for gaseous cleavage of compounds from solid phase supports.

At the end of solid phase organic synthesis, products are cleaved from the insoluble solid support. Depending on the type of linker, various reagents have been employed to enable the release of synthesized compounds. Excellent reviews have recently appeared in the literature that describe linkers and cleavage procedures *(8,9)*. Statistically, more than 60% of recent solid phase organic synthesis publications used acid-cleavable linkers, Wang *(10)* and Rink linkers *(11)* being the most commonly reported *(9)*. Nucleophiles were used to cleave compounds from linkers in 28% of cases. Typical cleavage protocols for Wang and Rink linkers involve the use of liquid cleavage cocktails, trifluoroacetic acid (TFA) is the reagent of choice. Obviously, the cleavage cocktail has to be added to all reaction vessels and, after cleavage, the cocktail has to be removed. Synthesis of large compound arrays therefore requires a time consuming operation handling strongly acidic reagents that are not user friendly.

An alternative method for cleaving compounds from resin beads is to treat the solid support with gaseous reagents. Bray et al. *(12)* were the first to describe the use of a gaseous reagent, ammonia, to cleave an ester bond of peptides synthesized on pins. The method was later adapted for production-scale cleavage of

From: *Methods in Molecular Biology, Combinatorial Library Methods and Protocols*
Edited by: L. B. English © Humana Press Inc., Totowa, NJ

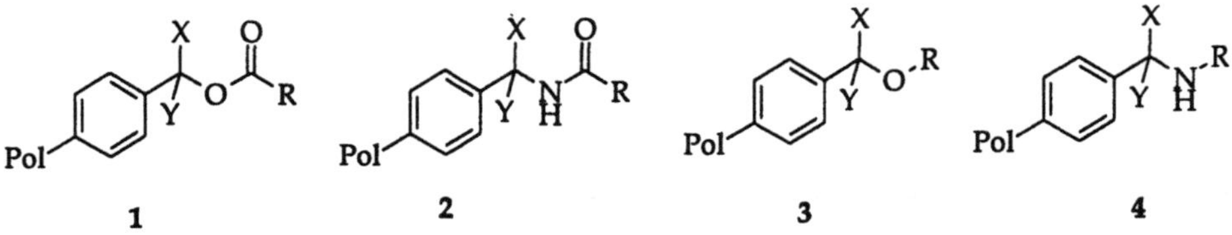

Series a: X = Y = H (benzyl)
Series b: X = H, Y = Ph (benzhydryl)
Series c: X = Y = Ph (trityl)

Fig. 1. Acid cleavable benzyl-type linkers (substitution at CH_2).

peptides from Merrifield resin *(13)*. Ammonia gas under pressure has also been described for deprotection and cleavage steps during the large-scale synthesis of oligonucleotides *(14)*.

The use of acidic gaseous reagents, HCl and TFA, for cleavage of acids, alcohols, and amines attached to trityl linker functionalized supports has been reported *(15)*. Jayawickreme et al. *(16)* used TFA for gradual cleavage of compounds from a solid support. The use of gaseous hydrogen fluoride (HF) has also been described for the release of compounds from the *p*-methylbenzhydrylamine (MBHA) *(17–20)* and dialkoxybenzylamine *(21–23)* linkers.

2. Materials and Methods

2.1. Acid-Labile Linkers

The most common linkers used in contemporary solid phase organic synthesis are acid-labile *(9)*. The majority of acid labile linkers can be viewed as a variation on the benzyl-type linkage (**Fig. 1**). Benzyl esters of type **1a** provide carboxylic acids upon cleavage from the resin, benzyl amides **2a** are cleaved to carboxamides, benzyl ethers **3a** provide alcohols, and benzylamines **4a** yield amines. Chloromethylated copoly(styrene-divinylbenzene) resin, referred to as Merrifield resin *(24)*, may be used to immobilize carboxylic acids via a benzyl ester linker **1a**. Attachments involving unsubstituted benzyl-type linkers, where X and Y are hydrogens, require harsh deprotection conditions, usually liquid HF, even in the case of the most labile type of compounds, the esters **1a**. Subsequently, the acid lability of linkers was increased by substitution at the CH_2 group (**Fig. 1**) and/or at the aromatic ring (**Fig. 2**). Replacing the X hydrogen with a phenyl or 4-methylphenyl group forms a benzhydryl linker **2b** *(25)* or *p*-methylbenzhydryl linker *(26)*, respectively. These linkers are widely used to prepare carboxamides, however, they still require HF cleavage. The trityl linker *(27,28)* replaces both X and Y with a phenyl ring, and derivatives of this type

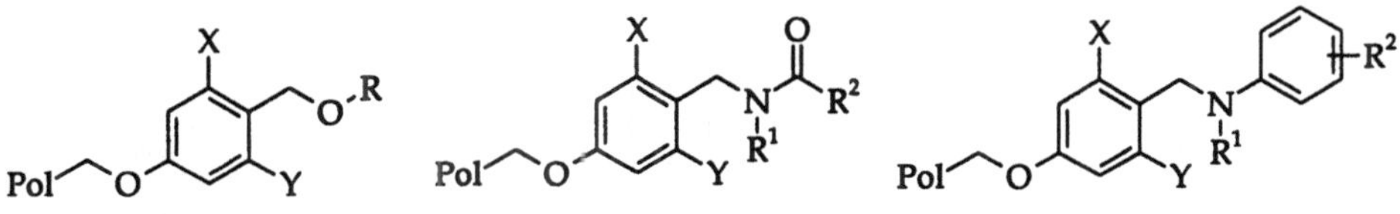

5 6 7

Series a: X = Y = H
Series b: X = H, Y = OMe
Series c: X = Y = OMe

Fig. 2. Acid cleavable benzyl-type linkers (ring substitution).

represent the most acid-labile linkers reported. It has been used to immobilize acids (**1c**), alcohols (**3c**), thiols, and amines (**4c**).

The ease of linker cleavage can be further fine-tuned by substitutions on the aromatic ring *(8,9)*. An electron-donating alkoxy group, as in linker **5**, increases acid lability and is used in Wang (**5a**) *(10)* and Sheppard *(29)* linkers. Additional methoxy groups further increase acid lability, as documented for Sasrin (**5b**) *(30)* or PAL (**5c**) *(31)* linkers. The linker **6b** having X = OMe and Y = H, referred to as AMEBA (acid-sensitive methoxybenzaldehyde) and linker **6c** *(32)*, X = Y = OMe, referred to as BAL (backbone amide linker) **6c** *(33)* allow target compounds to be cleaved by TFA. The acyl group was replaced by an aromatic ring in linker **7**. This facilitates cleavage by TFA and provides a route to nitrogen-containing heterocyclic compounds *(21,23,34)*.

2.2. Linkers Cleaved by Gaseous Reagents

Liquid TFA and HF are typical cleavage reagents for acid-cleavable linkers. Alternatively, cleavage may be effected by the use of gaseous HCl or HF. The efficiency of gaseous HCl cleavage can be compared to diluted TFA; the effectiveness of gaseous HF is comparable to liquid HF.

Among the commonly used acid-cleavable linkers, the esters of Merrifield resin **8** and carboxamides of the MBHA linker **9** (**Fig. 3**) are the most acid stabile. Gaseous HF was required to cleave acids from the Merrifield resin *(20)* and amides from the MBHA resin *(18,19,35)*. The acyl group illustrated on the MBHA linker **9** may be replaced with an aromatic group to form the resin-bound aniline **10** that can also be cleaved by gaseous HF. Cleavage of the C–N bond in **10** was used for a traceless synthesis of some nitrogen-containing heterocyclic compounds *(22)*.

The acid lability of both the Merrifield esters **8** and MBHA amides **10** was increased by introducing electron-donating groups on the aromatic ring of the linker. Thus the Wang resin **11** *(36)* and derivatized BHA linkers **12** *(17)* and

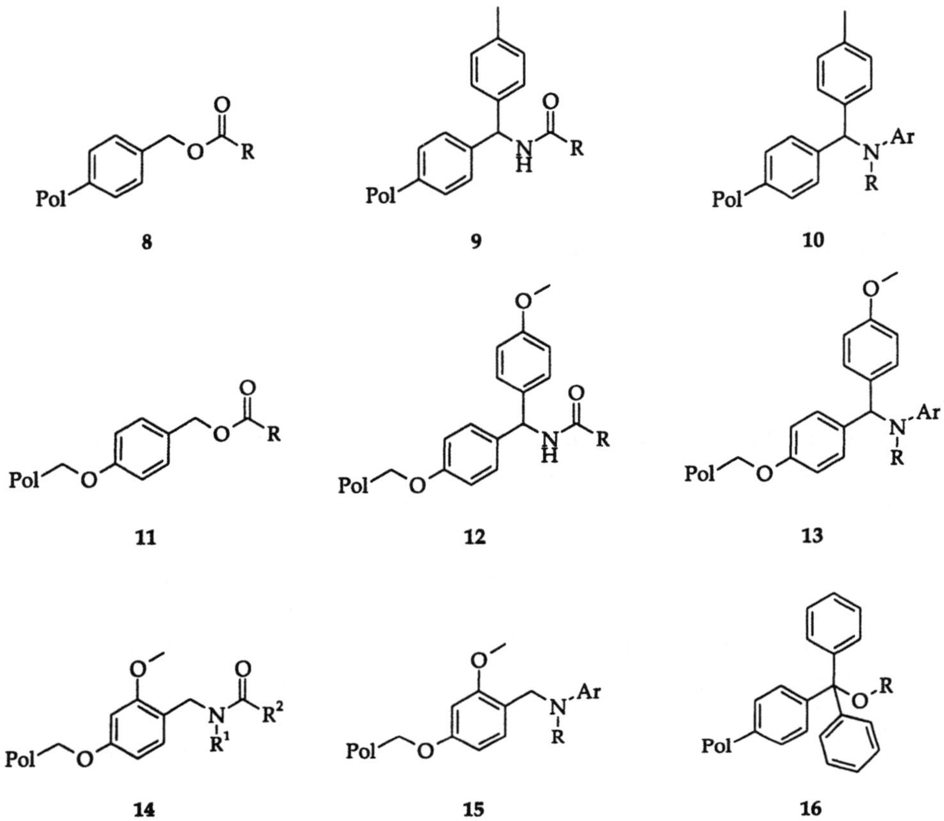

Fig. 3. Structure of derivatized linkers.

13 *(22)*, were cleavable by TFA and gaseous HCl. However, gaseous HCl treatment did not completely release model products from the resin **11**, and HF treatment was necessary *(37)*. The dialkoxybenzylamine linker **14**, originally developed for the synthesis of amides *(32,33,38,39)*, was found to be useful for synthesis of various target compounds, in particular, the synthesis of nitrogen-containing heterocyclic compounds (linker **15**). The first building block was attached to the aldehyde-derivatized resin using a reductive amination protocol (Scheme 1). At this stage, the bond between nitrogen and the methylene carbon of the benzyl group is stable to acid. Acylation (**14**) or arylation (**15**) of the nitrogen greatly reduces the acid stability and enables the target compounds to be cleaved by acidic reagents, including gaseous HCl *(21,23,37)*. Several target compounds that are cleavable from the dialkoxybenzylamine linker by gaseous HCl or HF are shown in the **Fig. 4**. The highly acid labile trityl linker **(16)** releases alcohols after treatment with HCl gas or TFA vapors **(15)**.

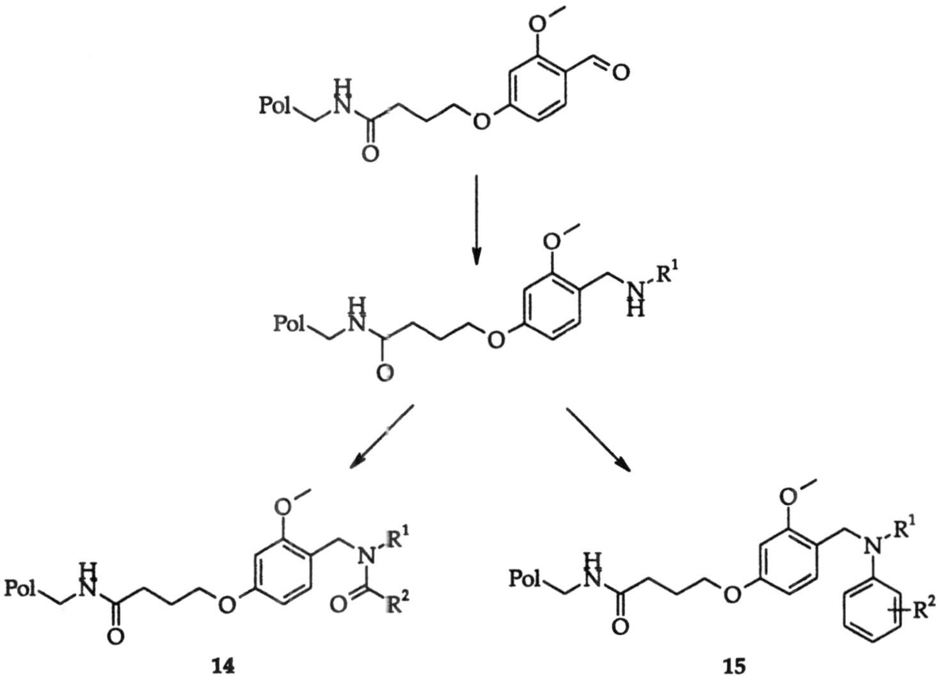

Scheme 1. Derivatization of an aldehyde linker.

2.3. Apparatus for Gaseous Cleavage

For reliable and safe cleavage of compounds from solid supports, a simple and user friendly apparatus was designed and built *(37)*. The apparatus consists of a reagent gas container (**Fig. 5**), where the resin-bound compounds are exposed to the reagent gas, and a control panel (**Fig. 6**) with valves and a series of four trap bottles for absorbing the reagent gas after the cleavage is finished (Torviq, Tucson, AZ, www.torviq.com).

The reaction gas container for gaseous cleavage is constructed from a polypropylene tube, enclosed by two polypropylene side-plates, and uses polypropylene foam as a seal (**Fig. 7**). The two side-plates also accommodate the inlet and outlet connectors for the reaction gas container. Two tightening screws are attached to the centers of the side-plates and allow both side-plates to firmly press against the tube and generate airtight seals. The entire reaction gas container is placed in a stainless-steel cage. The polypropylene tube is removed from the cage to load and unload reaction vessels containing resin-bound compounds (**Fig. 8**). The lengths of the tubes are 720 and 200 mm, the diameters are 143 mm. The larger tube accommodates thirteen 96-well 1.2 mL

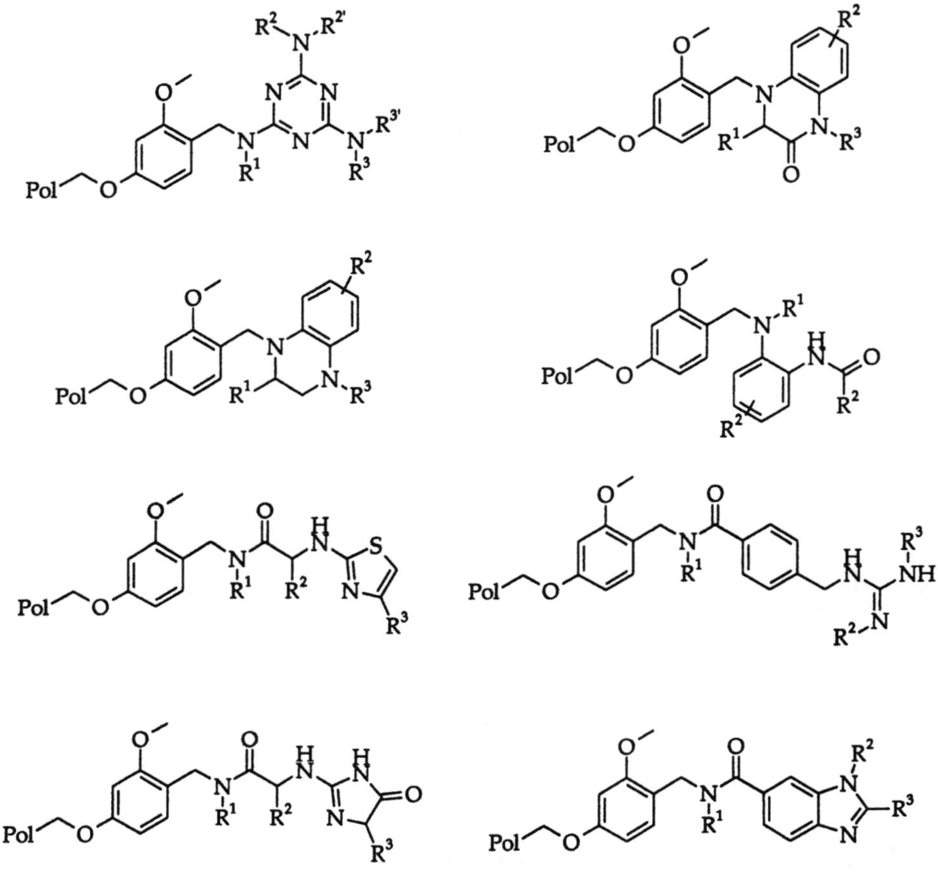

Fig. 4. Examples of compounds synthesized on the dialkoxybenzylamine linker that have been cleaved from the linker with HCl or HF gas.

plates, the smaller tube can be charged with three plates. Plates are arranged in two layers separated by a rigid polypropylene sheet.

The connections of the reaction gas container to the source of the gas, and to the trap bottles, are shown on **Fig. 9** (an arrangement with two reaction gas containers is shown). Wetted parts of the valves are made of Teflon, and 1/4" Teflon tubing is used for connections. Two ports of the three-port valve V1 are connected the source of an inert gas (nitrogen) and a reagent gas (HF, HCl, or NH_3). The third port is linked to the four-port valve V2. The outlet and inlet of the reaction gas container are connected to the neighboring ports of the four-port valve V2. A second four-port valve is used for the arrangement shown in **Fig. 9** that uses two reaction gas containers. A check valve is attached to the outlet of the reaction gas container. The last port is connected to the four-port

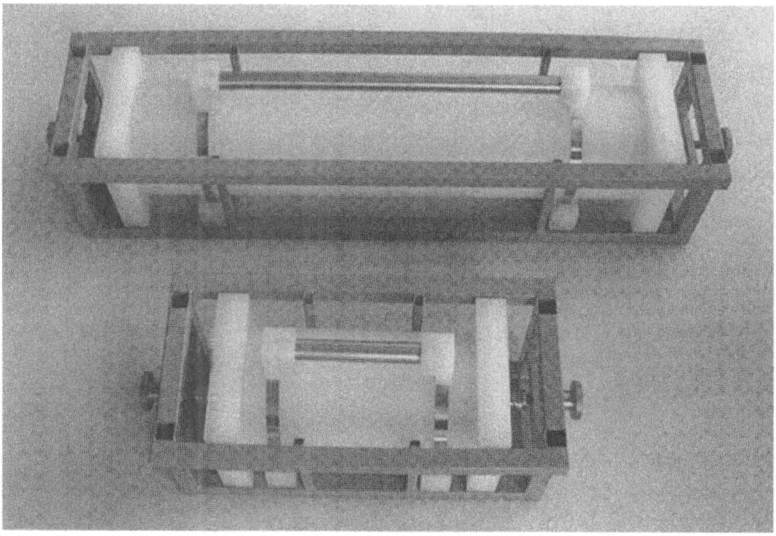

Fig. 5. Apparatus for gaseous cleavage.

valve V3. The V3 valve is connected to a vacuum pump. The suction port of the pump is connected to the V3 valve via the four-port valve V4. The remaining ports of the V4 valve are connected to the atmosphere. The last port of the V3 valve is connected to the first trap bottle. There are four 1-L polypropylene trap bottles, the first one empty, the next two filled with 0.5 L of water, and the last filled with 0.5 L of 10 m*M* aqueous NaOH solution and an acid–base indicator.

2.4. The Gas Cleavage Procedure

2.4.1. Sample Preparation

The resin-bound products are placed into either polypropylene 96-well plates or plastic syringes (with a frit and without a plunger). The reaction vessels (plates, syringes) are etched with a needle, razor blade, or engraver and all labels are removed. Hydrogen fluoride gas (*see* **Note 1**) destroys most labels made of paper or tape. The labels are not removed when using hydrogen chloride as a cleaving reagent.

2.4.2. Loading and Sealing the Reaction Gas Container

Syringes are placed in a rack or some other means of support and loaded into the reaction gas container (*see* **Notes 2** and **3**). The reaction gas container is sealed using polypropylene foam seals. The two seals are cut from polyethylene foam liner (VWR, Phoenix, AZ) with sides being 19.5 cm long. A hole

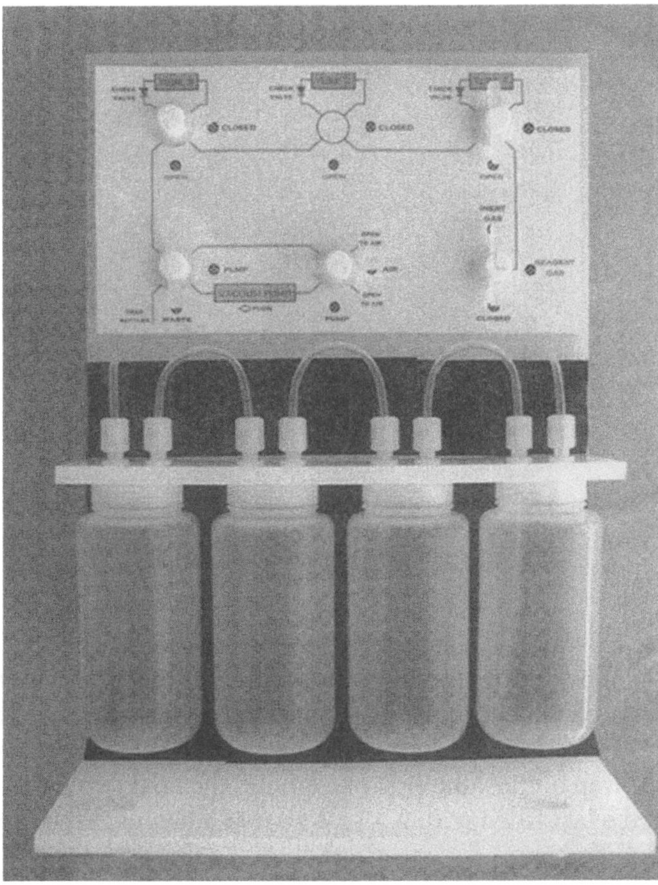

Fig. 6. Control panel.

with an approx 7 cm internal diameter is cut in the middle of each foam seal (the hole is absolutely necessary). Seals are placed on the ends of the tube and the tightening screw turned until a tight seal is achieved. Re-using foam seals is not recommended (*see* **Notes 4** and **5**).

2.4.3. Evacuation of the Reaction Gas Container

Since no vacuum gauge is used during the evacuation, the course of evacuation is monitored by the disappearance of bubbles in the trap bottles. Air bubbles through the traps very quickly initially. After a few minutes, as the rate of bubbling slows, the seals on the apparatus are re-tighten. Both the reaction gas container and the nitrogen input line are evacuated in this step (*see* **Notes 6–9**).

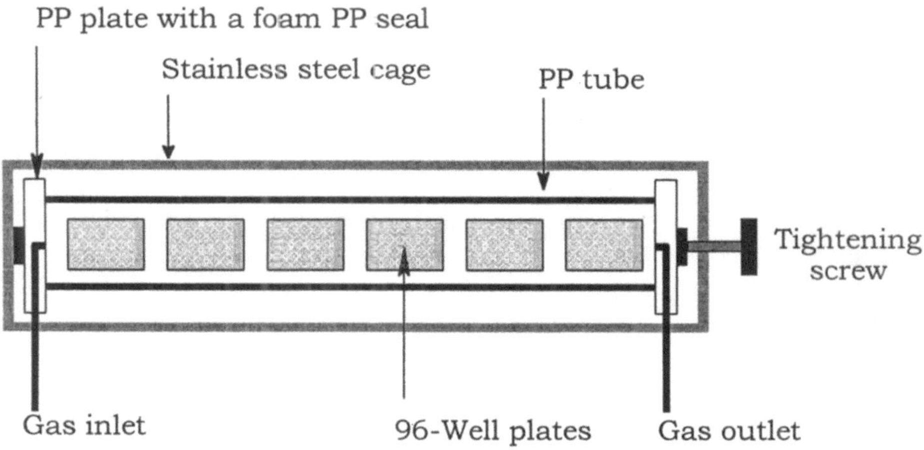

Fig. 7. Schematic drawing of the reaction gas container. PP, polypropylene.

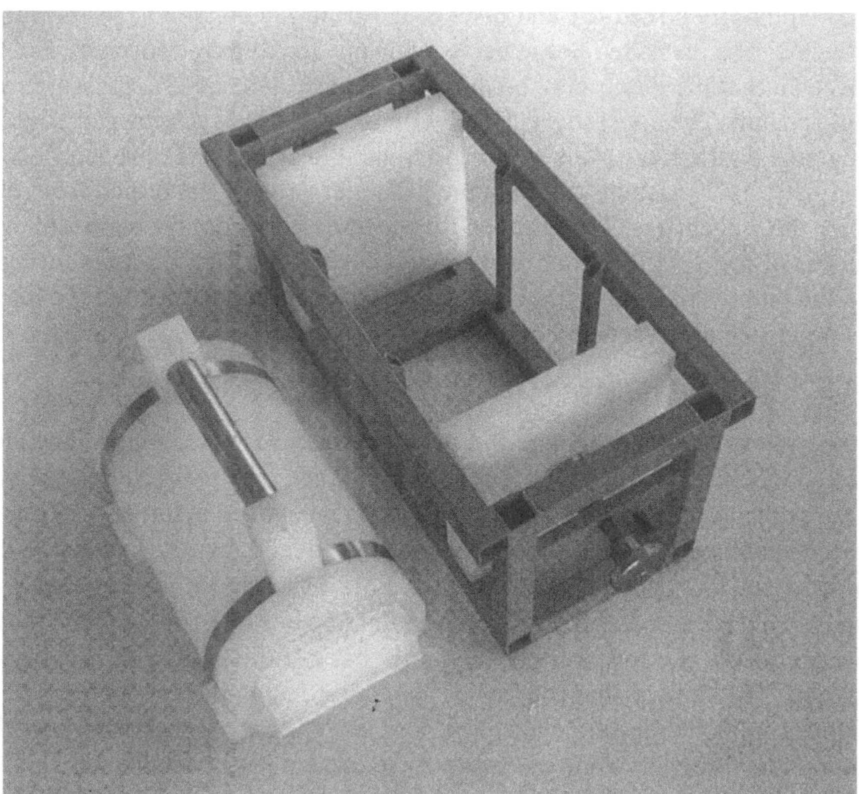

Fig 8. Loading the reaction gas container.

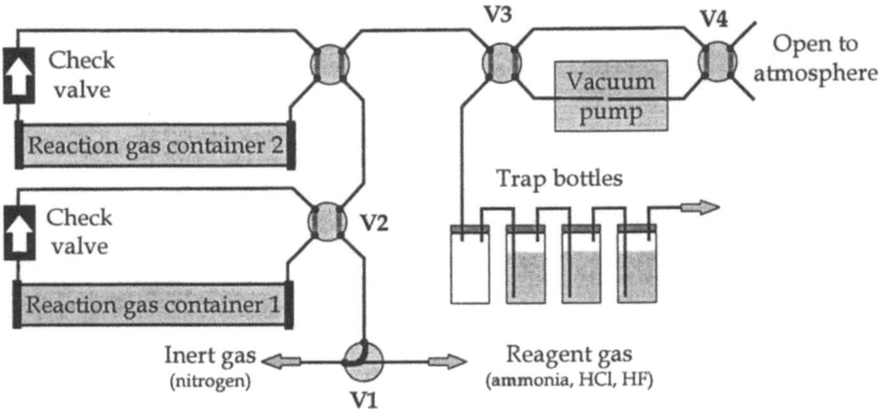

Fig. 9. Scheme of the apparatus used for gaseous cleavage.

2.4.4. Filling the Reaction Gas Container with Nitrogen and Evacuation

The pump is turned off and disengaged from the apparatus by turning the valve V3. The pressure release valve V4 is opened to the atmosphere and then closed so that the pump may be restarted against atmospheric pressure rather than vacuum. The apparatus is filled with inert gas such as nitrogen. Once the apparatus is filled, the traps will begin to bubble again. The nitrogen flow rate is adjusted to 2–3 L/min and then the gas input valve V1 is turned from nitrogen to the reagent gas. The nitrogen flow is turned off from the main tank while there is nitrogen still in the input line (*see* **Note 10**). The pump is turned on and the pump valve V3 is engaged. The reaction gas container and the reagent gas input line are evacuated in this step.

2.4.5. Filling the Reaction Gas Container with Reagent Gas

The pump is turned off and the pump valve V3 is disengaged. The check valve prevents air from coming into the container. The pressure release valve to the pump, V4, is opened and closed. The reagent gas is turned on from the main reagent gas tank and the reagent gas introduced into the evacuated reaction gas container via the valve V1 (*see* **Note 11**). It usually takes only a few minutes before bubbles start to appear in the trap bottles. This is a sign that the pressure inside the reaction gas container has equilibrated with atmospheric pressure. The flow of reagent gas is slowed but maintained at such a rate that bubbles continue to appear in the traps. This introduction of gas is maintained for 3 min. After 3 min of additional exposure to reagent gas, the container valve V2 is turned to disconnect the container from the flow of reagent gas. Reagent gas is now flowing directly into the traps and by-passing the container.

2.4.6. Flushing Lines with Nitrogen

The reagent gas remaining in the line has to be flushed out using nitrogen. The flow of nitrogen is turned on (flow rate: 2–3 L/min) from the main valve while reagent gas continues to flow. The gas selection valve V1 is turned from reagent gas to nitrogen and the flow of reagent gas is turned off at the source of the reagent gas. Reagent gases typically dissolve very quickly in water. If the flow of reagent gas were to be turned off without an inert gas to replace it, a vacuum would be formed in the traps and back-flow could occur. After a few minutes of flushing, the gas selection valve is turned to reagent gas and then the flow of nitrogen is turned off. This leaves the line under a slight nitrogen pressure.

2.4.7. Cleavage of Compounds

The compounds are exposed to reagent gas for a time necessary to cleave the resin-bound compounds (in most cases approx 2 h). During a hydrogen chloride cleavage, additional reagent gas is introduced into the container after 1 h of exposure. Hydrogen chloride is turned on to flow through the traps at about 2–3 L/min. The container valve V2 is turned to fill the container. Bubbling initially stops in the traps until atmospheric pressure is exceeded in the container. Once bubbling resumes, the flow is adjusted to a slower flow rate without stopping the bubbling, and this flow rate continues for 3 min. The container valve V2 is turned to disconnect the gas flow from the chamber. The flow of nitrogen is turned on and adjusted to 2–3 L/min. The gas selection valve V1 is turned to nitrogen and the reagent gas flow is turned off. After a few minutes of flushing, the flow of nitrogen is turned off.

2.4.8. Flushing the Container with Nitrogen

The flow of nitrogen is turned on at 2–3 L/min and the container is flushed for 30 min. The volume of nitrogen necessary to sufficiently flush the reaction gas container is dependent upon the size of the reaction gas container. The larger reaction gas container (approx 12 L volume) contains less than 0.6% residual reagent gas after flushing with nitrogen for 30 min at 2 L/min *(37)*. The container is evacuated and then refilled with nitrogen, before opening the container (*see* **Note 12**). Since the reaction vessels and compounds exposed to HF still contain residual adsorbed HF, the plates/syringes are left in the reaction gas container together with a dish of NaOH and evacuated (*see* **Notes 13–15**).

3. Conclusions

Cleavage of compounds from solid supports by gaseous reagents represents an alternative approach to the commonly used liquid cleavage cocktails. To safely perform the gaseous cleavage, an apparatus for gaseous cleavage of compounds from solid supports was described. The commercially available reaction gas container (Torviq, Tucson, AZ, www.torviq.com) provides secure airtight sealing and can be evacuated, ensuring fast and dependable filling with a reagent gas. Gaseous cleavage provides easy addition and removal of cleavage reagents. The advantages of gaseous cleavage are appreciated particularly during synthesis of large arrays of compounds or when the cleavage is performed on a routine basis.

4. Notes

1. HF is an extremely hazardous gas and appropriate care and proper handling is absolutely essential. All operations have to be done in a dedicated hood. Venting of reagent gas and disposal of concentrated acid waste are the primary safety concerns with HF gas cleavage. Gaseous HF very efficiently dissolves in water and three trap bottles filled with water securely adsorb the HF. The last bottle contains acid–base indicator to ensure that no acidic contaminant is leaving the traps. Procedures developed for disposal of the concentrated acid waste generated during cleavage have been described in detail elsewhere *(37)*.
2. The apparatus must be operated by trained personnel only.
3. The apparatus has to be installed in a well-vented hood.
4. The apparatus must not be pressurized with any gas.
5. The apparatus can be used at temperatures ranging from 25 to 35°C.
6. The material in contact with the gas is Teflon, polyethylene, and polypropylene. Gas not compatible with those materials must not be used in the apparatus.
7. The tightness of connectors to the reagent vessel and trap bottles must be checked before use.
8. The hood-sash has to be moved down before evacuating the container.
9. New polyethylene foam liner seals have to be used each time a cleavage process is performed to secure an airtight seal of the container.
10. The reagent gas and nitrogen have to be turned off at the source (cylinder), not by the V1 valve.
11. A mask, with filter for acidic vapors, and heavy duty gloves must be worn when working with HF or HCl.
12. Do not allow any part of the apparatus or of the vessels with HF-treated samples to come into contact with bare skin.
13. The valves have to be turned in the order described in the operating protocol.
14. Both side-plates and tube ends have to be kept clean and smooth. Scratches or contamination by particles (dust, resin beads, etc.) may compromise the airtight seal.
15. It is recommended to have calcium gluconate gel (Calgonate Corp., Rhode Island) on hand to treat accidental skin exposure to HF.

References

1. Frank, R., Heikens, W., Heisterberg-Moutsis, G., and Blocker, H. (1983) A new general approach for the simultaneous chemical synthesis of of large number of oligonucleotides. Segmental solid support. *Nucl. Acids Res.* **11,** 4365–4377.

2. Geysen, M. H., Meloen, R. H., and Barteling, S. J. (1984) Use of peptide synthesis to probe viral antigens for epitopes to a resolution of a single amino acid. *Proc. Natl. Acad. Sci. USA* **81,** 3998–4002.
3. Geysen, H. M., Barteling, S. J., and Meloen, R. H. (1985) Small peptides induced antibody with a sequence and structural requirements for binding antigen comparable to antibodies raised against the native protein. *Proc. Natl. Acad. Sci. USA* **82,** 178.
4. Frank, R. and Döring, R. (1988) Simultaneous multiple peptide synthesis under continuous flow conditions on cellulose paper discs as segmental solid supports. *Tetrahedron Lett.* **29,** 6031–6040.
5. Furka, A., Sebestyen, F., Asgedom, M., and Dibó, G. (1991) General method for rapid synthesis of multicomponent peptide mixtures. *Int. J. Pept. Protein Res.* **37,** 487–493.
6. Houghten, R. A., Pinilla, C., Blondelle, S. E., Appel, J. R., Dooley, C. T., and Cuervo, J. H. (1991) Generation and Use of Synthetic Peptide Combinatorial Libraries for Basic Research and Drug Discovery. *Nature* **354,** 84–86.
7. Lam, K. S., Salmon, S. E., Hersh, E. M., Hruby, V. J., Kazmierski, W. M., and Knapp, R. J. (1991) A new type of synthetic peptide library for identifying ligand-binding activity. *Nature* **354,** 82–84.
8. James, I. W. (1999) Linkers for solid phase organic synthesis. *Tetrahedron* **55,** 4855–4946.
9. Guillier, F., Orain, D., and Bradley, M. (2000) Linkers and cleavage strategies in solid-phase organic synthesis and combinatorial chemsitry. *Chem. Rev.* **100,** 2091–2157.
10. Wang, S. S. (1973) p-Alkoxybenzyl alcohol resin and p-alkoxybenzyloxycarbonylhydrazide resin for solid phase synthesis of protected peptide fragments. *J. Am. Chem. Soc.* **95,** 1328–1333.
11. Rink, H. (1987) Solid-phase synthesis of protected peptide fragments using a trialkoxydiphenyl-methylester resin. *Tetrahedron Lett.* **28,** 3787–3790.
12. Bray, A. M., Maeji, N. J., Jhingran, A. G., and Valerio, R. M. (1991) Gas Phase Cleavage of Peptides from a Solid Support with Ammonia Vapour—Application in Simultaneous Multiple Peptide Synthesis. *Tetrahedron Lett.* **32,** 6163–6166.
13. Flegel, M., Rinnova, M., Panek, Z., Lepsa, L., and Blaha, I. (1996) *Peptides: Chemistry,Structure and Biology, Proc.14.APS* (Kaumaya, P. T. P. and Hodges, R. S. eds.), Mayflower Scientific Ltd. Kingswinford, pp. 119–120
14. Iyer, R. P., Yu, D., Xie, J., Zhou, W., and Agrawal, S. (1997) The use of gaseous ammonia for the deprotection and cleavage steps during the solid-phase synthesis of oligonucleotides, and analogs. *Bioorg. Med. Chem. Lett.* **7,** 1443–1448.
15. Krchňák, V. and Weichsel, A. S. (1997) Polymer supported synthesis of diverse perhydro-1,4-diazepine-2, 5-diones. *Tetrahedron Lett.* **38,** 7299–7302.
16. Jayawickreme, C. K., Graminski, G. F., Quillan, J. M., and Lerner, M. R. (1994) Creation and Functional Screening of a Multi-Use Peptide Library. *Proc. Natl. Acad. Sci. USA* **91,** 1614–1618.

17. Smith, J., Gard, J., Cummings, W., Kaniszai, A., and Krchňák, V. (1999) Necklace-coded polymer-supported combinatorial synthesis of 2-arylaminobenzimidazoles. *J. Comb. Chem.* **1,** 368–370.
18. Lebl, M., Krchňák, V., Ibrahim, G., et al. (1999) Solid-phase synthesis of large tetrahydroisoquinolinone arrays by two different approaches. *Synthesis* 1971–1978.
19. Hebert, N., Hannah, A. L., and Sutton, S. C. (1999) Synthesis of oxadiazoles on solid support. *Tetrahedron Lett.* **40,** 8547–8550.
20. Lebl, M., Pires, J., Poncar, P., and Pokorny, V. (1999) Evaluation of Gaseous Hydrogen Fluoride as a Convenient Reagent for Parallel Cleavage from the Solid Support. *J. Comb. Chem.* **1,** 474–479.
21. Krchňák, V., Szabo, L., and Vagner, J. (2000) A solid phase traceless synthesis of quinoxalinones. *Tetrahedron Lett.* **41,** 2835–2838.
22. Krchňák, V., Smith, J., and Vagner, J. (2001) A solid phase traceless synthesis of 2-arylaminobenzimidazoles. *Tetrahedron Lett.* **42,** 1627–1630.
23. Krchňák, V., Smith, J., and Vagner, J. (2001) A solid phase traceless synthesis of tetrahydroquinoxalines. *Tetrahedron Lett.* **42,** 2443–2446.
24. Merrifield, R. B. (1963) Solid phase peptide synthesis. I. The synthesis of a tetrapeptide. *J. Am. Chem. Soc.* **85,** 2149–2154.
25. Tam, J. P., DiMarchi, R. D., and Merrifield, R. B. (1981) Design and synthesis of a multi-detachable benzhydrylamine resin for solid-phase peptide synthesis. *Tetrahedron Lett.* **22,** 2851–2854.
26. Matsueda, G. R. and Stewart, J. M. (1981) A *p*-methylbenzhydrylamine resin for improved solid-phase synthesis of peptide amides. *Peptides* **2,** 45–50.
27. Barlos, K., Gatos, D., Kallitsis, I., Papaioannou, D., and Sotiriou, P. (1988) Application of 4-polystyryltriphenylmethyl chloride to the syntheses of peptides and amino acid derivatives. *Liebigs Ann. Chem.* **1988,** 1079–1081.
28. Barlos, K., Chatzi, O., Gatos, D., and Stavropoulos, G. (1991) 2-Chlorotrityl chloride resin—Studies on Anchoring of Fmoc-Amino Acids and Peptide Cleavage. *Int. J. Pept. Protein Res.* **37,** 513–520.
29. Sheppard, R. C. and Williams, B. J. (1982) Acid-labile resin linkage agents for use in solid phase peptide synthesis. *Int. J. Pept. Protein Res.* **20,** 451–454.
30. Mergler, M. (1988) Peptide synthesis by combinaiton of solid-phase and solution methods: A new very acid-labile anchor group for the solid phase synthesis of fully protected fragments. *Tetrahedron Lett.* **29,** 4005
31. Albericio, F., Kneibcordonier, N., Biancalana, S., et al. (1990) Preparation and Application of the 5-(4-(9-Fluorenylmethyloxycarbonyl)aminomethyl-3,5-Dimethoxyphenoxy) valeric acid (Pal) handle for the solid-phase synthesis of C-terminal peptide amides under mild conditions. *J. Org. Chem.* **55,** 3730–3743.
32. Fivush, A. M. and Willson, T. M. (1997) AMEBA: An acid sensitive aldehyde resin for solid phase synthesis. *Tetrahedron Lett.* **38,** 7151–7154.
33. Jensen, K. J., Alsina, J., Songster, M. F., Vagner, J., Albericio, F., and Barany, G. (1998) Backbone amide linker (BAL) strategy for solid-phase synthesis of C-terminal modified and cyclic petides. *J. Am. Chem. Soc.* **120,** 5441–5452.

34. Smith, J. and Krchňák, V. (1999) A solid phase traceless synthesis of benzimidazoles with three combinatorial steps. *Tetrahedron Lett.* **40,** 7633–7636.
35. Lebl, M. and Krchňák, V. (1998) *Innovation & Perspectives in Solid Phase Synthesis & Combinatorial Libraries* (Epton, R. ed.), Mayflower Scientific Limited, Birmingham, UK, pp. 43–46.
36. Krchňák, V. (1998) Semi-automated high throughput combinatorial solid-phase organic synthesis. *Biotechnol. Bioeng. (Comb. Chem.)* **61,** 135–141.
37. Kerschen, A., Kaniszai, A., Botros, I., and Krchňák, V. (1999) Apparatus and method for cleavage of compounds from solid support by gaseous reagents. *J. Comb. Chem.* **1,** 480–484.
38. Boojamra, C. G., Burow, K. M., and Ellman, J. A. (1995) An expedient and high-yielding method for the solid-phase synthesis of diverse 1,4-benzodiazepine-2,5-diones. *J. Org. Chem.* **60,** 5742–5743.
39. Jensen, K. J., Songster, M. F., Vagner, J., Alsina, J., Albericio, F., and Barany, G. (1996) *Peptides: Chemistry, Structure and Biology* (Kaumaya, P. T. P. and Hodges, R. S. eds.), Mayflower Scientific Ltd. Birmingham, UK, pp. 30, 31.

6

Synthesis of DNA-Binding Polyamides

Robust Solid-Phase Methods for Coupling Heterocyclic Aromatic Amino Acids

Peter O. Krutzik and A. Richard Chamberlin

1. Introduction

Small molecules that recognize double-stranded DNA have the capacity to modulate various cellular processes, including DNA replication and repair, gene expression, cell cycle regulation, and growth, and therefore may serve as treatments for cancers or various genetic diseases. Currently, most DNA-binding therapeutics target short sequences of DNA with rather low selectivity, causing deleterious effects in both diseased and healthy cells *(1)*. Thus, the development of molecules that target longer, more cell-type-specific sequences of DNA is of great interest. In this chapter, a rapid solid-phase synthesis of hairpin polyamides, a promising class of minor groove binding agents developed by the Dervan group at Caltech, is described *(2)*. The robust coupling methods of aromatic amino acids outlined facilitate the application of combinatorial methods to polyamides and their further development as pharmaceutical reagents.

Hairpin polyamides comprised of N-methylpyrrole (Py) and N-methylimidazole (Im) aromatic amino acids bind in the minor groove of DNA and possess high DNA affinity, sequence specificity, target length, and apparent cell permeability *(3–7)*. Polyamides are one class of a series of molecules based on the naturally occurring compounds distamycin and netropsin, including lexitropsins and microgonotropens *(8–10)*. However, when synthesized with a flexible linker to allow hairpin formation, polyamides are able to read DNA by the side-by-side pairing of Py and Im monomers such that an Im-Py pair targets G-C, Py-Im targets C-G, and the Py-Py pair is degenerate for both A-T

From: *Methods in Molecular Biology, Combinatorial Library Methods and Protocols*
Edited by: L. B. English © Humana Press Inc., Totowa, NJ

and T-A base pairs *(2)*. Recently, polyamides were utilized in vivo to block HIV-1 replication in human lymphocytes *(11)*, transcription of the 5S RNA gene in cultured *Xenopus* kidney cells *(12)*, and to obtain gain- and loss-of-function phenotypes in the fruit fly, *Drosophila melanogaster (13,14)*. Conjugates of polyamides with the DNA alkylators chlorambucil and duocarmycin A have also been prepared and shown to have extremely high affinity and specificity for their target sequences *(15,16)*. With the potential uses of polyamides constantly expanding, a major limitation to the development of these molecules has been their somewhat laborious synthesis.

Solution-phase syntheses have been employed in the preparation of polyamides, but these typically allow only one coupling step to be performed per day *(16,17)*. Although fragment coupling methods in solution have made polyamides available in large quantities, they are limited in the number of different polyamides that can be produced *(18)*. Solid-phase syntheses using standard peptide chemistry have recently become more widespread and also provide the advantages inherent to resin-based chemistry, i.e., easy step-wise purification, high molar equivalents of reagents that drive reactions to completion, and high-throughput split-mix or parallel synthesis. *Tert*-butoxycarbonyl (Boc) and fluorenylmethoxycarbonyl (Fmoc) protection strategies in combination with benzotriazole (OBt) activation have both been utilized. The Fmoc-based syntheses are advantageous in their amenability to further peptide synthesis, and the wider availability of base-stable specialty resins and Fmoc-based peptide synthesizers *(19,20)*. However, the long coupling times necessary (3 h) make these methods laborious. Boc-OBt based strategies are more rapid, but still require more than an hour per cycle *(21)*, with products that often necessitate repeated high performance liquid chromatography (HPLC) purification to remove resin aminolysis reagents and side-products *(22)*. In addition, particular couplings require either double coupling or extended reaction times *(21,23)*.

This chapter describes an optimized synthesis of polyamides based on Boc protection and azabenzotriazole (OAt) activation that solves the problems of sluggish reactions and difficult purification (**Fig. 1**). It is divided into two sections: (1) the solid-phase synthesis of polyamides, and (2) resin aminolysis, purification, and characterization of the final product. Section 1 details the OAt-mediated couplings using either *O*-(7-azabenzotriazol-1-yl)-1,1,3,3-tetramethyluronium hexafluorophosphate (HATU) or dicyclohexylcarbodiimide (DCC)/1-hydroxy-7-azabenzotriazole (HOAt) *(24–28)*. Diisopropylethylamine (DIEA) is used both to catalyze the formation of the OAt ester when employing HATU and to deprotonate the amine-trifluoroacetate salt formed after Boc deprotection. With OAt activation, coupling time is reduced to 20 min per step, making nine residue polya-

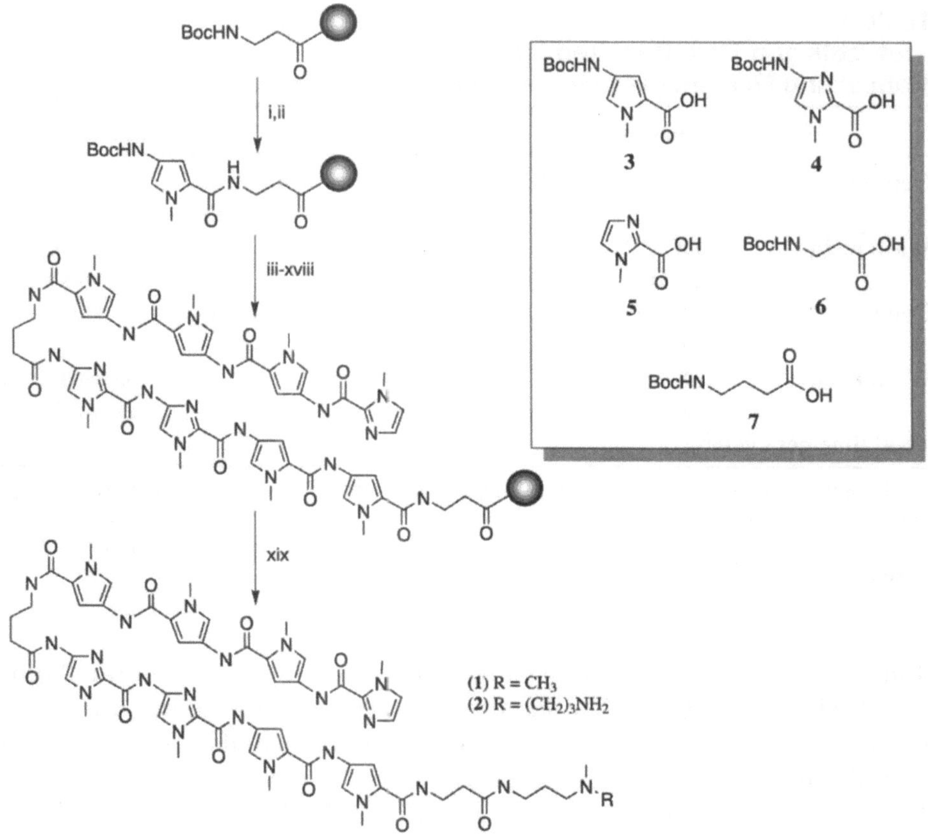

Fig. 1. Solid-phase synthesis of polyamides. Outlined is the synthesis of Im-Py-Py-Py-γ-Im-Im-Py-Py-β-Dp *(1)* and its amine derivative, Im-Py-Py-Py-γ-Im-Im-Py-Py-β-BAPMA *(2)* on Boc-β-Ala-Pam resin: (i) TPW mixture (92.5% TFA, 5% phenol, 2.5% water); (ii) Boc-Py-OH, HATU, DIEA; (iii) TPW; (iv) Boc-Py-OH, HATU, DIEA; (v) TPW; (vi) Boc-Im-OH, DCC/HOAt, DIEA; (vii) TPW, extended; (viii) Boc-Im-OH, DCC/HOAt, DIEA; (ix) TPW, extended; (x) Boc-γ-Abu-OH, HATU, DIEA; (xi) TPW; (xii) Boc-Py-OH, HATU, DIEA; (xiii) TPW; (xiv) Boc-Py-OH, HATU, DIEA; (xv) TPW; (xvi) Boc-Py-OH, HATU, DIEA; (xvii) TPW; (xviii) Im-OH, DCC/HOAt, DIEA; (xix) DMPA for **1** or BAPMA for **2**, 55°C, 14–16 h. (Inset) Monomers for solid phase synthesis: Boc-Py-OH *(3)*, Boc-Im-OH *(4)*, Im-OH *(5)*, Boc-β-Ala-OH *(6)*, Boc-γ-Abu-OH *(7)*.

mides obtainable in approx 5 h (**Table 1**). Naturally occurring amino acids may be substituted at any step in the preparation of peptide-polyamide libraries to further develop their DNA binding properties. Also, to promote

Table 1
Protocols and Time Required for Manual and Machine-Assisted Solid-Phase Synthesis

	Reagent	Manual	Machine-Assisted
Deprotect[a]	TFA, phenol, water (92.5:5:2.5)	1 min mix, 2 × 2 min mix	
Wash resin	DCM	30 s flow	2× wash
	DMF	1 min flow	3× wash
Couple	OAt ester, DIEA, DMF/NMP	20 min mix	
Wash resin	DMF	1 min flow	3× wash
	DCM	30 s flow	2× wash
Total time per cycle[b]		28 min	35 min

[a]Deprotection times are for aliphatic and pyrrole monomers, Boc-imidazole requires an additional 10 min deprotection period.

[b]The total time per cycle is longer in machine-assisted syntheses because of slower resin washing.

combinatorial techniques, the synthesis has been optimized on the Argonaut Technologies Quest 210 Parallel Synthesizer, capable of performing 20 parallel reactions. Because of possible *t*-butylation of the aromatic amino acids during Boc deprotection, a mixture of trifluoroacetic acid (TFA) with the scavengers phenol and water is used ***(29–31)***. This eliminates the malodorous thiophenol used previously and produces comparable products.

Upon completion of solid-phase synthesis, Section 2 outlines cleavage of the polyamide from the solid support, as well as purification and characterization. Aminolysis of the resin-bound polyamide with the primary aliphatic amine, dimethylaminopropylamine (DMPA), yields a product with a positively charged tail in neutral solutions, aiding solubility ***(21)***. A diamine, such as bis-aminopropylmethylamine (BAPMA), can be substituted for DMPA to produce a final product with a reactive amino handle for further modification or library production. The polyamide is then precipitated from the amine solution and purified by preparative reversed-phase HPLC. Once isolated, the products are characterized by a combination of matrix-assisted laser desorption-ionization (MALDI) or electrospray (ES) mass spectrometry (MS), HPLC, UV-Vis spectroscopy, and proton nuclear magnetic resonance (^{1}H-NMR) ***(32,33)***. These methods complement one another in assessing purity and identity. Relying on one method too heavily may result in fallacious assumptions concerning the polyamide synthesized (*see* **Note 1**).

2. Materials

2.1. Solid-Phase Synthesis of Polyamides

2.1.1. Equipment

1. Machine-assisted synthesis: Quest 210 Parallel Synthesizer with Auto-Solvent Wash (ASW) from Argonaut Technologies (San Carlos, CA). However, any peptide synthesizer that is TFA compatible can be used.
2. Manual synthesis: Glass 10 mL peptide synthesis vessel with a three-way stopcock and vacuum adapter (Chemglass, Vineland, NJ).

2.1.2. Monomers, Coupling Reagents, and Resin

1. Boc-β-alanine-(4-carbonylaminomethyl)-benzyl-ester-copoly(styrene-divinylbenzene)resin, (Boc-β-Ala-Pam resin). Optimal loading level of 0.2–0.25 mmol/g (Peptides International, Louisville, KY).
2. 4-[(*tert*-Butoxycarbonyl)amino]-1-methylpyrrole-2-carboxylic acid (Boc-Py-OH) and 4-[(*tert*-butoxycarbonyl)amino]-1-methylimidazole-2-carboxylic acid (Boc-Im-OH) can either be prepared as described ***(22)*** or purchased from Bachem (Torrance, CA). Store at room temperature (RT) with desiccation or at –20°C (*see* **Note 2**).
3. 1-Methylimidazole-2-carboxylic acid (Im-OH) was prepared as described ***(35)*** and stored desiccated at RT.
4. Boc-β-alanine (Boc-β-Ala-OH). Store at RT.
5. Boc-γ-aminobutyric acid (Boc-γ-Abu-OH). Store at RT.
6. Dicyclohexylcarbodiimide (DCC). Store at RT. Caution: DCC is a toxic reagent, avoid contact or inhalation.
7. 7-aza-1-hydroxybenzotriazole (HOAt) (PerSeptive Biosystems, Framingham, MA). Store at –20°C.
8. O-(7-azabenzotriazol-1-yl)-1,1,3,3-tetramethyluronium hexafluorophosphate (HATU) (PerSeptive Biosystems). Store at –20°C.

2.1.3. Solvents and Solutions

1. Dimethylformamide (DMF). Reagent-grade DMF is adequate for resin washing; however, it should be stored over activated 4 Å molecular sieves for at least 24 h prior to use as the coupling reaction solvent to remove water and dimethylamine. For manual synthesis, fill a 500 mL squirt bottle with DMF for rapid resin washing.
2. Dichloromethane (DCM). For manual synthesis, fill a 500 mL squirt bottle as for DMF.
3. 1-Methyl-2-pyrrolidone (NMP), 99+% grade.
4. Diisopropylethylamine (DIEA). ACS reagent grade. Appearance is clear and colorless. Aliquot 5–10 mL in a vial and cap with a rubber septum prior to each synthesis to minimize oxidation of the stock. Distill or discard if yellow (*see* **Note 3**).

5. TPW Boc deprotection mixture: 92.5% TFA, 5% phenol, 2.5% water (*see* **Note 4**) *(31)*. Caution: TFA is extremely corrosive and volatile! It should be handled in a fume hood at all times. TFA (99% grade) can be used without further purification. Also, phenol is highly toxic, and care should be taken to avoid skin contact and inhalation. Use in a fume hood as well. The cleavage mixture should be prepared fresh before each synthesis, and kept under nitrogen with an inert, syringe-accessible seal.
6. Methanol (MeOH).
7. Diethyl ether.

2.2. Resin Aminolysis, Product Purification, and Characterization

1. Dimethylaminopropylamine (DMPA), 99%. Store at RT.
2. *Bis*-aminopropylmethylamine (BAPMA), 96% (Lancaster, Windham, NH). Store at RT under nitrogen when not in use.
3. HPLC Buffer A: 0.1% v/v TFA in water. Filter through a 0.4 µm filter and degas prior to use.
4. HPLC Buffer B: acetonitrile. Prepare as for Buffer A.
5. HPLC columns: Reversed-phase C_{18} columns are used for both analytical and preparative HPLC. In these studies, Varian Microsorb MV, 5 µm, 100 Å, 250 × 4.6 mm, and Waters NovaPak, 6 µm, 60 Å, 100 × 25 mm, columns were employed, respectively.
6. MALDI or ES mass spectrometer. In these studies a PerSeptive Biosystems Voyager Elite STR MALDI was employed (*see* **Note 5**).
7. Matrix solution for MALDI: saturated (10–20 mg/mL) α-cyano-4-hydroxycinnamic acid (CHCA) in 50% acetonitrile, 50% water, 0.3% v/v TFA. Centrifuge and use supernatant. Prepare fresh prior to each use.
8. Deuterated dimethylsulfoxide (d_6-DMSO), 99.9%, and deuterium oxide (D_2O), 100%.
9. UV-Vis spectrophotometer.

3. Methods

3.1. Solid-Phase Synthesis of Polyamides

3.1.1. Presynthesis Preparation

1. Weigh 4 equivalents (relative to resin loading) of the necessary Py and Im monomers and 8 equivalents of γ-Abu and β-Ala monomers into separate glass vials (5 mL capacity).
2. Add 3.6 equivalents of HATU to Py vials, 7.2 equivalents of HATU to γ-Abu and β-Ala vials, and 4 equivalents of DCC and HOAt to Im monomers (*see* **Note 6**).
3. Weigh out and place the Boc-β-Ala-Pam resin into the peptide synthesis vessel or Quest reaction vessel. Typical syntheses are performed on 200–400 mg of resin, i.e., 0.05–0.1 mmol scale (to give 10–30 mg final product).
4. Swell the resin by rinsing twice with DCM, then mixing for 15 min with DCM (*see* **Note 7**).

Table 2
Reagent Reference Chart

Monomer	FW[a]	RW[b]	Activation method	Time[c]	Coupling reagent	FW	Density
Boc-Py-OH	240.3	122	HATU	>3 min	DCC	206.3	—
Boc-γ-Abu-OH	203.2	85	HATU	>3 min	HOAt	136.1	—
Boc-β-Ala-OH	189.2	71	HATU	>3 min	HATU	380.2	—
Im-OH	126.1	108	DCC/HOAt	1 h	DIEA	129.3	0.74
Boc-Im-OH	241.3	123	DCC/HOAt	2 h			

[a]FW, formula weight.
[b]RW, residue weight for mass spectrometry analysis.
[c]Time required for activation to occur. Longer times are not deleterious, and are somewhat beneficial for DCC/HOAt activations.

5. Rinse the resin twice with DMF, then mix for 15 min in DMF.
6. Finally, wash the resin three times with DCM.

3.1.2. TFA Deprotection and Coupling Reaction

1. Prior to beginning the first deprotection, the activation of the upcoming monomers must be planned according to the times listed in **Tables 1** and **2**. The time required per cycle can be estimated as 30 min.
2. Py, β-Ala, γ-Abu activation: Dissolve the monomers in 0.75 mL 1:1 DMF:NMP per 100 mg of resin. This is typically done at the end of the previous 20 min coupling cycle, or before synthesis is begun for the first step. Then, add 12 equivalents of DIEA to the solution approx 3 min (or greater, up to 10 min) before it will be added to the resin in step 11, i.e., during the third TFA deprotection period (**Subheading 3.1.2., step 7**).
3. Im activation: Dissolve the Im-OH and Boc-Im-OH monomers and DCC/HOAt in 0.75 mL 1:1 DMF:NMP per 100 mg of resin 1 and 2 h, respectively, before they are needed. Stir the solution vigorously with a magnetic stir bar. Thus, Im-OH is dissolved two cycles prior to its use, and Boc-Im-OH four cycles ahead of time. Activation is clearly indicated by the formation of dicyclohexylurea (DCU) precipitate. Immediately prior to addition to the resin in **Subheading 3.1.2., step 11**, the DCU should be filtered, and 12 equivalents DIEA added.
4. Boc deprotection: to the resin, add approx 0.4 mL of the TPW mixture per 100 mg of resin and mix for 1 min.
5. Drain the reaction vessel.
6. Add TPW as above, mix for 2 min, and drain the reaction vessel.
7. Repeat **step 6** of **Subheading 3.1.2., step 6**.
8. For deprotection of Boc-Im *only*, add TPW, mix for 10 min, and drain.
9. Wash resin (manual). Rinse resin with DCM for 30 s with vacuum assistance (approx 15–20 mL). Follow by washing with DMF for 1 min (approx 15–20 mL). Use DMF to spray resin that adheres to the walls of the vessel down to the bot-

tom. Also, spray vigorously into the resin to produce a suspension and promote thorough washing (*see* **Note 8**).

10. Wash resin (automated). Measure the time it takes for the Quest ASW to fill the reaction vessels with approximately 0.75 mL of solvent per 100 mg of resin. Create a wash program as follows: Set "drain/purge" function to drain. Set line 1 to wash twice with DCM, for the above fill time, with 10 s of agitation, and a 10 s drain time. Set line 2 to wash three times with DMF, for the above fill time, with 20 s of agitation, and a 20 s drain time (*see* **Note 9**).
11. Add the proper monomer solution from **Subheading 3.1.2., steps 2** or **3** to the resin.
12. Mix for 20 min with complete resin suspension. Clumps should not be visible.
13. Drain the reaction vessel.
14. Wash the resin with an inversion of the previous wash method, i.e., wash first with DMF, then with DCM. Create a separate wash program on the Quest ASW for this process.
15. If desired, take a small resin sample (3–5 mg) for monitoring synthesis progress (*see* **Subheading 3.1.3.**).
16. Repeat from **Subheading 3.1.2., step 2** until synthesis is complete.
17. Upon completion, wash the resin thoroughly with DMF, DCM, MeOH, and finally, ether. A typical final wash includes 3 × DMF, 3 × DCM, 2 × DMF, 2 × DCM, 3 × MeOH, 3 × DCM, 3 × MeOH, 2 × DCM, 5 × ether (*see* **Note 10**).
18. Dry the resin under vacuum overnight (or at least for 30 min).
19. Weigh the dried resin. Calculate a crude yield (*see* **Note 11**).

3.1.3. Stepwise Reaction Monitoring

1. Place the resin sample in a small test tube.
2. Add 200 µL of DMPA and heat at 100°C for 10 min (longer times are required for longer polyamides, up to 30 min).
3. Filter the solution.
4. For ES-MS, add approximately 30 µL of the amine solution to 1 mL of acetonitrile, and subject to MS.
5. For analytical HPLC, add 100 µL of 0.1% TFA/water to the amine solution and inject 20 µL, running a gradient of 1–2% Buffer B/min from 30 to 60% B.

3.2. Resin Aminolysis, Product Purification, and Characterization

3.2.1. Resin Aminolysis

1. Heat a sand or oil bath to 50–55°C.
2. Weigh the dried resin into a small round bottom flask or vial with a Teflon or polypropylene cap (*see* **Note 12**).
3. Add 0.75 mL of DMPA or BAPMA per 100 mg of resin, blanket with nitrogen or argon, and cap with an inert seal.
4. Stir with a small stir bar for 14–16 h.
5. Cool the vessel to room temperature.

6. Filter the resin and amine through a polypropylene syringe tip filter into a 50 mL centrifuge tube.
7. Add an equivalent volume of DCM (relative to the amine used) to the resin remaining, and mix.
8. Using the same syringe, swell the resin present in the syringe and the filter with the DCM just added.
9. Filter into the centrifuge tube.
10. Repeat the DCM rinse two more times (*see* **Note 13**).
11. Add 5–8 vol of diethyl ether (relative to the volume of DCM-amine mixture) with mixing. A thick white precipitate should form immediately upon addition of 1 vol.
12. Vortex to mix.
13. Cool to –20°C for 30 min.
14. Centrifuge in any table top centrifuge at full speed for 3–5 min to pellet the solid and then decant the supernatant.
15. Add approximately the same volume of DCM that was used in washing the resin to resuspend the solid (not necessarily dissolve) with vigorous mixing.
16. Add 5–8 vol of ether and cool to –20°C for 15 min to precipitate.
17. Centrifuge as above.
18. Dry the crude solid under vacuum for 1 h or until dry and powdery.

3.2.2. HPLC Purification

1. Dissolve the crude product in 15% acetonitrile/water, 0.1% v/v TFA, using approx 3 mL per 100 mg of resin (*see* **Note 14**). Vigorous mixing, sonication, and warming to 40°C may be necessary to obtain product solution. Check the pH of the solution. If basic, add 10% TFA/H_2O in 50 μL portions until the pH is 2–4. This will also increase product solubility. Keep a 0.1 mL aliquot for crude analytical HPLC and mass spectrometry (**Fig. 2**).
2. Filter through a 0.4 μm filter, preferably PTFE.
3. Equilibrate the preparatory reversed phase HPLC column in 85% Buffer A:15% Buffer B with detection at 254 nm.
4. Inject the polyamide mixture and hold at 15% B for 5 min, then ramp to 25% B over 60 min. The products typically elute at 21–23% Buffer B.
5. Collect 1 min fractions from 20–25% Buffer B.
6. After analysis by mass spectrometry, pool the appropriate fractions and lyophilize. Store the product at –20°C (*see* **Note 15**).

*3.2.3. Product Characterization (see **Note 1**)*

1. For MALDI-MS, spot 1 μL from each HPLC fraction together with 1 μL of matrix solution onto a target plate. Also prepare dilutions of product-containing fractions to obtain concentrations of 1–10 pmol/μL to obtain accurate analysis (**Fig. 2, Table 2**, *see* **Note 5**).
2. For ES-MS, inject HPLC fractions directly.

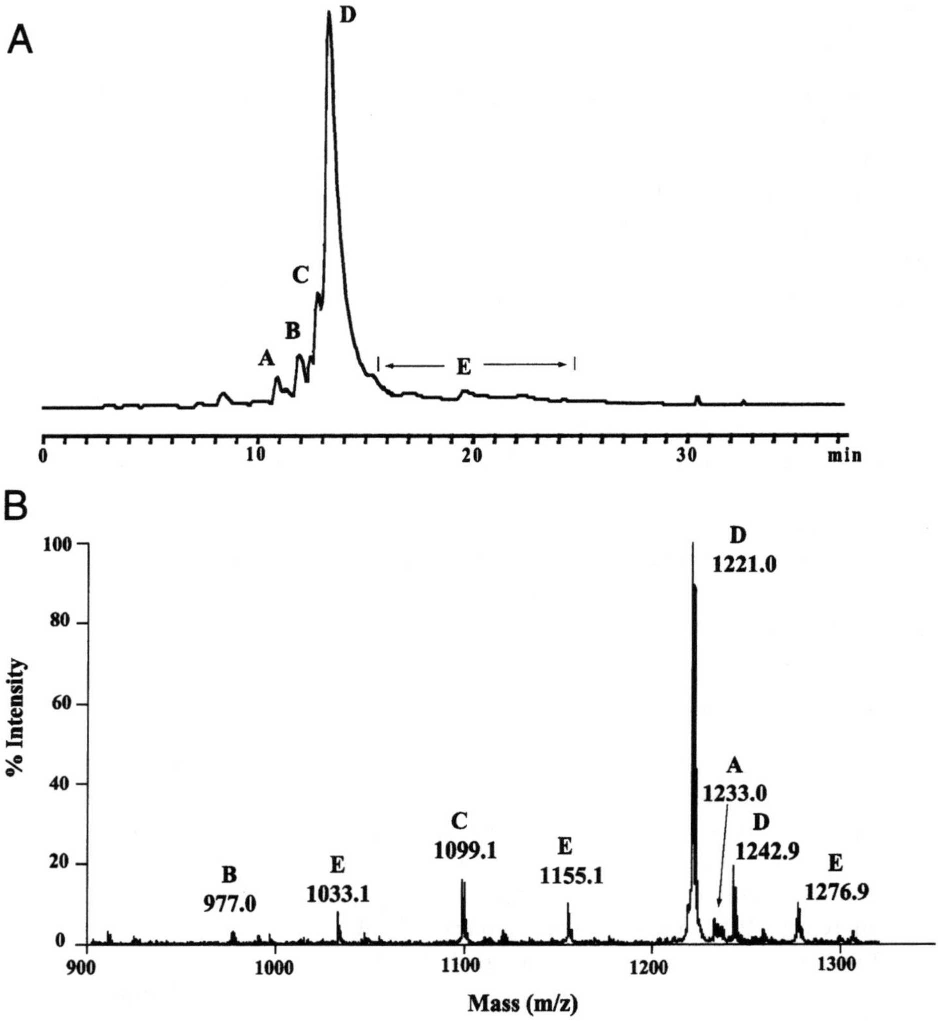

Fig. 2. HPLC and MALDI-MS analysis of a crude synthesis product. (A) Analytical HPLC trace of crude Im-Py-Py-Py-γ-Py-Py-Py-Py-β-Dp after cleavage from resin and ether precipitation. Immediately prior to preparative HPLC, 25 μL of the sample was subjected to analytical HPLC with a linear gradient of 20–50% Buffer B over 30 min (detection at 254 nm). Peaks labeled **A–E** correspond to the masses observed in the MALDI spectrum using CHCA as matrix (B). **A**, unknown impurity of +12 amu as well as oxidation products; **B**, double Py deletion product; **C**, single Py deletion; **D**, full length polyamide and its sodium adduct; **E**, *tert*-butyl adducts of **B**, **C**, and **D**. Note that the *t*-butyl adducts elute in a very broad band on the analytical HPLC, and are often overlooked as insignificant impurities.

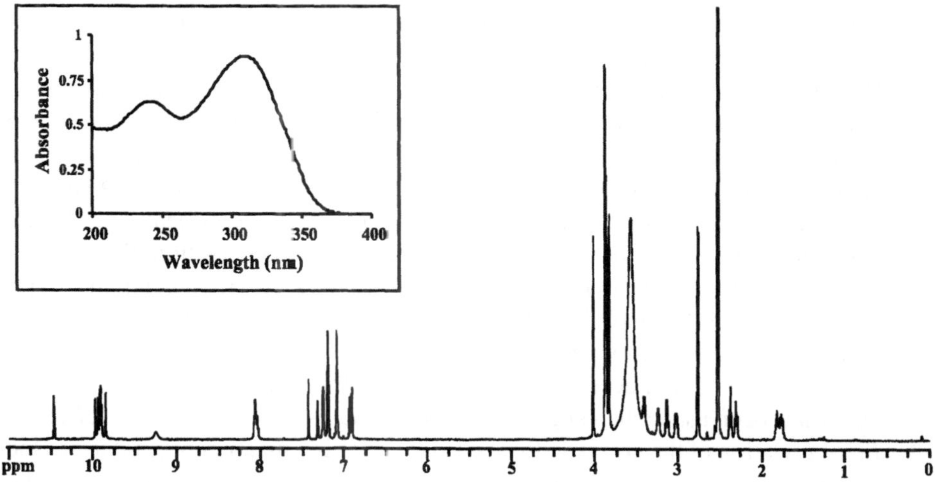

Fig. 3. Proton NMR and absorbance spectra of polyamides. The spectra shown are of purified Im-Py-Py-Py-γ-Py-Py-Py-Py-β-Dp. The NMR spectrum is of a 2 m*M* solution of polyamide in d_6-DMSO. Of interest are the amide protons at 8 and 10 ppm, the aromatic protons at 7 ppm, the N-methyl peaks at 4 ppm, and the aliphatic region of 1.5–3.5 ppm. Note that the N-methyl protons of imidazole appear slightly downfield of those of pyrrole. Also, the large peak at 3.3–3.6 ppm corresponding to water can easily be shifted to reveal the aliphatic region by addition of D_2O. (Inset) Absorbance spectrum in water, showing the characteristic maximum at 300–315 nm. Contaminated products often show significantly more absorbance near 200 nm.

3. For analytical HPLC, inject 10 µL of a 1 mg/mL solution of polyamide in 0.1% TFA, and run a linear gradient from 15–45% Buffer B over 30 min with detection at 254 nm. The products typically elute at 35–40% Buffer B (**Fig. 2**).
4. For ^{1}H-NMR, dissolve 1–3 mg of product in d_6-DMSO. After obtaining the DMSO spectrum, add 0.5 vol of deuterium oxide to shift the HDO peak to 4 ppm or greater to reveal the aliphatic region (**Fig. 3**).
5. For UV-Vis spectroscopy, dissolve polyamides in water, and use the approximate extinction coefficient of 8333 L/mol/cm per aromatic residue at the absorbance maximum (between 300 and 315 nm) (**Fig. 3**). Do not include the β-Ala or γ-Abu residues in calculation of the extinction coefficient.

4. Notes

1. Analysis of polyamides is not a trivial process, and care must be taken to be thorough. Each analytical technique described has its own merit. Mass spectrometry gives product identity and is a sensitive technique for determining the presence of monomer dele-

tion products. NMR is critical in determining the presence of low-molecular-weight residual aminolysis reagents that often co-elute with the product (present at 2–3 ppm). However, a clean NMR spectrum does not indicate a clean product, since a 50/50 mixture of product and a single monomer deletion has been observed to produce a spectrum nearly identical to pure product. HPLC can give product purity, but, because deletion products are very difficult to separate, HPLC *must* be coupled to mass spectrometry.

2. In these studies, the monomers were all synthesized as described. Boc-Py-OH has been stored at RT for 6 mo with no deleterious decomposition. Boc-Im-OH decomposed approx 25% over 1 yr at RT, with decarboxylation being the major side-product. Im-OH has also been stored at RT for 6 mo with little decomposition, but is also prone to decarboxylation. Storage at –20°C with desiccation is recommended but not necessary for storage less than 2 mo.
3. DIEA has been used when slightly yellow without observable negative effects.
4. *Tert*-butylation of the aromatic Py and Im residues is a major concern during Boc deprotection of polyamides, and necessitates the presence of scavengers in the deprotection mixture. Different scavenger mixtures have been used, with those containing phenol and water being as effective as thiophenol, and eliminating the stench associated with this reagent ***(21)***. Triisopropylsilane, often used in combination with phenol and water (known as reagent B), does not appear to have great benefit, and is therefore excluded. None of the scavenger mixtures used has completely eliminated formation of *t*-butyl adducts.
5. MALDI- and ES-MS are essentially interchangeable, and agree closely on product purity. However, MALDI is slightly favorable over ES because of the small sample volume needed, and the prevalence of $(M+H)^+$ ions in MALDI spectra. This allows for more clear product analysis and is not reliant upon flight characteristics of fragmented ions. Typical side-products include monomer deletions (*see* **Table 2** and **Fig. 2**), *t*-butylation (+56), and oxidation (+16). Exercise caution by preparing dilutions of MALDI samples to avoid erroneous results due to ion suppression. *t*-Butylated products fly poorly in concentrated samples. ES spectra typically show singly and doubly charged ions, as well as minor fragmentation.
6. The DCC/HOAt activation method is substituted for HATU for Im monomers because of their slow activation (activating for longer than 1–2 h is beneficial, and Boc-Im-OH may be activated overnight prior to use). If residual HATU is present in the monomer solution being added to the resin, it will immediately react with the free amines to produce guanidine adducts and terminate the synthesis. Imidazole monomers may be activated by HATU, however, if 4 equivalents of 2,4,6-collidine are used instead of DIEA, and the activation is allowed to proceed for 1 h. DIEA is then added immediately before addition of the monomer to the resin. Residual HATU is easily identified when DIEA is added by brown color development. With full activation and consumption of the reactive HATU, monomer solutions turn bright yellow with DIEA, and remain this way throughout the coupling period. Changes to red or brown are indicative of side reactions. It is also important to note that although OAt mediated couplings are robust, the coupling of Py to the imidazole amine is problematic, and necessitates the

synthesis of the Boc-Py-Im-OH dimer in solution ***(21)***. However, Py-Im can often be substituted by β-Ala-Im, which can be synthesized stepwise, without loss of binding or sequence specificity ***(2)***. Utilization of dimer blocks, such as Boc-Py-Im-OH, Boc-γ-Abu-Im-OH, or Boc-β-Ala-Im-OH is beneficial as it avoids slow imidazole deprotection, and is recommended if their synthesis is possible in existing facilities.

7. Rinsing refers to a vacuum-assisted flow wash in manual synthesis, or to solvent addition, brief agitation to obtain suspension, and vessel draining in machine-assisted synthesis. Mixing entails bubbling nitrogen through the manual reaction vessel or to agitation on the Quest.
8. Because polystyrene resins adhere to glass, it is important to wash all of the resin to the bottom of the reaction vessel prior to continuing. Also, prior to addition of the monomer solution, DMF should be used to vigorously suspend the resin (or added and nitrogen bubbled through to mix), then drained.
9. Agitation and drain times may need to be adjusted depending on the volumes added and the amount of resin present. The times listed work well for 200 mg of resin or less. In some instances, the stir bar gets stuck in the resin and requires a longer time to begin agitation. Make sure that full suspension is obtained before the vessel is drained or overfilling may occur, in essence ruining the synthesis.
10. The final resin wash protocol can be modified as one sees fit. Though it may seem excessive, the final wash is an opportunity to completely remove any reagents that may remain from the synthesis or side-products that may be adhering to the resin. Thorough washing promotes smoother aminolysis and cleaner products. A powdered resin should result. Clumps indicate the presence of non-volatile solvents such as DMF.
11. New resin loading can be calculated as follows: $L_{new}(\text{mmol/g}) = L_{old}/[1 + L_{old}(W_{new} - W_{old}) \times 10^{-3}]$. L is loading and W is molecular weight of the compound attached to the resin. The Boc-β-Ala-PAM resin has a W_{old} of 171.
12. Although the aminolysis reaction can be performed in Quest reaction vessels, it is problematic when small amounts of resin (200 mg or less, with less than 1.5 mL of amine) are being cleaved. Too much amine is condensed near the top of the reaction vessels, and the resin may be lodged on top of the agitator bars, causing it to be heated without solvent present. At least 1.5–2.0 mL of aminolysis reagent need to be used to avoid these problems. Otherwise, use conventional heating techniques.
13. It is critical to re-swell the resin thoroughly with the first DCM wash, as polyamides are highly soluble in the aminolysis reagents, but not in DCM.
14. Polyamides are not highly soluble in water (approx 1 mg/mL) at neutral or basic pH (7–10), with larger polyamides being less soluble than smaller ones. Residual aminolysis reagent may cause the pH to be above neutral and have deleterious effects to the reversed phase HPLC column. Make sure that enough TFA is added to produce a pH of 2–4. Acetonitrile may be added to 20–25%, but larger percentages lead to product loss and a decrease in resolution due to band broadening.
15. Yields for typical 7–9 residue polyamides are 20–40%. After lyophilization, polyamides appear as white to very light yellow solids, typically amorphous.

Presence of a deep yellow color, or oils, is indicative of residual amines and requires repurification. Polyamides are stable at RT, but 0.1 mg aliquots should be prepared in microcentrifuge tubes and stored at –20°C. The best way to make the aliquots is by dissolving a larger amount of polyamide into water, measuring the absorbance, determining the concentration, and pipetting an appropriate amount into the tubes. The solutions are then re-lyophilized to produce solids.

References

1. Lawley, P. D. and Phillips, D. H. (1996) DNA adducts from chemotherapeutic agents. *Mutat. Res.* **355,** 13–40.
2. Dervan, P. B. and Burli, R. W. (1999) Sequence-specific DNA recognition by polyamides. *Curr. Op. Chem. Biol.* **3,** 688–693, and references therein.
3. Trauger, J. W., Baird, E. E., and Dervan, P. B. (1996) Recognition of DNA by designed ligands at subnanomolar concentrations. *Nature* **382,** 559–561.
4. Swalley, S. E., Baird, E. E., and Dervan, P. B. (1997) Discrimination of 5'-GGGG-3', 5'-GCGC-3', and 5'-GGCC-3' sequences in the minor groove of DNA by eight-ring hairpin polyamides. *J. Am. Chem. Soc.* **119,** 6953–6961.
5. Trauger, J. W., Baird, E. E., and Dervan, P. B. (1998) Recognition of 16 base pairs in the minor groove of DNA by a pyrrole-imidazole polyamide dimer. *J. Am. Chem. Soc.* **120,** 3524–3535.
6. Kielkopf, C. L., Baird, E. E., Dervan, P. B., and Rees, D. C. (1998) Structural basis for G-C recognition in the DNA minor groove. *Nature Struct. Biol.* **5,** 104–109.
7. Hawkins, C. A., de Clairac, R. P., Dominery, R. N., Baird, E. E., White, S., Dervan, P. B., and Wemmer, D. E. (2000) Controlling binding orientation in hairpin polyamide DNA complexes. *J. Am. Chem. Soc.* **122,** 5235–5243.
8. Satz, A. L. and Bruice, T. C. (1999) Synthesis of a fluorescent microgonotropen (FMGT-1) and its interactions with the dodecamer d(CCGGAATTCCGG). *Bioorg. Med. Chem. Lett.* **9,** 3261–3266.
9. Filipowsky, M. E., Kopka, M. L., Brazil-Zison, M., Lown, J. W., and Dickerson, R. E. (1996) Linked lexitropsins and the in vitro inhibition of HIV-1 reverse transcriptase RNA-directed DNA polymerization: A novel induced-fit of 3,5 m-pyridyl bis-distamycin to enzyme-associated template primer. *Biochemistry* **35,** 15,397–15,410.
10. For review of distamycin and netropsin analogues: Bailly, C. and Chaires, J. B. (1998) Sequence-specific DNA minor groove binders. Design and synthesis of netropsin and distamycin analogues. *Bioconjugate Chem.* **9,** 513–538.
11. Dickinson, L. J., Gulizia, R. J., Trauger, J. W., Baird, E. E., Mosier, D. E., Gottesfeld, J. M., and Dervan, P. B. (1998) Inhibition of RNA polymerase II transcription in human cells by synthetic DNA-binding ligands. *Proc. Natl. Acad. Sci. USA* **95,** 12,890–12,895.
12. Gottesfeld, J. M., Neely, L., Trauger, J. W., Baird, E. E., and Dervan, P. B. (1997) Regulation of gene expression by small molecules. *Nature* **387,** 202–205.
13. Janssen, S., Cuvier, O., Muller, M., and Laemmli, U. K. (2000) Specific gain- and loss-of-function phenotypes induced by satellite-specific DNA-binding drugs fed to *Drosophila melanogaster. Mol. Cell* **6,** 1013–1024.

14. Janssen, S., Durussel, T., and Laemmli, U. K. (2000) Chromatin opening of DNA satellites by targeted sequence-specific drugs. *Mol. Cell* **6,** 999–1011.
15. Wurtz, N. R. and Dervan, P. B. (2000) Sequence specific alkylation of DNA by hairpin pyrrole-imidazole polyamide conjugates. *Chemistry and Biology* **7,** 153–161.
16. Tao, Z.-F., Fujiwara, T., Saito, I., and Sugiyama, H. (1999) Rational design of sequence-specific DNA alkylating agents based on duocarmycin A and pyrrole-imidazole hairpin polyamides. *J. Am. Chem. Soc.* **121,** 4961–4967.
17. Nishiwaki, E., Tanaka, S., Lee, H., and Shibuya, M. (1988) Efficient synthesis of oligo-N-methylpyrrolecarboxamides and related compounds. *Heterocycles* **27,** 1945–1952.
18. Xiao, J., Yuan, G., Huang, W., Chan, A. S. C., and Lee, K.-L. D. (2000) A convenient method for the synthesis of DNA-recognizing polyamides in solution. *J. Org. Chem.* **65,** 5506–5513.
19. Vazquez, E., Caamano, A. M., Castedo, L., and Mascarenas, J. L. (1999) An Fmoc solid phase approach to linear poly-pyrrole peptide conjugates. *Tet. Lett.* **40,** 3621–3624.
20. Wurtz, N. R., Turner, J. M., Baird, E. E., and Dervan, P. B. (2001) Fmoc solid phase synthesis of polyamides containing pyrrole and imidazole amino acids. *Org. Lett.* **3,** 1201–1203.
21. Baird, E. E. and Dervan, P. B. (1996) Solid phase synthesis of polyamides containing imidazole and pyrrole amino acids. *J. Am. Chem. Soc.* **118,** 6141–6146.
22. Pitie, M., Van Horn, J. D., Brion, D., Burrows, C. J., and Meunier, B. (2000) Targeting the DNA cleavage activity of copper phenanthroline and Clip-phen to A-T tracts via linkage to a poly-N-methylpyrrole. *Bioconjugate Chem.* **11,** 892–900.
23. Sharma, S. K., Tandon, M., and Lown, J. W. (2001) A general solution and solid phase synthetic procedure for incorporating three contiguous imidazole moieties into DNA sequence reading polyamides. *J. Org. Chem.* **66,** 1030–1034.
24. Schnolzer, M., Alewood, P., Jones, A., Alewood, D., and Kent, S. B. (1992) *In situ* neutralization in Boc-chemistry solid phase peptide synthesis. *Int. J. Pept. Res.* **40,** 180–193.
25. Albericio, F. and Carpino, L. A. (1997) Coupling reagents and activation. *Methods Enzymol.* **289,** 104–126.
26. Alewood, P., Alewood, D., Miranda, L., Love, S., Meutermans, W., and Wilson, D. (1997) Rapid in situ neutralization protocols for Boc and Fmoc solid-phase chemistries. *Methods Enzymol.* **289,** 14–29.
27. Miranda, L. P. and Alewood, P. F. (1999) Accelerated chemical synthesis of peptides and small proteins. *Proc. Natl. Acad. Sci. USA* **96,** 1181–1186.
28. Carpino, L. A. and El-Faham, A. (1999) The diisopropyl/1-hydroxy-7-azabenzotriazole system: segment coupling and stepwise peptide assembly. *Tetrahedron* **55,** 6813–6830.
29. Lundt, B. F., Johansen, N. L., Volund, A., and Maskussen, J. (1978) Removal of *t*-butoxycarbonyl protecting groups with trifluoroacetic acid. *Int. J. Pept. Prot. Res.* **12,** 258–268.

30. Pearson, D. A., Blanchette, M., Baker, M. L., and Guindon, C. A. (1989) Trialkylsilanes as scavengers for the trifluoroacetic acid deblocking of protecting groups in peptide synthesis. *Tet. Lett.* **30,** 2739–2742.
31. Sole, N. A. and Barany, G. (1992) Optimization of solid-phase synthesis of [Ala^8]-dynorphin A. *J. Org. Chem.* **57,** 5399–5403.
32. Moore, W. T. (1997) Laser desorption mass spectrometry. *Methods Enzymol.* **289,** 520–542.
33. Burdick, D. J. and Stults, J. T. (1997) Analysis of peptide synthesis products by electrospray ionization mass spectrometry. *Methods Enzymol.* **289,** 499–519.
34. Wade, W. S., Mrksich, M., and Dervan, P. B. (1992) Design of peptides that bind in the minor groove of DNA at 5'-(A/T)G(A/T)C(A/T)-3' sequences by a dimeric side-by-side motif. *J. Am. Chem. Soc.* **114,** 8783–8794.

7

The Preparation of Phenyl-stilbene Derivatives Using the Safety Catch Linker

Amy Lew and A. Richard Chamberlin

1. Introduction

Very few diseases are directly caused by ion channel mutations, yet many drugs for various diseases work by modulating ion channel activity. This is because ion channels are cellular "gatekeepers," monitoring intracellular concentrations of Na^+, Ca^+, K^+, or Cl^- to control cardiac pacemaking, membrane potential, neurotransmitter release, hormone secretion, cell proliferation, cell volume regulation, and lymphocyte differentiation. One ion channel that controls lymphocyte differentiation is the highly T cell specific Kv1.3 channel. Since the discovery of Kv1.3 channels in T_c cells, these lymphocyte channels have been targeted in developing immunomodulatory drugs that might exhibit lower toxicities than those currently in use *(1,2)*.

The Kv1.3 channel, cloned in the lab of George Chandy, is a voltage-gated K^+ channel, 400–600 of which are expressed in each T_c cell *(3)*. Blocking these ion channels depolarizes the cell and ultimately halts interleukin-2 (IL-2) production and T cell proliferation in vitro *(4,5)* and in vivo *(6)*, providing an effective means of immunosuppression. Current Kv1.3 channel blockers, such as the peptidyl toxins Margatoxin *(7,8)*, Charybdotoxin *(9–11)*, and ShK toxin *(12,13)*, as well as the natural products Correolide *(14–17)* and Candelalides *(18,19)*, demonstrate that there are two available sites to induce Kv1.3 channel blockage—the outer vestibule and the intracellular S5 pocket (**Fig. 1**). Virtually no structural data exists for the S5 pocket, making rational ion channel modulator design difficult. However, models of the Kv1.3 channel's outer vestibule have been developed from site-directed mutagenesis of rigid picomolar affinity scorpion toxins that bind to the outer vestibule of the Kv1.3 channel. For example, Chandy and Aymar have identified nine pairs of toxin-channel

From: *Methods in Molecular Biology, Combinatorial Library Methods and Protocols*
Edited by: L. B. English © Humana Press Inc., Totowa, NJ

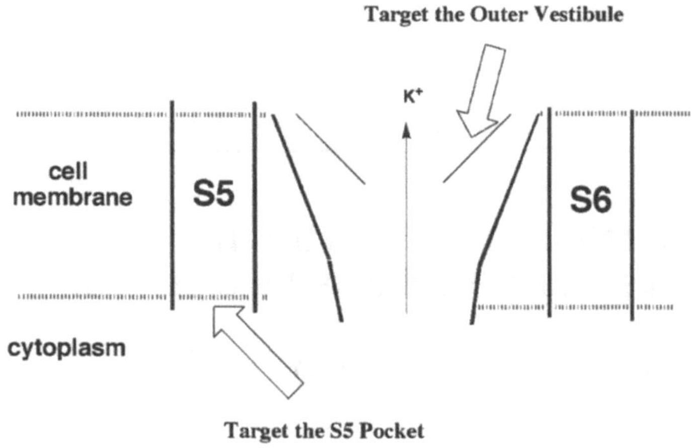

Fig. 1. The Kv1.3 channel can be blocked at two general sites–either the outer vestibule or the intracellular S5 pocket.

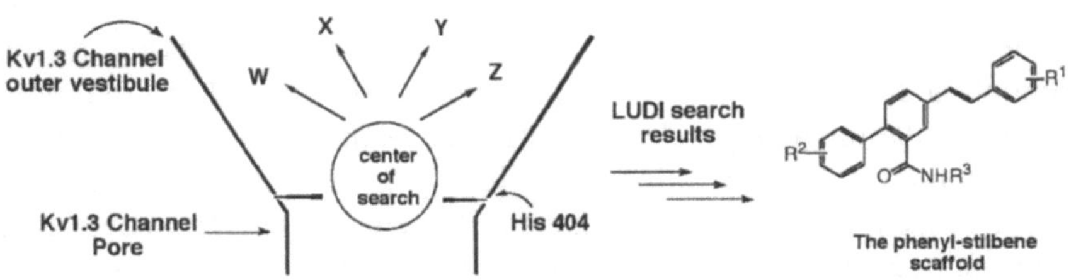

Fig. 2. A Protocol for LUDI Searches with the center of search chosen around the critically unique His 404 of the postulated Kv1.3 channel model. LUDI calculations predicted that a phenyl-stilbene scaffold with various R^1, R^2, and R^3 functionalities would fit in the outer vestibule of the channel.

interactions to develop a model ***(20,21)*** of the Kv1.3 vestibule based on Guy and Dunnel's 1994 model ***(22,23)*** of a Shaker K^+ channel.

With the advent of more powerful computer programs and the increasing number of available three-dimensional protein structures in the public domain, computer-aided rational drug design has become a popular tool in medicinal chemistry ***(24)***. We took advantage of the available Kv1.3 channel coordinates and a computer modeling algorithm LUDI ***(25,26)*** to design *de novo* Kv1.3 channel blockers (**Fig. 2**). Through these studies, we designed, screened, and assayed to discover low micromolar phenyl-stilbene derivatives that blocked the outer vestibule of the Kv1.3 channel ***(27)***. In this chapter, we will describe the detailed steps in preparing libraries of phenyl-stilbene derivatives on solid phase (**Fig. 3**). In this synthesis, the Wittig salt **1** was linked to solid support

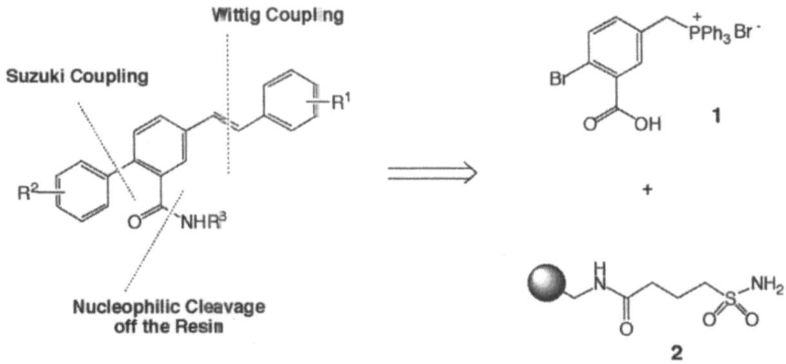

Fig. 3. A combinatorial strategy to synthesize the phenyl-stilbene derivatives.

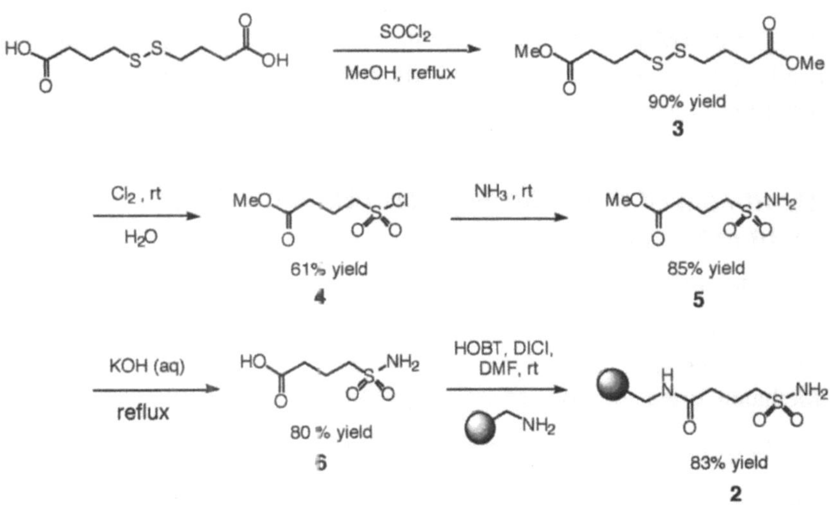

Fig. 4. Synthesis of the alkyl sulfonamide resin.

via activation of the carboxylic acid and then coupled to Ellman's alkyl sulfonamide resin **2** *(28–30)*. Once on solid support, variations in R^1 were made through a solid-phase Wittig coupling *(31–35)* with various commercially available aryl aldehydes. The R^2 substituent was varied using a solid-phase Suzuki coupling *(36,37)* with various commercially available aryl boronic acids. Lastly, variations in R^3 were introduced by nucleophilic cleavage off of Ellman's modified Kenner's safety catch linker with various commercially available amines. This chapter is divided into four parts: (1) synthesis of the solid support, (2) synthesis of the Wittig salt, (3) synthesis of the Suzuki catalyst, and (4) synthesis of the phenyl-stilbenes on solid support (**Figs. 4–6**).

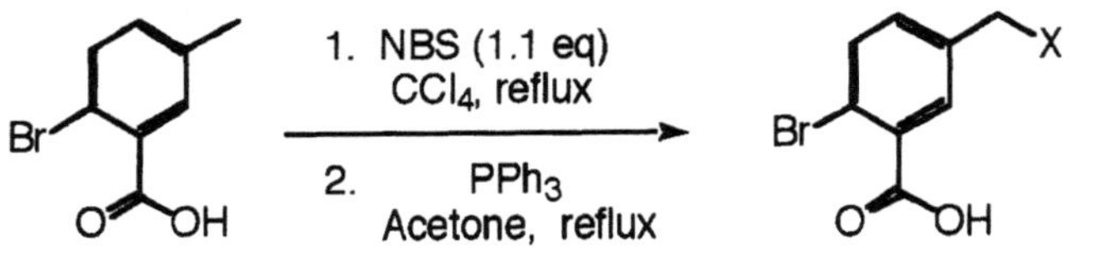

Fig. 5. Synthesis of the Wittig Salt 1.

2. Materials

All reagents and solvents employed in the reactions, the reaction work-ups, and the solid phase washes were purchased from Fisher Scientific (Pittsburg, PA) or Aldrich (Milwaukee, WI) and used without further purification unless specified. Specific reagents and handling methods required for the synthesis of each compound are listed below under the name of each corresponding compound synthesized.

2.1. Synthesis of the Solid Support (Ellman's Alkyl Sulfonamide Resin)

2.1.1. Dimethyl 4,4'-Dithiobutyrate (3)

1. 4,4'-Dithiobutyric acid.
2. Thionyl chloride.
3. Methanol, dried and distilled over 3 Å molecular sieves.

2.1.2. Methyl 4-(Chlorosulfonyl)Butyrate (4)

1. Chlorine gas, 99.999% pure lecture bottle (Matheson, Montgomeryville, PA).

2.1.3. Methyl 4-Sulfamoylbutyrate (5)

1. Diethyl ether, dried and distilled over potassium-benzophenone ketyl.
2. Ammonia gas, 99.999% pure lecture bottle (Matheson, Montgomeryville, PA).

2.1.4. Sulfonamide Linker (2)

1. Aminomethylated resin, 0.8 meq/g, stored at room temperature (RT). Beads must be 0.8 meq/g or lower in loading; otherwise yields will be low due to low coupling in each step (Advanced ChemTech, Louisville, KY).
2. Hydroxy 1-hydroxybenzotriazole (HOBT), stored at 4°C, but warmed to RT before use.
3. 1,3-Diisopropyl carbodiimide (DICI), stored in a dessicator at RT.
4. THF, dried and distilled over CaH_2.
5. Reagents for the Kaiser Ninhydrin test were purchased from Aldrich and Fisher and used without further purification: phenol, ninhydrin, KCN, pyridine.

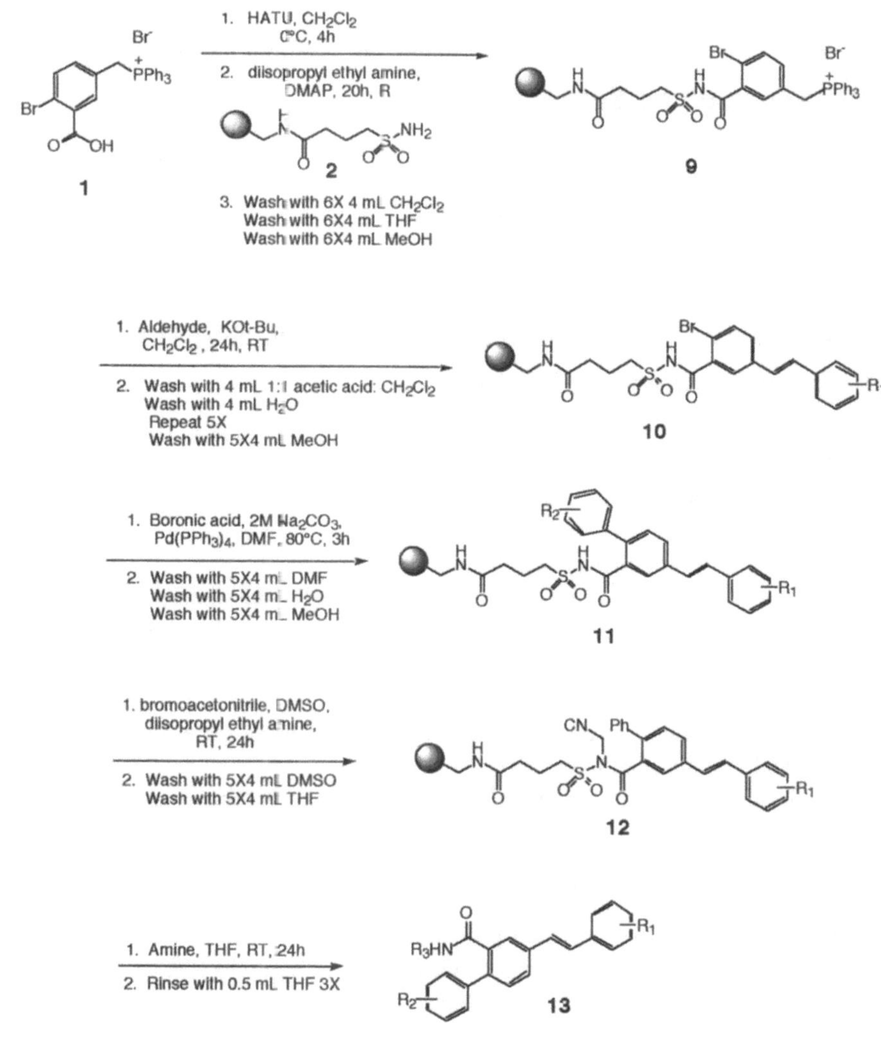

Fig. 6. Combinatorial synthesis of LUDI hits.

2.2. Synthesis of the Wittig Salt

2.2.1. Monobrominated (8)

1. 2-Bromo-5-methyl benzoic acid (Transworld Chemicals, Rockville, MD).
2. Carbon tetrachloride; avoid vapors and skin contact, as CCl_4 is a potent carcinogen.
3. N-bromosuccinimide (NBS), stored at RT in the absence of light.

2.2.2. Wittig Salt (1)

1. Acetone, dried by treating with K_2CO_3 at RT, then filtering off the solids, and distilling into a dry round bottom flask.

2. Triphenyl phosphine.
3. Diethyl ether for the washes, dried and distilled over potassium-benzophenone.

2.3. Synthesis of the Suzuki Catalyst

1. $PdCl_2$, stored at RT under Argon. Compound should be a rusty red brown color (Alfa Aesar, Ward Hill, MA).
2. DMSO, dried over 3Å molecular sieves.
3. Triphenyl phosphine.
4. Hydrazine hydrate.

2.4. Synthesis of the Phenyl-Stilbenes on Solid Support

2.4.1. Solid Supported Scaffold (9)

1. *N*-[dimethylamino)-1H-1,2,3-triazolo[4,5-b] pyridin-1-ylmethylene]-*N*-methyl-methanaminium hexafluorophosphate N-oxide (HATU), stored at –20°C and warmed to RT before use. The reagent should be a brownish, dry, light powder (PerSeptive Biosystems, Framingham, MA).
2. CH_2Cl_2, dried and distilled over CaH_2.
3. *N,N*-dimethyl amino pyridine (DMAP), stored in a RT dessicator.
4. Diisopropyl ethyl amine, distilled over CaH_2 and stored over 3 Å molecular sieves.

2.4.2. Wittig Coupling of Various Aldehydes

1. Potassium *t*-butoxide, stored in a plastic bag within a sealed metal container at RT.
2. THF, dried and distilled over CaH_2.
3. All aryl aldehydes were purchased from Aldrich and Fischer and used without further purification.

2.4.3. Suzuki Coupling of Various Boronic Acids

1. DMF, degassed by two repeated procedures of cooling the DMF in a dry ice/acetone bath under high vacuum followed by warming up to RT.
2. $Pd(PPh_3)_4$, can be purchased from Aldrich, or made as described in this chapter under **Subheading 3.3.**
3. All boronic acids were purchased from Aldrich, stored in a RT dessicator and used without further purification.

2.4.4. Alkylation with Bromoacetonitrile

1. DMSO, dried and stored over 3 Å molecular sieves.
2. Diisopropyl ethyl amine, distilled over CaH_2 and stored over 3 Å molecular sieves.
3. Bromoacetonitrile, stored at 4°C, warmed to RT before use. Avoid skin contact as bromoacetonitrile is a potent carcinogen.

2.4.5. Cleavage with Various Nucleophiles

1. All amines were purchased from Aldrich and used without further purification.

3. Methods

3.1. Synthesis of the Solid Support (Ellman's Alkyl Sulfonamide Resin) (28,29,38) *(see Note 1)*

3.1.1. Dimethyl 4,4'-dithiobutyrate (3)

1. To a flame dried 200-mL round bottom flask, equipped with a magnetic stir bar and reflux condenser, add 4,4'-dithiobutyric acid (21.0 g, 88.2 mmol), thionyl chloride (15.0 mL, 200 mmol), and MeOH (200 mL) under anhydrous conditions (*see* **Note 2**).
2. Heat the mixture to reflux under nitrogen for 4 h and then cool slowly to 0°C before adding 33.3 mL of water dropwise to the stirring mixture.
3. Extract twice with 200 mL of CH_2Cl_2.
4. Combine the organic extracts and wash three times with 200 mL of 1 *M* $NaHCO_3$ (*see* **Note 3**).
5. Transfer the organic extracts to an Erlenmeyer flask and dry with anhydrous Na_2SO_4 for 45 min.
6. Decant into a clean round bottom flask. Rinse the Erlenmeyer flask twice with 10 mL of CH_2Cl_2 and add these rinses to the round bottom flask.
7. Concentrate the organic extracts with a rotary evaporator to obtain 7.53 g of a crude brown oil.
8. Filter through a plug of silica to obtain 7.01 g *(90%)* of **3** as a yellow oil (*see* **Note 4**). IR (neat) 1728, 2951 cm^{-1}; ^{1}H-NMR (300 MHz, $CDCl_3$) δ 3.70 (s, 6 H), 2.75 (t, $J = 7.2$ Hz, 4H), 2.45 (t, $J = 7.2$, 4 H), 2.05 (quintet, $J = 7.2$ Hz, 4 H); ^{13}C-NMR (500 MHz, $CDCl_3$) δ 173.3, 37.7, 32.3, 24.1; HRMS (CI) calcd for $C_{10}H_{18}O_4S_2$ (M^+) 266.0647, found 266.0645.

3.1.2. Methyl 4-(Chlorosulfonyl)Butyrate (4)

1. To a 100-mL three-necked round bottom flask equipped with a magnetic stir bar, add 50 mL of water and 7.36 g (27.7 mmol) of **3**.
2. Cool the heterogeneous mixture to 0°C. While stirring vigorously with a magnetic stir bar, slowly bubble in chlorine gas that has been passed through a saturated solution of Na_2CO_3. Trap any excess Cl_2 exiting the reaction with a saturated solution of sodium meta bisulfite (*see* **Note 5**).
3. After 3 h at 0°C, the reaction should turn bright yellow. When the solution turns bright yellow, continue the slow bubbling of Cl_2 and vigorous stirring for another 0.5–1 h at 0°C until the reaction is homogeneous.
4. Stop bubbling in the chlorine gas. Instead, bubble nitrogen into the reaction via a glass Pasteur pipet with vigorous stirring for 1 h at 0°C to remove excess Cl_2 (*see* **Note 6**).
5. Extract the reaction three times with 50 mL of CH_2Cl_2.
6. Combine the organic layers and dry them with anhydrous Na_2SO_4.
7. Concentrate under reduced pressure to obtain 6.79 g (61%) of **4** as a crude yellow oil. The crude methyl 4-(chlorosulfonyl)butyrate is sufficiently pure for the next step. IR (neat) 1162, 1370, 1736, 2358–3630 cm^{-1}; ^{1}H-NMR (500 MHz, $CDCl_3$)

δ 3.81 (t, $J = 7.0$ Hz, 2 H), 3.72 (s, 3 H), 2.60 (t, $J = 7.0$ Hz, 2 H), 2.37 (quintet, $J = 7.0$ Hz, 2 H); ^{13}C-NMR (500 MHz, $CDCl_3$) δ 172.1, 64.0, 52.0, 30.9, 19.7; HRMS (CI) calcd for $C_5H_9ClO_4S$ [$(M - Cl)^+$] 165.0221, found 165.0222.

3.1.3. *Methyl 4-Sulfamoylbutyrate* (5)

1. To a 500-mL round bottom flask equipped with a magnetic stir bar, add 6.79 g (33.9 mmol) of crude **4** and 95.0 mL of dry ether.
2. Cool the mixture with an ice water bath and bubble in gaseous ammonia via a glass Pasteur pipet while vigorously stirring the mixture. A white precipitate should form within 30 minutes. Continue bubbling in NH_3 for 3–4 h at 0°C to ensure that the reaction reaches completion (*see* **Note 7**).
3. Stop bubbling in the ammonia gas, and concentrate the slurry down to approx 10 mL in volume before dissolving in 19.0 mL of water.
4. Extract the aqueous layer five times with 50 mL of hot ethyl acetate.
5. Dry the organic extracts with Na_2SO_4 and concentrate under reduced pressure to obtain 5.2 g (85%) of 5 as a pale yellow oil, which is sufficiently pure for the next step. IR (neat) 1149, 1322, 1722, 3255, 3351 cm^{-1}; ^{1}H-NMR (500 MHz, d_6-DMSO) δ 6.82 (s, 2 H), 1.93 (quintet, $J = 7.5$ Hz, 2 H), 3.60 (s, 3 H), 3.00 (t, $J = 7.5$, 2 H), 2.48 (t, $J = 7.5$ Hz, 2 H); ^{13}C-NMR (500 MHz, d_6-DMSO) δ 173.1, 53.9, 51.8, 31.8, 19.6; HRMS (CI) calcd for $C_5H_{11}NO_4S$ [$(MH)^+$] 182.0493, found 182.0487.

3.1.4. *3-Carboxypropane Sulfonamide* (6)

1. To a 100-mL round bottom flask equipped with a stir bar and reflux condenser, add 4.67 g (25.8 mmol) of crude **5** and 30.0 mL of 1.8 *M* KOH (*see* **Note 8**).
2. Stir under reflux for 1 h. Then cool the reaction to RT, then 0°C before slowly adding in 30.0 mL of 1.8 *M* HCl.
3. Concentrate the neutral mixture under reduced pressure to a volume of 2–2.5 mL and add in an equivalent amount of ACS grade acetone to precipitate out the KCl salts.
4. Filter this mixture through a coarse fritted glass funnel to remove the precipitated KCl and rinse with a minimum amount of acetone:water (1:1, v/v).
5. Concentrate the filtrate and dry overnight in a P_2O_5 dessicator to obtain 3.1 g (72%) of the crude translucent solid **6** that could be carried directly to the next step.

3.1.5. *Sulfonamide Linker* (2)

1. In a flame dried 200-mL round bottom flask, preswell 2.5 g of amino methylated resin (0.8 meq/g) with 20.0 mL of dry THF.
2. Add 13 mL of THF, 0.336 g (2.67 mmol) of DICI, 0.35 g (2.67 mmol) of HOBT, and 0.52 g (2.67 mmol) of crude 3-carboxypropane sulfonamide **6** under anhydrous conditions (*see* **Note 9**).
3. Cap the round bottom flask with a polypropylene cap and shake on a shaker at RT for 12 h (*see* **Note 10**).

4. Filter the beads through a fritted glass funnel and wash them three times with 25 mL of CH_2Cl_2. Dry the beads under vacuum to obtain 3.82 g (0.125 mmol/g) of alkyl sulfonamide resin **2** (*see* **Note 11**).
5. Confirm the presence of the sulfonamide on solid support with the standard Kaiser ninhydrin test *(39)* outlined below:

   For Solution A, mix together 8 g of phenol, 0.5 g of ninhydrin, and 10 mL of EtOH, and store the solution in a scintillation vial. In a separate scintillation vial, mix together 0.013 g of KCN and 10 mL of pyridine. This is Solution B. In a small test tube, place several beads (approx 1 mg) along with approx 0.1 mL of Solution A and 0.1 mL of Solution B. Mix and heat the contents of the test tube to 70°C with a sand bath. A positive ninhydrin test for the nonreacted amino methylated resin gives a deep blue color while the sulfonamide linker **2** gives a pale red color.
6. Resin loading is determined by exhaustively coupling a known amount of resin with the activated Wittig salt **1** as discussed in **Subheadings 3.4.1.**, **3.4.4.**, and **3.4.5.** to determine the amount of recovered Wittig salt. For amino methylated resin loading of 0.8 meq/g, you should expect about a 0.125 meq/g loading for the sulfonamide linker.

3.2. Synthesis of the Wittig Salt

3.2.1. Monobrominated (8)

1. To a flame-dried round bottom flask equipped with a stir bar and reflux condenser, add 8.56 g (40 mmol) of 2-bromo-5-methyl benzoic acid **(7)**, 320.0 mL of CCl_4, 7.20 g (42.3 mmol) of N-bromosuccinimide (NBS), and 0.031 g of benzoyl peroxide under anhydrous conditions (*see* **Note 12**).
2. Reflux the reaction under nitrogen for 6 h. Cool the reaction to RT, and then 0°C.
3. Filter the yellow-white solid through a fritted glass funnel.
4. Remove the succinimide by-product by rinsing the solid twice with 10 mL of H_2O followed by washing twice with 10 mL of pentane.
5. Transfer the solid yellow powder to a round bottom flask and dry on vacuum at RT overnight in the absence of light to obtain the crude monobrominated **8**, which can be carried directly to the next step (*see* **Note 13**).

3.2.2. Wittig Salt (1)

1. In a flame-dried round bottom flask, add 10.6 g (36 mmol) of crude **8**, 9.73 g (37.1 mmol) of triphenyl phosphine, and 334 mL of dry acetone under anhydrous conditions.
2. Reflux the reaction under nitrogen for 2 h; cool it to RT, and then 0°C (*see* **Note 14**).
3. Filter the resulting white precipitate through a fritted glass funnel, and wash twice with 100 mL of dry ether.
4. Collect the filtrate and concentrate it to one third of its volume. Heat the concentrated filtrate until all the solids are dissolved. Then allow the mixture to cool slowly to 0°C to collect a second crop of pure Wittig salt **1**. Also rinse this second crop of salt twice with 25 mL of dry ether.

5. Dry the combined crops of Wittig salt **1** under vacuum at 60°C overnight to obtain 15.56 g (overall yield, 70%) of pure white powdery salt.
   mpt: 260–261°C; IR (KBr pellet) 1714, 2300–4000 cm^{-1}; ^{1}H-NMR (500 MHz, d_6-DMSO) δ 13.38 (bs, 1 H), 7.70–8.00 (m, 15 H), 7.60 (d, *J*= 8.3 Hz, 1 H), 7.34 (s, 1 H), 7.09 (d, *J* = 8.3 Hz, 1 H), 5.28 (d, *J*= 15.8 Hz, 2H); ^{13}C-NMR (500 MHz, d_6-DMSO) δ 167.1, 135.9, 135.2, 135.1, 134.6, 134.3, 133.8, 130.9, 128.6, 121.0, 118.4, 117.7; HRMS (CI) calcd for $C_{26}H_{21}O_2PBr$ (M^+) 475.0463, found 475.0476.

3.3. Synthesis of the Suzuki Catalyst (40) (see Note 15)

1. To a flame dried three-necked round bottom flask equipped with a magnetic stir bar and swivel frit, add 0.886 g (5.0 mmol) of $PdCl_2$, 6.55 g (25.0 mmol) of triphenyl phosphine, and 60 mL of dry DMSO under nitrogen.
2. Heat the stirring solution under nitrogen with an oil bath until all of the solid has dissolved (approx 140°C for 15 min). Note that the reaction turns from yellow to orange when all the solids are in solution.
3. Remove the reaction from the oil bath and continue stirring for 15 min to produce an orange precipitate.
4. Make sure that the system is set up to allow pressure relief through a nitrogen bubbler. Add in hydrazine hydrate (0.917 mL, 20.0 mmol) rapidly via a needle and syringe over a period of 1 min. Vigorous evolution of nitrogen and crystallization of the bright yellow $Pd(PPh_3)_4$ catalyst should occur.
5. Cool the reaction to RT with a water bath and filter with the swivel frit under N_2. It is important to keep the catalyst under N_2 and away from air throughout the filtrations and washes, as air oxidizes the catalyst, turning it from bright yellow to dark orange.
6. Wash the yellow catalyst twice with 2.5 mL of dry EtOH, and twice with 2.5 mL of dry ether before drying with a stream of N_2. Store the dried $Pd(PPh_3)_4$ away from light and under N_2 at –20°C.

3.4. Synthesis of the Phenyl-Stilbenes on Solid Support

The following are procedures to prepare 40 solid-phase reactions at a time using Advanced ChemTech's ReacTech. However, these reactions can also be performed on any other parallel solid-phase reactor with heating capabilities, mode of bead agitation, and wells of at least 5 mL capacity, i.e., Argonaut's Quest. The inside of the vessels should preferably be coated with Teflon.

3.4.1. Solid Supported Scaffold (9)

1. Fill each of 40 Teflon-coated reaction vessels with 0.2 g (0.025 mmol) of dry sulfonamide linker **2** (0.125 mmol/g) and 0.03 g (0.24 mmol) of DMAP and seal the vessels under nitrogen with Teflon-coated rubber septa (*see* **Note 16**).
2. Cool a flame dried round bottom flask to 0°C and add 160 mL of dry CH_2Cl_2, 4.4 g (7.92 mmol) of dry Wittig Salt **1**, and 3.8 g (10.0 mmol) of HATU under anhydrous conditions.

3. Stir this reaction for 3 h under nitrogen at 0°C before delivering 4.0 mL (0.20 mmol, 8.0 eq) of this mixture via a dry syringe and needle to each of 40 sealed wells already filled with sulfonamide resin **2** and DMAP (*see* **Note 17**),
4. When all of the slurry is added to each reaction vessel, add distilled dry diisopropyl ethyl amine (0.5 mL, 3.12 mmol) to each well.
5. Program the reaction block to shake at 450 rpm for 20 h at RT (*see* **Note 18**),
6. Empty the reaction vessels using positive N_2 pressure, and wash with 6× 4 mL of CH_2Cl_2 , 6× 4 mL of THF, and 6× 4 mL of MeOH. Each wash step entails addition of solvent, mixing for 2 min at 450 rpm, and emptying under a positive N_2 pressure for 2 min.
7. Dry the beads in each of the 40 reactions with a continuous stream of N_2 for 6 min.

3.4.2. Wittig Coupling of Various Aldehydes (10)

1. To each of the sealed 40 vessels containing the above resin **9**, deliver via a dry syringe and needle, 3.0 mL (0.27 mmol, 10.8 eq) of a solution of 0.09 *M* potassium *t*-butoxide in dry THF (*see* **Note 19**).
2. Deliver a unique aryl aldehyde to each sealed reaction vessel with a dry syringe and needle; liquid aldehydes (0.1 mL, approx 0.7 mmol, 28 eq) , solid aldehydes (0.7 mL of a 1 *M* solution in THF, 0.7 mmol, 28 eq).
3. Program the reaction vessel to shake at 450 rpm for 25 h at RT (*see* **Note 20**).
4. Empty the reaction vessels with positive N_2 pressure, and then wash each reaction alternatingly with 4 mL of H_2O and 4 mL of 1:1 acetic acid: CH_2Cl_2 for a total of six times. Each wash step entails addition of solvent, mixing for 2 min at 450 rpm and emptying under a positive N_2 pressure for 2 min (*see* **Note 21**).
5. Finally, wash each of the 40 reactions with 6× 4 mL of MeOH and dry over a stream of N_2 for 6 min.

3.4.3. Suzuki Coupling of Various Boronic Acids (11)

1. To each of the 40 sealed vessels containing the above resin **10**, add 4.0 mL of a 0.00135 *M* solution of $Pd(PPh_3)_4$ in degassed DMF (0.005 mmol) via a syringe and needle. Again, the resin **10** does not need to be preswollen (*see* **Note 22**).
2. After shaking at RT for 5 min, add in 0.5 mL of 2 *M* Na_2CO_3 in H_2O and 0.5 mL (0.3 mmol, 12 eq) of a 1 *M* solution of the appropriate boronic acid in degassed DMF to each reaction well.
3. Program the reaction block to shake at 450 rpm for 3 h at 80°C. Then allow the block to cool to RT for 1.5 h (*see* **Note 23**).
4. Empty the vessels using positive N_2 pressure and wash with 6× 4 mL of DMF, 6× 4 mL of H_2O, and 6× 4 mL of MeOH. Each wash step entails addition of solvent, mixing for 2 min at 450 rpm, and emptying under a positive N_2 pressure for 2 min.
5. Dry the 40 reactions under a stream of N_2 for 6 min.

3.4.4. Alkylation with Bromoacetonitrile (12)

1. To each of the sealed 40 reactions, add 4.0 mL of dry DMSO, 0.2 mL of dry diisopropyl ethyl amine, and 0.2 mL of bromoacetonitrile via a dry syringe and needle (*see* **Note 24**).
2. Program the reaction block to shake at 450 rpm for 20 h at RT. Empty the vessels using positive N_2 pressure, and wash with 6× 4 mL of DMSO and 6× 4 mL of THF. Each wash step entails addition of solvent, mixing for 2 min at 450 rpm and emptying under a positive N_2 pressure for 2 min.
3. Dry the beads under a stream of N_2 for 6 min.

3.4.5. Cleavage with Various Nucleophiles (13)

1. Set the parallel synthesizer up for product collection.
2. To each of the 40 vessels, add 4.0 mL of dry THF and 0.1 mL of a 3 *M* solution of the appropriate nucleophile via a syringe and needle.
3. Program the reaction block to shake at 450 rpm for 1 h at RT and then empty the product into the their respective cleavage vials.
4. Rinse the beads for each reaction twice with 0.5 mL of THF and collect the rinses in the same corresponding vials.
5. Filter the crude product from each reaction through a plug of silica to remove excess amines to obtain between 13 and 20 mg of purified product before sending them off for biological testing.

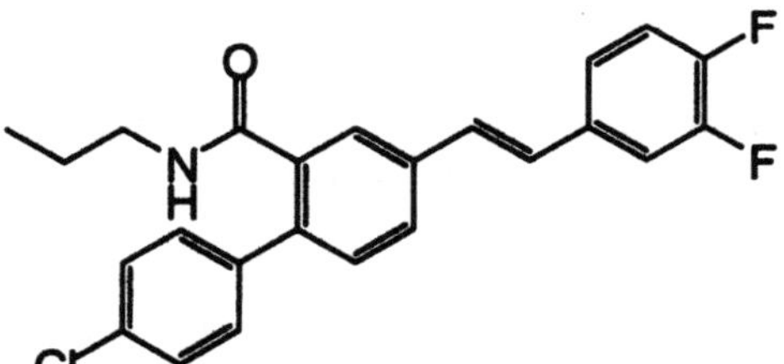

Scheme 1. Spectral data for one example of a phenyl-stilbene derivative made by these procedures.

IR (neat) υ_{max} 3335, 1643 cm^{-1}; ^{1}H-NMR (500 MHz, $CDCl_3$) δ 7.85 (s, 1H), 7.63 (dd, *J* = 10.14, 2.44 Hz, 1H), 7.47–7.37 (m, 5 H), 7.26–7.20 (m, 3 H), 7.10 (apparent q, *J* = 19.8 Hz, 2 H), 5.3 (bt, 1 H), 3.24 (q, *J* = 8.7 Hz, 2 H), 1.37 (sextet, *J* = 9.11 Hz, 2 H), 0.79 (t, *J* = 9.2 Hz, 3 H), HRMS (CI) calcd for $C_{24}H_{20}ClF_2NO$ (MH^+) 412.1280, found 412.1287.

4. Notes

1. This resin is commercially available, but is relatively expensive. If large quantities are needed follow the procedures in Section 3.1, which are procedures adapted from **refs.** ***29*** and ***38***.

2. Thionyl chloride should be added with a disposable metal needle. Although the thionyl chloride should be added dropwise, prolonged exposure to the metal needle corrodes the needle and contaminates the reaction leaving a green brown tint.
3. Caution. Addition of 1 *M* $NaHCO_3$ results in the release of high amounts of CO_2. Do not cap the separatory funnel until bubbling ceases.
4. Dimethyl 4,4'-dithiobutyrate (**3**) is not heat stable. Attempts to distill the crude product without the use of a high vacuum results in decomposition. Running the crude product through a plug of silica gel will yield product that is sufficiently pure for the next step.
5. The reaction must be vigorously stirred as the dimethyl 4,4'-dithiobutyrate (**3**) is insoluble in water. In addition, chlorine must be bubbled in through a glass pipet, as bubbling in via a metal needle leads to contamination of the reaction due to metal corrosion. Passing the Cl_2 gas through a solution of saturated Na_2CO_3 removes any HCl present. Finally, failure to keep the reaction at low temperatures and prolonged reaction times leads to methyl ester hydrolysis and other uncharacterized side products. The reaction is done when it turns bright yellow and becomes homogeneous. If the reaction separates into two layers, the reaction is not complete.
6. Again, the reaction has to be kept at 0°C during the nitrogen bubbling to prevent hydrolysis and side reactions. Once the yellow color of the reaction disappears, the reaction needs to be worked up as soon as possible.
7. After the white precipitate forms, the reaction can also be monitored by TLC using a 1:1 EtOAc:hexanes as the eluent, and staining with a ninhydrin-based stain.
8. The hydrolysis reaction is kept as concentrated as possible to prevent intramolecular cyclization. However, even under these reaction conditions, approx 20% cyclized product is commonly observed.
9. Dicyclohexyl carbodiimide (DCC) should not be used as the insoluble dicyclohexyl urea (DCU) by-product is very difficult to remove, leaving residual DCU that can contaminate future reactions. Also, no dimerization of the HOBT ester of 3-carboxypropane sulfonamide **6** occurs as the sulfonamide is nonreactive toward HOBT esters. This was also observed by Ellman *(38)*.
10. The reaction cannot be stirred with a magnetic stir bar as the integrity of the beads will deteriorate. Instead, agitate the beads on a shaker or use an overhead mechanical stirrer, making sure that the stirring paddle does not contact the bottom of the flask.
11. If you had chosen to use DCC in the coupling step instead of DICI, wash the beads until no more of the white precipitate (dicyclohexyl urea) remains. This may take more than the three washes.
12. The benzoyl peroxide is used as a free radical initiator. The reaction can be done without the peroxide, but longer reaction times will be required. The reaction is complete when the succinimide floats to the top in hot CCl_4. Small aliquots of the reaction should still be taken and monitored by ^1H-NMR to ensure that the reaction is complete before a work up is done.

13. Analysis by ^{1}H-NMR analysis shows that the crude product contains approx 5% of the dibrominated species.
14. Depending on the purity of the triphenylphosphene, the refluxing reaction may turn purple in color after 30 min, but disappear later in the reaction. However, this does not seem to affect the purity and yield of the final product.
15. This procedure is adapted from the already excellent procedure in ref. ***40***.
16. The sulfonamide resin **2** need not be preswollen.
17. The HATU-activated Wittig salt is a white slurry. Deliver the slurry with a thick 16 gauge needle, making sure to mix the slurry periodically during the delivery.
18. Coupling of this hindered activated Wittig salt **1** is difficult, and long reaction times are needed. Several other activation methods such as the corresponding symmetrical anhydride and pyBOP ester produced even lower coupling yields. If the yield is still low despite the 20 h reaction with the HATU ester, repeat the coupling by repeating steps 1–7 for a second coupling of the HATU activated Wittig salt **1**.
19. The solid supported scaffold **9** was not preswollen.
20. This Wittig coupling protocol gives a mixture of the *cis/trans* alkene in a ratio of 20:80 for a majority of the aldehydes used.
21. This wash sequence is to ensure the complete removal of potassium t-butoxide. Open the reaction vessels at the end of the nitrogen drying step to inspect for the presence of a white precipitate. If any white precipitate is detected, repeat the water and acetic acid/CH_2Cl_2 wash steps 4–6.
22. DMF must be degassed prior to addition of $Pd(PPh_3)_4$. Otherwise, the palladium catalyst turns a dark orange brown color in solution and is inactive.
23. DMF was used in this Suzuki coupling because the boiling point of DMF is at least 25°C higher than the required reaction temperature of 80°C. If your parallel synthesizer is equipped with a reflux condenser, a lower boiling solvent may be used in place of DMF.
24. Bromoacetonitrile activation of the sulfonamide linker gives better amine cleavage yields than activation with diazomethane.

References

1. Chandy, K. G., Strong, M., Aiyar, J., and Gutman, G. A. (1997) Structural and biochemical features of the Kv1.3 potassium channel: an aid to guided drug design. *Cellular Physiol. Biochem.* **7,** 135–147.
2. Lew, A. and Chamberlin, A. R. (2000) Human T cell Kv1.3 potassium channel blockers: new strategies for immunosuppression. *Exp. Opin. Ther. Patents* **10(6),** 905–915.
3. Chandy, K. G. and Gutman, G. A. (1995) Voltage-gated potassium channel genes. In Handbook of Receptors and Channels: Ligand and Voltage Gated ion Channels (North, A., ed.), CRC Press, Boca Raton, FL.
4. Chandy, K. G., DeCoursey, T. E., Cahalan, M. D., McLaughlin, C., and Gupta, S. J. (1984) Voltage-gated channels are required for human T lymphocyte activation. *J. Exp. Med.* **160,** 369–385.

5. DeCoursey, T. E., Chandy, K. G., Sudhir, G., and Cahalan, M. D. (1984) Voltage gated potassium channels in human T lymphocytes: A role in mitogenesis? *Nature* **307,** 465.
6. Koo, G. C., Blake, J. T., Talento, A., et al. (1997) Blockade of the voltage-gated potassium channel Kv1.3 inhibits immune responses in vivo. *J. Immunol.* **22,** 5120–5128.
7. Garcia-Calvo, M., Leonard, R., Novick, J., et al. (1993) Purification, characterization, and biosynthesis of Margatoxin, a component of Centruroides-margaritatus venom that selectively inhibits voltage-dependent potassium channels. *J. Biol. Chem.* **25,** 18,866–18,874.
8. Helms, L., Felix, J., Bugianesi, R., et al. (1997) Margatoxin binds to a homomultimer of Kv1.3 channels in Jurkat cells. Comparison with Kv1.3 expressed in CHO cells. *Biochem.* **36,** 3737–3744.
9. Price, M., Lee, S. C., and Deutsch, C. (1989) Charybdotoxin inhibits proliferation and interleukin-2 production in human peripheral blood lymphocytes. *Proc. Natl. Acad. Sci. USA* **86,** 10,171–10,175.
10. Deutsch, C., Price, M., Lee, S., King, F., and Garcia, M. L. (1991) Characterization of high affinity binding sites for Charybdotoxin in human T lymphocytes: evidence for association with the voltage-gated K^+ channel. *J. Biol. Chem.* **266,** 3668.
11. Leonard, R. J., Garcia, M. L., Slaughter, R. S., and Reuben, J. P. (1992) Selective blockers of voltage-gated K^+ channels depolarize human T lymphocytes: mechanism of antiproliferative effect of Charybdotoxin. *Proc. Natl. Acad. Sci. USA* **89,** 10,094–10,098.
12. Tudor, J. E., Pallaghy, P. K., Pennington, M. W., and Norton, R. S. (1996) Solution structure of ShK toxin, a novel potassium channel inhibitor from a sea anemone. *Nature Struc. Biol.* **3(4),** 317–320.
13. Kalman, K., Pennington, M. W., Lanigan, M. D., et al. (1998) ShK-Dap22, a potent Kv1.3-specific immunosuppressive polypeptide. *J. Biol. Chem.* **273(49),** 32,697–32,707.
14. Goetz, M. A., Hensens, O. D., Zink, D. L., Bet al. (1998) Potent nor-triterpenoid blockers of the voltage-gated potassium channel Kv1.3 from spachea correae. *Tetrahedron Lett.* **39,** 2895–2898.
15. Hanner, M., Schmalhofer, W. A., Green, B., et al. (1999) Binding of correolide to the Kv1 family potassium channels. *J. Biol. Chem.* **274(36),** 25,237–25,244.
16. Felix, J. P., Bugianesi, R. M., Schmalhofer, W. A., et al. (1999) Identification and biochemical characterization of a novel nortriterpene inhibitor of the human lymphocyte voltage-gated potassium channel, Kv1.3. *Biochemistry* **38,** 4922–4930.
17. Koo, G. C., Blake, J. T., Shah, K., et al. (1999) Correolide and derivatives are novel immunosuppressants blocking the lymphocyte Kv1.3 potassium channels. *Cellular Immunol.* **197,** 99–107.
18. Singh, S. B., Zink, D. L., Dombrowski, A. W., et al. (2001) Candelalides A-C: novel diterpenoid pyrones from fermentations of sesquicillium candelabrum as blockers of the voltage-gated potassium channel Kv1.3. *Organic Lett.* **3(2),** 247–250.

19. Goetz, M. A., Zink, D. L., et al. (2001) Diterpenoid pyrones, novel blockers of the voltage-gated potassium channel Kv1.3 from fungal fermentations. *Tetrahedron Lett.* **42(7),** 1255–1257.
20. Aiyar, J., Withka, J. M., Rizzi, J. P., et al. (1995) Topology of the pore-region of a K+ channel revealed by the NMR-derived structures of scorpion toxins. *Neuron* **15,** 1169.
21. Aiyar, J., Rizzi, J. P., Gutman, G. A., and Chandy, K. G. (1996) The signature sequence of voltage-gated potassium channels projects into the external vestible. *J. Biol. Chem.* **271(49),** 31,013.
22. Guy, H. R. and Durell, S. R. (1994) Using homology in modeling the structure of voltage-gated ion channels. In *Molecular evolution of physiological processes* (Farmbrough, D. M., ed.), Rockefeller University Press, New York.
23. Guy, H. R. and Durell, S. R. (1995) Structural model of Na^+, Ca^{2+}, and K^+ channels. In *Ion channels and genetic diseases* (Dawson, D., ed.), Rockefeller University Press, New York.
24. Bohm, H. and Stahl, M. (2000) Structure based library design: molecular modelling merges with combinatorial chemistry. *Current Opin. Chem. Biology* **4(3),** 283–286.
25. Bohm, H. J. (1992) The computer program LUDI: a new method for the de novo design of enzyme inhibitors. *J. Comp. Aided Molec. Design* **6,** 61–78.
26. Bohm, H. J. (1992) LUDI: rule-based automatic design of new substituents for enzyme inhibitor leads. *J. Comp. Aided Molec. Design* **6,** 593–606.
27. Lew, A. and Chamberlin, A. R. (1999) Blockers of human T cell Kv1.3 potassium channels using de novo design and solid phase parallel combinatorial chemistry. *Biorg. Med. Chem. Lett.* **9,** 3267–3272.
28. Backes, B. J., Virgilio, A. A., and Ellman, J. A. (1996) Activation method to prepare a highly reactive acylsulfonamide “safety-catch” linker for solid-phase synthesis. *J. Am. Chem. Soc.* **118(12),** 3055.
29. Backes, B. J. and Ellman, J. A. (1999) An alkanesulfonamide “safety-catch” linker for solid phase synthesis. *J. Org. Chem.* **64,** 2322–2330.
30. Kenner, G. W., McDermott, J. R., and Sheppard, R. C. (1971) The safety catch principle in solid phase peptide synthesis. *J. Chem. Soc. Chem. Commun.* 636–637.
31. Camps, F., Castells, J., Font, J., and Vela, F. (1971) Organic synthesis with functionalized polymers: II. Wittig reaction with polystyryl-p-diphenylphosphoranes. *Tetrahedron Lett.* **20,** 1715–1716.
32. Frechet, J. M. and Schuerch, C. (1971) Solid-phase synthesis of oligosaccharides. I. Preparation of the solid support Poly[p-(1-propen-3-ol-1-yl)styrene]. *J. Am. Chem. Soc.* **93,** 492–496.
33. Chen, C., Ahlberg Randall, L. A., Miller, B. R., Jones, D. A., and Kurth, M. J. (1994) “Analogous” organic synthesis of small-compound libraries: validation of combinatorial chemistry in small-molecule synthesis. *J. Am. Chem. Soc.* **116,** 2661–2662.
34. Leznoff, C. C., Fyles, T. M., and Weatherson, J. (1977) The use of polymer supports in organic synthesis. 8. Solid phase syntheses of insect sex attractants. *Canad. J. Chem.* **55,** 1143–1153.

35. Leznoff, C. C., Fyles, T. M., and Weatherson, J. (1978) Bifunctionalized resins. applications to the synthesis of insect sex attractants. *Canad. J. Chem.* **56,** 1031.
36. Frenette, R. and Friesen, R. W. (1994) Biaryl synthesis via suzuki coupling on a solid support. *Tetrahedron Lett.* **35,** 9177–9180.
37. Backes, B. J., and Ellman, J. A. (1994) Carbon-carbon bond-forming methods on solid support. Utilization of Kenner's "safety-catch" linker. *J. Am. Chem. Soc.* **116,** 11,171–11,172.
38. Beck, M. L. (1968) Synthesis of 3-sulfopropionimide. *J. Org. Chem.* **33(2),** 897–899.
39. Kaiser, E., Colescot, R. L., Bossinger, C. D., and Cook, P. I. (1970) *Anal. Biochem.* **34,** 595.
40. Coulson, D. R. (1972) Tetrakis(triphenylphosphine) palladium (0). *Inorg. Syn.* 121–124.

8

Host–Guest Chemistry

Combinatorial Receptors

Brian R. Linton

1. Introduction

At the heart of host–guest chemistry is the design and construction of artificial receptors. While great progress has been achieved in the specific recognition of a multitude of substrates, rational design has still failed to equal the degree of association strength and substrate specificity observed in nature. Evolutionary selection has allowed biopolymers to take advantage of their large size to position varied functionality in the rigid context of tertiary structure, but the limitations of chemical synthesis dictate that artificial receptors will be smaller, less structurally defined, and, as a result, less effective. Combinatorial chemistry offers a new approach to the task of receptor design by quickly creating a library of combinatorial receptors containing a variety of binding or catalytic groups, and subsequently determining the most active component of that library ***(1–5)***. Combinatorial receptors also provide the possibility for using substrate binding to selectively create a receptor from a dynamic library ***(6–8)***. To be effective, however, library synthesis should require minimal synthetic expenditure and the entire library must be screened efficiently.

The approach detailed in this chapter uses metal-templated self-assembly to create combinatorial receptors through the dimerization of separate binding units ***(9)***. Nature follows a similar strategy in the creation of antibodies. The vast diversity of antigen recognition is accomplished by the variety within a single protein sequence, but is also greatly enhanced by the dimerization of the light and heavy protein chains to create one dimeric binding site. An analogous approach can be employed in small-molecule receptors, where a series of

From: *Methods in Molecular Biology, Combinatorial Library Methods and Protocols*
Edited by: L. B. English © Humana Press Inc., Totowa, NJ

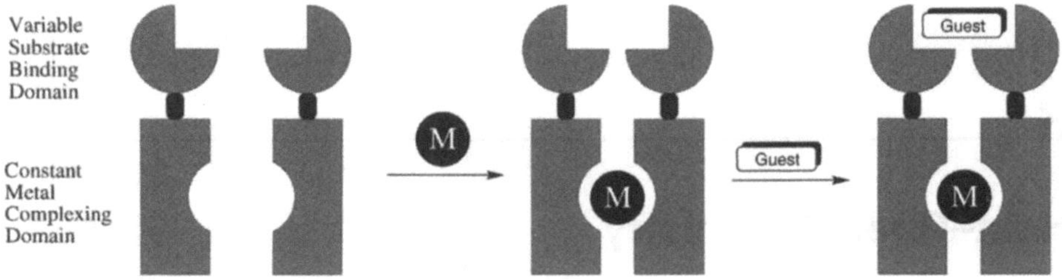

Fig. 1. Formation of dimeric receptor by metal-templated self-assembly.

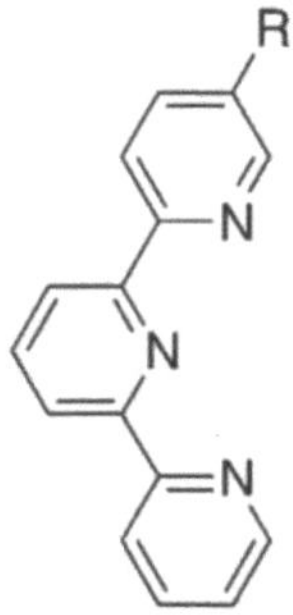

R =
A: CH_3
B: $CH_2NHC(S)NHCH_2CH_2CH_2CH_3$
C: $CH_2OCH_2CH_2OH$
D: CH_2OCHPh_2
E: $CH_2N(C_2H_4OC_2H_4OC_2H_4)_2O$
F: CH_2Br
G: CO_2H
H: $C(O)NHCH_2CH_2$(4-imidazole)

Fig. 2. Terpyridine monomers.

monomers can be oligomerized to create a larger, more functionally diverse binding cleft. **Figure 1** illustrates the strategy for the creation of dimeric receptor libraries through metal-templated self-assembly. This requires a series of molecules containing both a metal complexing domain, and a variable substrate binding domain. Addition of a metal recruits two monomers to form a dimeric receptor, and orients both substrate binding domains to form one binding cleft (10,11). With this facile dimerization, a library of receptors can be formed from a smaller number of monomers with minimal synthetic expenditure. Subsequent association with a guest is aided by the accumulation of binding domains, as well as the variety of binding groups possible in a receptor library.

For the approach detailed above to be successful, the metal binding domain must form robust complexes, yet retain synthetic flexibility. Although substituted phenanthroline and bipyridine easily form self-assembled receptors ***(12)***, the substitution stability of terpyridine (**Fig. 2**) makes this ligand more suited for the creation of libraries. The choice of metal also permits some flexibility in library creation. Iron complexes form easily through the addition of ammonium iron(II) sulfate hexahydrate, but can be cleaved by the addition of chelating guests ***(11,13)***. Cobalt complexes are also easily formed using cobalt(II)

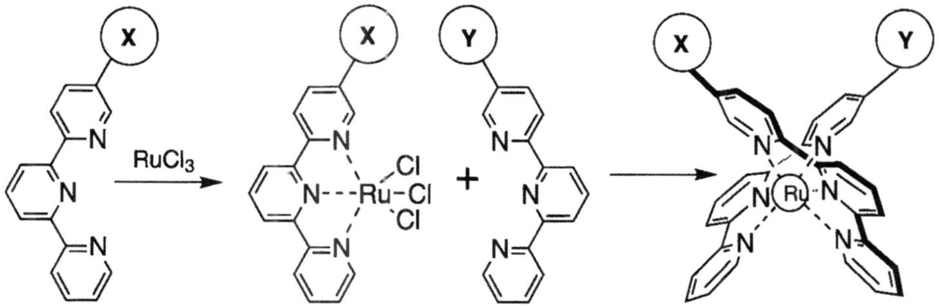

Fig. 3. Stepwise synthesis of asymmetric ruthenium complex.

chloride and subsequent oxidation to cobalt(III) renders these complexes substitution inert *(14)*. The most robust complexes are formed using ruthenium trichloride; once formed, these are inert to ligand substitution by an added guest *(15,16)*. The use of ruthenium also permits the step-wise formation of asymmetric complexes, as shown in **Fig. 3**. Addition of metal and terpyridine forms an intermediate complex, followed by the subsequent addition of a second ligand. With the other metals described, only mixtures of symmetric and asymmetric complexes are formed with more than one terpyridine monomer.

All terpyridine monomers began with 5-methyl-terpyridine *(17)*. Conversion to either bromomethyl (F) or carboxylic acid (G) provided an attachment for additional binding or catalytic groups. From these simple derivatives, a series of terpyridine monomers was quickly created. These initial substrate binding domains include no binding group (A), hydrogen bonding groups (B and C), hydrophobic group (D), and crown ether (E). These five simple monomers were easily assembled into a 15-membered spatially addressed library using ruthenium trichloride and tested for binding to guests using several methods *(15)*. **Figure 4** illustrates the calorimetric response of each member of the library to the addition of bis(tetrabutylammonium) pimelate, with the greatest heat production correlating to the largest association. The receptor formed (B•B) with two thiourea derivatives indicated the strongest association, and the complex is shown in **Fig. 4**. Association with pentane-diammonium dipicrate salt was determined using solid–liquid extraction. This salt is insoluble in organic solvents, but is made soluble through association with an appropriate guest. Receptor E•E demonstrated the greatest ability to extract this guest into solution, indicating the interaction between both crown ethers and the diammonium salt. Binding can also be visually assayed by covalently linking the guest to a bead, then adding a solution of each ruthenium or iron complex

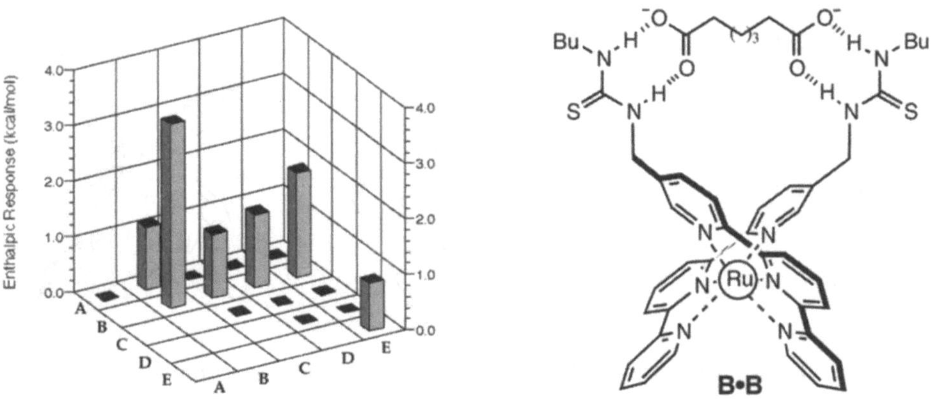

Fig. 4. Response of the library to TBA pimelate, and strongest binding complex **B•B**.

to the resin. Association is indicated by the amount of color retained by the bead after washing.

This combinatorial approach permits the rapid creation of a large number of multimeric receptors from a collection of monomers. In addition, subsequent generations can be generated through the modification of initial leads. For example, 10 monomers containing potentially catalytic groups were quickly dimerized with cobalt(II) chloride to form a 55-member library. Ten solutions contained only symmetric complexes while 45 contained mixtures of symmetric and asymmetric complexes. Each component mixture was combined with *bis*(2,4-dinitrophenyl)phosphate and the rate of phosphodiester cleavage was determined by measuring the change in absorbance at 400 nm over time. One component of the library, a symmetric complex containing an amide and an imidazole (H•H), was significantly more potent than the other members of the library. Using this initial lead, 12 additional monomers were prepared with variations of the active functional groups, and assembled into a second-generation 91-member library. Several complexes performed significantly better than the initial lead and the complex in **Fig. 5** demonstrated the greatest rate of cleavage from the second-generation receptor library. The creation of such a molecule is not only efficient using metal-templated self-assembly, but also would probably not have been conceived through a process of rational design.

This metal-templated self-assembly approach to combinatorial receptors is presented below, starting from functionalized terpyridine derivatives. Methods include the creation of monomers, dimerization with several metals to form dimeric receptors, and several procedures used to gauge association strength and catalytic activity of library components.

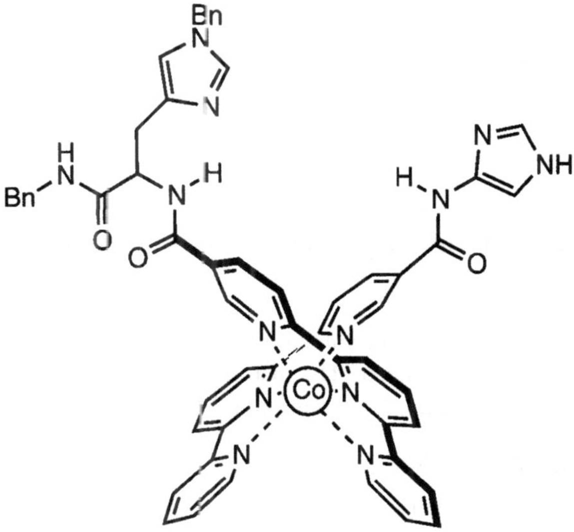

Fig. 5. Second-generation catalytic complex.

2. Materials

2.1. Creation of Terpyridine Monomers

2.1.1. Monomers via 5-Bromomethyl-Terpyridine

1. 5-Methyl-terpyridine.
2. N-bromo-succinimide.
3. Benzoyl peroxide.
4. Additional alcohol or amine with attached substrate binding domain (*see* **Note 1**).
5. Carbon tetrachloride.
6. Tetrahydrofuran (THF).
7. Hexanes.

2.1.2. Monomers via Terpyridine-5-Carboxylic Acid

1. 5-Methyl-terpyridine.
2. Potassium permanganate.
3. Dichloromethane.
4. Concentrated hydrochloric acid.
5. Thionyl chloride.
6. Additional alcohol or amine with attached substrate binding domain (*see* **Note 1**).

2.2. Preparation of Dimeric Receptors

1. Functionalized terpyridine (from above).
2. Ammonium iron(II) sulfate hexahydrate or cobalt(II) chloride or ruthenium trichloride trihydrate (*see* **Note 2**).

3. Ferrocenium hexafluorophosphate (if cobalt is chosen above).
4. N-ethyl-morpholine (if ruthenium is chosen above).
5. Saturated aqueous solution of potassium hexafluorophosphate.
6. Flash silica.
7. Ammonium hexafluorophosphate.
8. Dichloromethane, acetonitrile, and methanol.

2.3. Methods for Determining Binding Affinity and Catalytic Activity

2.3.1. Isothermal Calorimetry

1. Dimethylsulfoxide (DMSO) or solvent of choice for binding assay.
2. Isothermal titration calorimeter. The instrument used for these studies was purchased from Microcal, Inc., Northhampton, MA.

2.3.2. Solid–Liquid Extraction

1. Solvent for binding assay (in this case 5% acetonitrile/chloroform).
2. Picrate salt of desired guest.
3. UV/visible spectrophotometer.

2.3.3. Solid-Phase Colorimetric Detection

1. Functionalizable resin (Wang resin was used in this example).
2. Desired guest to be immobilized on the resin.
3. Equimolar solutions of each member of the library.
4. Mechanical shaker.
5. Chloroform for washing.

2.3.4. Catalysis of Nitrophenol Cleavage

1. Catalytic substrate [in this case bis(2,4-dinitrophenyl)phosphate].
2. UV/visible spectrophotometer.

3. Methods

3.1. Creation of Terpyridine Monomers

3.1.1. Monomers via 5-Bromomethyl-Terpyridine

1. Dissolve 5-methyl-terpyridine (roughly 1 g) in 20 mL carbon tetrachloride.
2. Add an equimolar amount of N-bromo-succinimide and 50 mg benzoyl peroxide.
3. Reflux the solution for 3 h, then concentrate to dryness.
4. Recrystalize the yellow solid with hexanes (*see* **Note 3**).
5. Dissolve the 5-bromomethyl-terpyridine in THF with an equimolar amount of desired variable alcohol or amine (*see* **Note 1**) and reflux overnight.
6. Concentrate to dryness and recrystallize with hexanes.

3.1.2. Monomers via Terpyridine-5-Carboxylic Acid

1. Dissolve 5-methyl-terpyridine (roughly 1 g) in 300 mL water.

2. Add three molar equivalents of aqueous potassium permanganate (*see* **Note 4**).
3. After refluxing for 16 h, filter the solution while still hot.
4. Wash with dichloromethane to remove unreacted terpyridine.
5. Remove the water via lyophilization, and dissolve the resulting white solid in 50 mL water.
6. Acidify the solution with aqueous HCl until pH = 3 or until a precipitate starts to form (*see* **Note 5**).
7. Collect the solid by vacuum filtration, and dry under high vacuum.
8. Suspend the residue in 50 mL dichloromethane, and add 50 mL thionyl chloride.
9. Reflux for 12 h.
10. Filter any insoluble materials, then concentrate to dryness under reduced pressure.
11. Suspend the residue in 10 mL dichloromethane with an equimolar amount of the desired variable amine (*see* **Note 1**) and reflux for 4 h.
12. Concentrate to dryness and recrystallize with hexanes.

3.2. Preparation of Dimeric Receptors

3.2.1. Bis(terpyridine)–Iron Complex

1. Dissolve 50 mg of the functionalized terpyridine in 2 mL acetonitrile. If a mixture of asymmetric complexes is desired, use more than one ligand (*see* **Note 6**).
2. Add one half of an equivalent of ammonium iron(II) sulfate hexahydrate dissolved in 2 mL water, and stir for 2 h.
3. Pour the solution above into 50 mL saturated potassium hexafluorophosphate, causing the immediate formation of a precipitate (*see* **Note 7**).
4. Collect the violet solid by vacuum filtration (*see* **Note 8**).

3.2.2. Bis(terpyridine)–Cobalt Complex

1. Dissolve the functionalized terpyridine in 15 mL acetonitrile (*see* **Note 6**).
2. Add one equivalent of cobalt(II) perchlorate hexahydrate and stir the solution for 1 h.
3. If desired, the crude cobalt(II) complex can be isolated as a red solid by removing the solvent under reduced pressure. These complexes are air stable and produce acceptable, if widely dispersed NMR spectra.
4. Dissolve the crude cobalt(II) complex in 15 mL acetonitrile. Add one equivalent of ferrocenium hexafluorophosphate *(18)* and stir the solution for 2 h (*see* **Note 9**).
5. Reduce the solvent volume to 2 mL under reduced pressure and add to a solution of saturated aqueous potassium hexafluorophosphate. The yellow precipitate was collected by filtration (*see* **Note 8**).

3.2.3. Asymmetric Bis(terpyridine)–Ruthenium Complex

1. Dissolve one functionalized terpyridine in 10 mL absolute ethanol.
2. Add an equimolar amount of ruthenium trichloride trihydrate and heat the solution at reflux for 2 h.
3. After cooling to room temperature, use vacuum filtration to collect the brown solid that is formed.

4. Dissolve this mono-terpyridine ruthenium complex in 20 mL absolute ethanol.
5. Add a solution containing one equivalent each of a second functionalized terpyridine and N-ethyl-morpholine, and heat the solution to reflux for 4 h.
6. After cooling to room temperature, pour the solution into saturated potassium hexafluorophosphate.
7. Collect the red/black solid precipitate using vacuum filtration.
8. Recrystallize from acetone/ethyl ether.
7. Further purification was accomplished by silica column chromatography (*see* **Note 10**). The silica must first be saturated with ammonium hexafluorophosphate before the sample is eluted with dichloromethane/acetonitrile/methanol (*see* **Note 8**).

3.3. Methods for Determining Binding Affinity and Catalytic Activity

3.3.1. Isothermal Calorimetry

1. Prepare a solution (roughly 1 m*M*) of each member of the receptor library in DMSO (*see* **Note 11**).
2. Rinse the cell of the isothermal titration calorimeter once with this receptor solution, then fill the cell with a second aliquot of the same solution.
3. Prepare a second solution in DMSO containing the guest at roughly 6–7 times the concentration of the receptor. Fill the injection syringe with the guest solution and place it into the calorimeter cell. Equilibrate to a constant baseline over several minutes.
4. Inject 10 μL of the guest solution into the receptor solution and measure the change in the heat needed to maintain the baseline.
5. Determine the heat of dilution by repeating **step 4**, but substituting pure solvent for the receptor solution. Subtract the value of the change in heat observed during the dilution experiment from that obtained during complex formation.
6. The binding affinity is indicated by the relative amount of heat produced from association (*see* **Note 12**).

3.3.2. Solid–Liquid Extraction

1. Prepare a solution of each metal-templated receptor of equal concentration (roughly 0.25 m*M* was used for these procedures). For these protocols 5% acetonitrile/chloroform was most effective, although other solvents are possible (*see* **Note 11**).
2. Add each solution to a vial containing an equivalent amount of the solid guest, as the picrate salt (*see* **Note 13**). One microgram of the picrate salt was sufficient to observe binding. To ease setup, a solution of the guest can be dispersed, and subsequently evaporated to dryness under reduced pressure.
3. Subject all vials to sonication for 1 h to aid in dissolution of guest.
4. Cool each sample to room temperature and allow each to settle for 1 h.
5. Obtain a ultraviolet–visible (UV-Vis) spectra for each solution. If available, a microplate reader is recommended for the assay of large libraries.
6. Determine UV-Vis absorbance of a stock solvent containing no metal complex and subtract to determine the amount of picrate extracted into solution by the receptor.

3.3.3. Solid-Phase Colorimetric Detection

1. Immobilize the guest on a solid resin support. For this study attach an amino acid (glutamate *t*-butyl ester) to alkoxybenzyl alcohol (Wang) resin using carbonyldiimidazole. Deprotect both esters using 50% trifluoroacetic acid/dichloromethane, and treat with 1 *M* tetrabutylammonium hydroxide in methanol to generate the dianion.
2. Prepare a 1.0 m*M* solution of each bis(terpyridine)–ruthenium receptor in 5% acetonitrile/chloroform. Dilute a 400 µL aliquot of each solution with 2.00 mL chloroform (*see* **Note 14**).
3. Add 10 mg of the resin containing the guest to the receptor solution, and agitate on a mechanical shaker overnight.
4. Decant the solution from the resin, and wash twice with chloroform to remove any residual receptor solution.
5. Remove residual liquid under reduced pressure. Visually evaluate the color of each resin compared to a control containing no receptor. The degree of color change is consistent with the degree of binding by the metal-templated receptors.

3.3.4. Catalysis of Nitrophenol Cleavage

1. Prepare mixtures of bis(terpyridine)–cobalt(II) complexes by the method in **Subheading 3.2.2.** (*see* **Note 2**).
2. Add a solution of bis(2,4-dinitrophenyl)phosphate to each mixture, and monitor the change in absorbance at 400 nm due to the release of 2,4-dinitrophenoxide. The rate of change is proportional to the rate of phosphodiester cleavage.
3. Correct the rates of the asymmetric complexes by subtracting the contributions by the symmetric complexes. Each mixture contains 25% of each symmetric complex and 50% of the asymmetric complex (*see* **Note 6**).
4. Determine the most active components of the library.
5. Assemble a second-generation library by creating a second collection of terpyridine monomers with modifications of the initial lead, and repeat from **step 1**.

4. Notes

1. Functionalized terpyridines can be created to contain a variety of binding and catalytic functional groups *(15)*. Bromomethyl derivatives can be reacted with the sodium salts of alcohols to form ethers, or reacted with the sodium salts of amines to form substituted amines. Each sodium salt was formed by reaction with sodium hydride. Gabriel synthesis leads to the creation of an aminomethyl derivative, which can be converted into amide with acyl chlorides, ureas with isocyanates, or thioureas with isothiocyanates. Terpyridine-5-carboxylic acid is converted to the acyl chloride with thionyl chloride and can be coupled with a variety of amines to form amide derivatives.
2. The choice of metal does not noticeably affect the structure of the bis(terpyridine) complex, only its stability. Complexes between iron and terpyridine are easily cleaved by weakly basic guests such as carboxylate anions. Cobalt(II) complexes

are also substitution labile, but can be easily oxidized to the substitution inert cobalt(III) complex. Once they are created, ruthenium complexes are the most stable, but require more strenuous conditions for formation. While cobalt presents the best overall characteristics for library formation and screening, ruthenium is optimal for complete characterization of asymmetric complexes.

3. This initial residue contains mainly bromomethyl-terpyridine, but also the dibromo derivative as well as unreacted methyl-terpyridine. Proceeding with the next step without purification does not lead to reduced yields.
4. Maximal results were obtained by slow addition of portions of the aqueous potassium permanganate. Each portion was added only when the purple color had faded from the previous addition.
5. After collecting the initial precipitate, further addition of acid resulted in the formation of more white solid. In this fashion several portions of the product were collected.
6. While the use of one terpyridine monomer results in only one dimeric complex, the use of more than one monomer leads to a mixture of complexes. Two monomers (**X** and **Y**) will lead to 25% dimeric **X•X**, 25% dimeric **Y•Y**, and 50% mixed complex **X•Y**; determined by high-pressure liquid chromatography (HPLC) of reaction products. Complexes (**X•Y**) and (**Y•X**) are degenerate and only one need be considered. Each bis(terpyridine)•metal complex exists as a pair of stereoisomers due to the two possible arrangements around the metal center, and no attempt was made to distinguish or separate these stereogenic pairs. The total number of components of these combinatorial libraries must include both symmetric and asymmetric complexes. If n terpyridine monomers are combined, the same number of symmetric complexes (**X•X**) are formed. The number of asymmetric complexes (**X•Y**) formed is $n!/[(n-2)!2!]$ or better represented as $n(n-1)/2$. Thus the total number of dimeric complexes is the sum of n and $n(n-1)/2$, which is $n(n+1)/2$. For example, a library with 5 monomers will form 15 dimers, 10 monomers will produce 55 dimers, and 20 monomers will produce 210 dimers.
7. Purification of bis(terpyridine)–iron complexes can be complicated by the potential for subsequent loss of terpyridine ligand. Precipitation as the insoluble hexafluorophosphate salt provides an efficient purification without risking ligand substitution.
8. Purity and identity are determined using nuclear magnetic resonance (NMR) and mass spectra. Complexes formed from iron(II), cobalt(III), and ruthenium(III) are diamagnetic and exhibit traditional NMR spectra. Cobalt(II) complexes are paramagnetic and produce broadened spectra with downfield shifts in all signals. With increased sweep widths, a series of singlets are observed from 0 to 100 ppm. Mass spectra using fast atom bombardment (FAB) ionization were successful for cobalt(III) and ruthenium(III) complexes , with each producing a molecular ion signals for the metal complex without counterions. More substitution labile iron(II) and cobalt(II) complexes did not display similar molecular ion peaks.
9. A variety of oxidants can transfer an electron from the cobalt center including bromine, selenium oxide, and lead oxide. Ferrocenium hexafluorophosphate provided the most mild conditions and no purification difficulties.

10. Separation of bis(terpyridine)–ruthenium complexes using flash silica chromatography proved difficult due to tight binding of these ion pairs to the silica. Elution efficiency was increased by presaturating the silica with ammonium hexafluorophosphate. Silica was allowed to stand in a dichloromethane solution of ammonium hexafluorophosphate for 1 h followed by removing the salt solution and washing the silica with several volumes of dichloromethane.
11. All combinatorial receptors in this account are created as spatially addressed libraries with each solution containing either one symmetric complex or mixtures of at most three symmetric and asymmetric complexes. These methods also permit the formation of large mixed libraries where a variety of monomers are mixed with one half an equivalent of the metal.
12. Calorimetric analysis requires some prior knowledge of the nature of the interaction. The formation of hydrogen-bonded host-guest complexes with dicarboxylates in DMSO was found to be exothermic ***(19)***. Endothermic association would require a reversal in the treatment of binding effectiveness.
13. Picrate salts are chosen due to their strong UV-Vis absorption and insolubility in organic solvents.
14. These solvent conditions depend on the degree of association. Initial bis(terpyridine) complexes are insoluble in pure chloroform but association is weak in solvent mixtures containing too much of the polar acetonitrile. Stronger binding is usually observed in more nonpolar solvent conditions.

References

1. DeMiguel, Y. R. and Sanders, J. K. M. (1997) Generation and screening of synthetic receptor libraries. *Curr. Op. Chem. Biol.* **2,** 417–421.
2. Chen, C.-T., Wagner, H., and Still, W. C. (1998) Fluorescent, sequence-selective peptide detection by synthetic small molecules. *Science* **279,** 851–853.
3. Haberhauer, G., Somogyi, L., and Rebek, J. (2000) Synthesis of a second-generation pseudopeptide platform. *Tetrahedron Lett.* **41,** 5013–5016.
4. Calama, M. C., Hulst, R., Fokkens, R., Nibbering, N. M. M., Timmerman, P., and Reinhoudt, D. N. (1998) Libraries of non-covalent hydrogen-bonded assemblies; combinatorial synthesis of supramolecular systems. *Chem. Commun.* 1021–1022.
5. Shimizu, K. D., Snapper, M. L., and Hoveyda, A. H. (1998) High-throughput strategies for the discovery of catalysis. *Chem. Eur. J.* **4,** 1885–1889.
6. Lehn, J.-M. and Eliseev, A. V. (2001) Dynamic combinatorial chemistry. *Science* **291,** 2331–2332.
7. Epstein, D. M., Choudhary, S., Churchill, M. R., Keil, K. M., Eliseev, A. V., and Morrow, J. R. (2001) Chloroform-soluble schiff-base Zn(II) or Cd(II) complexes from a dynamic combinatorial library. *Inorg. Chem.* **40,** 1591–1596.
8. Calama, M. C., Timmerman, P., and Reinhoudt, D. N. (2000) Guest-templated selection and amplification of a receptor by noncovalent combinatorial synthesis. *Angew. Chem. Int. Ed.* **39,** 755–757.
9. Linton, B. and Hamilton, A. (1997) Formation of artificial receptors by metal-templated self-assembly. *Chem. Rev.* **97,** 1669–1680.

10. Goodman, M. S., Weiss, J., and Hamilton, A. D. (1994) A self-assembling receptor for dicarboxylic acids. *Tetrahedron Lett.* **35,** 8943–8946.
11. Goodman, M. S., Jubian, V., and Hamilton, A. D. (1995) Metal-templated receptors for the effective complexation of dicarboxylates. *Tetrahedron Lett.* **36,** 2551–2554.
12. Goodman, M. S., Hamilton, A. D., and Weiss, J. (1995) Self-assembling chromogenic receptors for the recognition of dicarboxylic acids. *J. Am. Chem. Soc.* **117,** 8447–8455.
13. Constable, E. C. and Thompson, A. M. W. C. (1995) Strategies for the assembly of homo- and hetero-nuclear metalosupramolecules containing 2,2':6',2"-terpyridine metal-binding domains. *J. Chem. Soc. Dalton Trans.* 1615–1627.
14. Potts, K. T., Usifer, D. A., Guadalupe, A., and Abruna, H. D. (1987) 4-Vinyl, 6-vinyl, and 4'vinyl-2,2':6',2"terpyridinyl ligands: Their synthesis and the electrochemistry of their transition-metal coordination complexes. *J. Am. Chem. Soc.* **109,** 3961–3967.
15. Goodman, M. S., Jubian, V., Linton, B., and Hamilton, A. D. (1995) A combinatorial library approach to artificial receptor design. *J. Am. Chem. Soc.* **117,** 11610–11611.
16. Sauvage, J.-P., Collin, J.-P., Chambron, J.-C., Guillerez, S., and Coudret, C. (1994) Ruthenium(II) and osmiun(II) bis(terpyridine) complexes in covalently-linked multicomponent systems: Synthesis, electrochemical behavior, absorption spectra, and photochemical and photophysical properties. *Chem. Rev.* **94,** 993–1019.
17. Constable, E. C., Ward, M. D., and Corr, S. (1988) A convenient high yield synthesis of terpyridine. *Inorg. Chim. Acta.* **141,** 201–203.
18. McCollum, D. G., Yap, G. P. A., Rhinegold, A. L., and Bosnich, B. (1996) Bimetallic reactivity—Synthesis of bimetallic complexes of macrocyclic binucleating ligands containing 6-coordinate and 4-coordinate sites and their reactivity with dioxygen and other oxidants. *J. Am. Chem. Soc.* **118,** 1365–1379.
19. Linton, B. and Hamilton, A. D. (1999) Calorimetric investigation of guanidinium-carboxylate interactions. *Tetrahedron* **55,** 6027–6038.

9

Automated Structure Verification of Small Molecules Libraries Using 1D and 2D NMR Techniques

Gérard Rossé, Peter Neidig, and Harald Schröder

1. Introduction

The purity control and the structure verification of compound collections from automated synthesis and combinatorial chemistry play an essential role in the success of medicinal chemistry programs. High performance liquid chromatography (HPLC), mass spectrometry (MS), and liquid chromatography–mass spectrometry (LC-MS) techniques are generally accepted as the most appropriate means of characterization *(1,2)*. While these analytical methods are fast and easy to automate, they do not provide sufficient structural and quantitative data about the desired products.

Nuclear magnetic resonance (NMR) spectroscopy is the most informative analytical technique and is widely applied in combinatorial chemistry. However, an automated interpretation of the NMR spectral results is difficult *(3,4)*. Usually the interpretation can be supported by use of spectrum calculation *(5–18)* and structure generator programs *(8,12,18–21)*. Automated structure validation methods rely on ^{13}C NMR signal comparison using substructure/subspectra correlated databases or shift prediction methods *(8,15,22,23)*. We have recently introduced a novel NMR method called AutoDROP (Automated Definition and Recognition of Patterns) to rapidly analyze compounds libraries *(24–29)*. The method is based on experimental data obtained from the measured 1D or 2D $^{1}H,^{13}C$ correlated (HSQC) spectra.

The focus of this chapter is on the application of AutoDROP toward structure validation of every member of a substituted 4-phenylbenzopyrans library **1**. The results of the NMR interpretation are compared with electrospray-ionization mass spectrometry (ESIMS) and HPLC analysis. A detailed description of

From: *Methods in Molecular Biology, Combinatorial Library Methods and Protocols*
Edited by: L. B. English

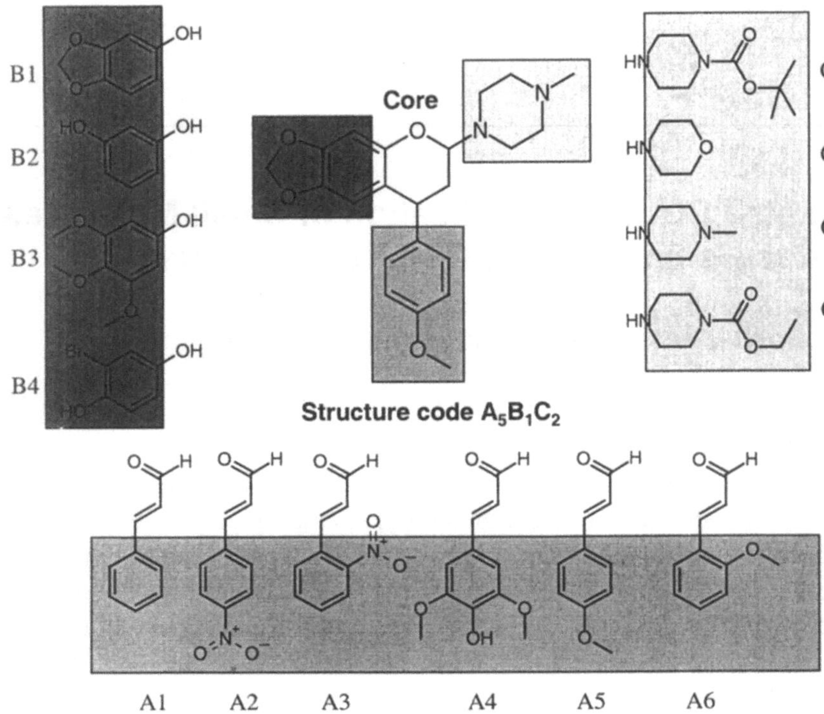

Fig. 1. Ninty-six 4-phenylbenzopyrans generated in a three component reaction. For $x = 6$, $y = 4$, and $z = 4$, $x \times y \times z = 96$ different compounds with the substructure codes $A_xB_yC_z$ are obtained. Library **1** is characterized by $x + y + z$ + core = 15 different structural fragments and six from the 96 compounds contain all structural fragments (e.g., $A_1B_1C_1$, $A_2B_2C_2$, $A_3B_3C_3$, $A_4B_4C_4$, $A_5B_1C_3$, and $A_6B_2C_4$).

AutoDROP has been published elsewhere *(24–29)*, so we will review only its basic principles in this chapter.

1.1. Modular Structural Elements of a Combinatorial Library

In combinatorial chemistry large numbers of compounds are synthesized by systematic combination of a relatively small number of molecules. For example, a three-component reaction may involve linking three classes of molecules (building blocks) A, B, C to form a product denoted ABC (**Fig. 1**). Each class may contain several species (A_1, A_2 . . . , A_i; B_1, B_2 . . . , B_i; C_1, C_2 . . . , C_i). With just 10 building blocks in each class, 1000 different products can be formed. Thus, the structures of the synthesized products can be formally represented as a combination of individual molecular fragments (substructures) with one fragment coming from each class of building block. In many cases a nonvariable region (core) occurring in all molecules can be observed. A substructure code $A_xB_yC_z$ defined by the synthesis can be assigned to each product. Both spectroscopic and chromatographic data can be regarded as the sum of data belonging to the substructures of a molecule.

1.2. Structure Verification Using AutoDROP Based on 2D NMR Spectra

Since synthesized products can be formally represented as a combination of individual molecular fragments, 2D NMR spectra can be regarded as the sum of spectra of substructures. The key idea of AutoDROP is to systematically examine 2D C,H correlated NMR spectra and to derive subspectra of the individual molecular fragments. The subspectra are managed as spectral patterns in AutoDROP. For the example described in **Fig. 1**, a total of 14 molecular fragments of the class A, B, and C and the core fragment were used to generate 96 different compounds of the form $A_xB_yC_z$. Each of the molecular fragments occurs repetitively in all structures (A_x 16-fold, B_y 24-fold, and C_z 24-fold). The repetitive occurrence of the corresponding signals in the total set of spectra is used by AutoDROP to automatically isolate and define the spectral patterns of specific molecular fragments. A linear combination of spectra is thereby used in a systematic way (**Fig. 2**) *(28)*.

Once the spectral patterns of all individual substructures have been defined, all available spectra can be tested for the presence of a particular substructure in the synthesized compounds. The proposed structure is verified (true) if all expected fragments are found (**Fig. 3A, Table 1**). If at least one of the expected patterns is not found, then the spectrum is not verified (false, **Fig. 3B**). Spectra with a low signal-to-noise ratio or with large amounts of impurities were automatically assigned an "unclear" category and should be checked manually. In the simplest case the verification procedure is based on the integration of spectral patterns and comparison to an automatically detected noise level. Better results are obtained if a signal (e.g., from the core) can be defined as an internal reference signal to normalize all integrals. Then a reference spectrum is defined for each substructure pattern. The corresponding integrals of the reference spectrum are defined as 100% and corresponding integral values of all other spectra are rescaled accordingly. During the verification it is then possible to apply an additional threshold that expresses the minimum signal intensity of identified patterns. Example: a spectrum related to the structure code $A_1B_1C_1$ would be classified as true if A_1, B_1, and C_1 are identified and at least each integral exceeds 30%.

1.3. Structure Verification Using AutoDROP Based on 1D NMR Spectra

The basic principles and strategies remain the same as described for AutoDROP based on 2D NMR. The 1D NMR spectra are translated into

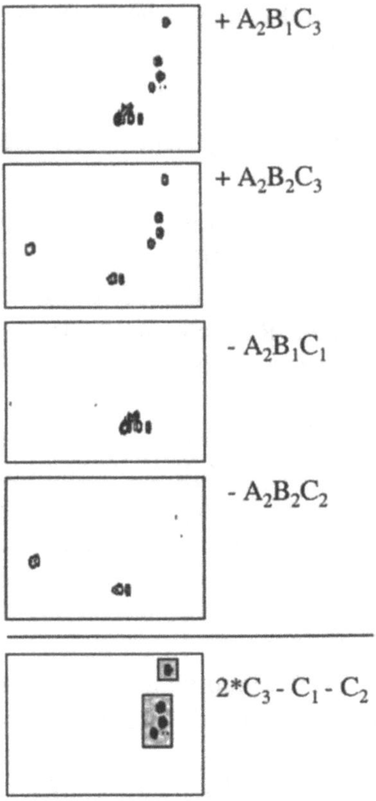

Fig. 2. Linear combination of spectra to extract pattern box C_3. Signals are peak picked and transformed into peak areas. Overlapping peak areas of spectra containing the structural fragment code C_3 are added (counted) and peak areas of spectra not containing C_3 are subtracted. The threshold is adjusted so that only peak areas of C_3 remain, and after a clustering step boxes are defined for each remaining peak area.

one-dimensional peak lists and clusters. It was necessary to make changes in various calculation techniques such as peak picking, cluster analysis, spectral pattern definition, noise level estimation, and pattern integration *(28)*. Again, individual structural fragments yield signals in different spectral regions and spectra are classified true if all requested spectral patterns could be verified (**Fig. 4**). Although there is usually more overlap of signals in the 1D spectra, good verification results have been obtained in many cases.

1.4. Aquisition of the NMR Spectra

The 4-phenylbenzopyran library **1** was synthesized using a multicomponent reaction *(30–32)* by the combination of phenols, unsaturated aldehydes, and

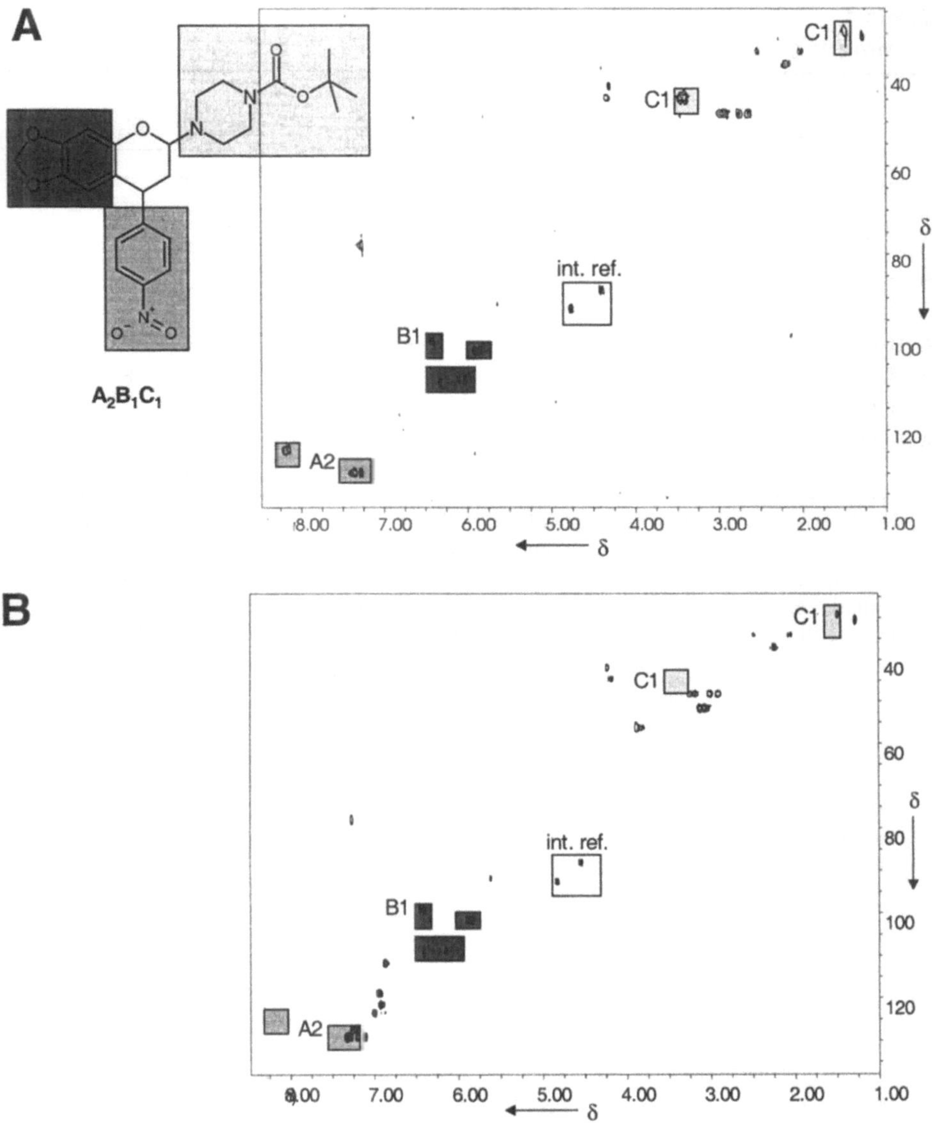

Fig. 3. Decomposition of a 2D HSQC spectrum of a compound into subspectra corresponding to each of the three molecular fragments A_2, B_1, and C_1. The width and height of the boxes indicate the expected range of chemical shift for the signals of a given fragment. A spectral pattern is defined by the combination of the corresponding boxes. **(A)** The spectral patterns of each fragment are found and the structure of the expected compound $A_2B_1C_1$ is therefore validated. **(B)** Structure of compound $A_2B_1C_1$ is not verified because the spectral patterns of both A_2 and C_1 are missing.

Table 1
Summary of the Verification Results

	NMR result	A_1	A_2	A_3	A_4	A_5	A_6	B_1	B_2	B_3	B_4	C_1	C_2	C_3	C_4
$A_1B_1C_1$	+	+	–	–	–	–	–	+	–	–	–	+	–	–	–
$A_1B_2C_1$	+	+	–	–	–	–	–	–	+	–	–	+	–	–	–
$A_1B_3C_1$	–	+	–	–	–	–	–	–	–	–	–	+	–	–	–
$A_1B_4C_1$	?	+	–	–	–	–	–	–	–	–	+	–	–	–	–
$A_1B_1C_2$	+	+	–	–	–	–	–	+	–	–	–	–	+	–	–
$A_1B_2C_2$	+	+	–	–	–	–	–	–	+	–	–	–	+	–	–
$A_1B_3C_2$	–	+	–	–	–	–	–	–	+	–	–	–	+	–	–
$A_1B_4C_2$	–	+	–	–	–	–	–	–	–	–	+	–	–	–	–
$A_2B_1C_1$	+	–	+	–	–	–	–	+	–	–	–	+	–	–	–
$A_2B_2C_1$	+	–	+	–	–	–	–	–	+	–	–	+	–	–	–

[a] In the columns labeled with fragment codes, the "+" and "–" entries indicate whether or not the corresponding spectral pattern is identified in a given spectrum. The column labeled NMR results indicates whether the structure is verified (+), false (–) or unclear (?). For example, for compound $A_1B_3C_2$ pattern A_1, B_2, and C_2, was identified and the compound was assigned false. In this case the sample has been exchanged and the correct structure code would be $A_1B_2C_2$.

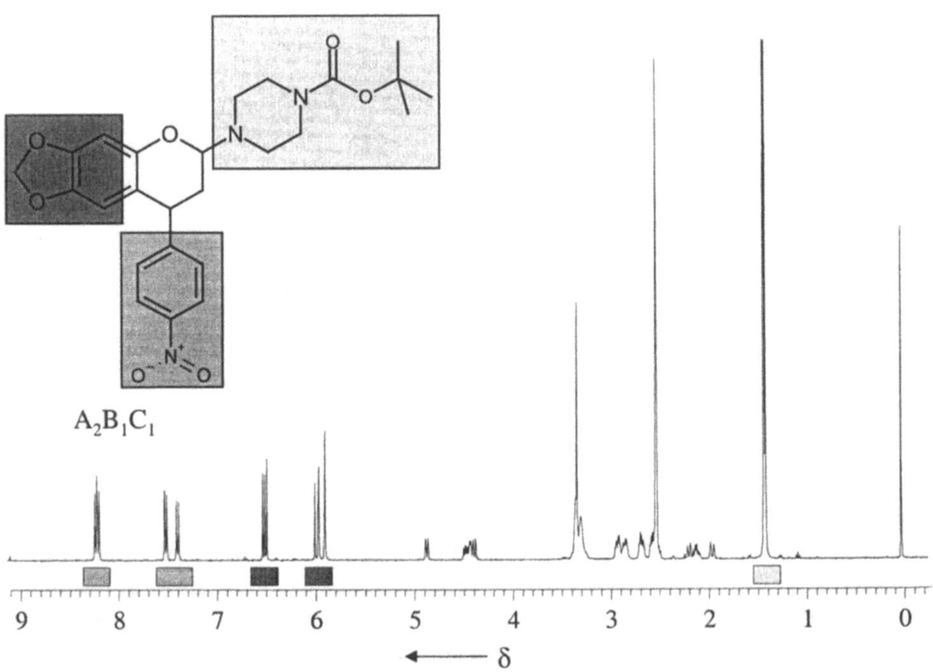

Fig. 4. 1D spectrum of a synthesis product. Different signals are related to different molecular fragments.

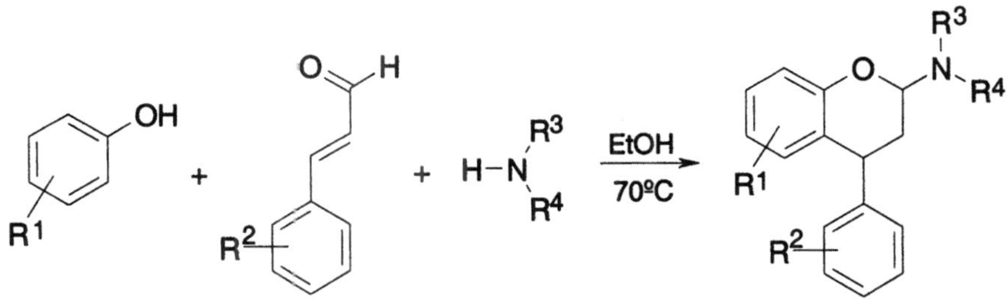

1

Fig. 5. Synthesis of 4-phenylbenzopyran library **1**.

secondary amines (**Fig. 5**). The products were purified before analysis. The ^1H NMR and 2D HSQC spectra of the 96 4-phenylbenzopyrans were measured using standard NMR probes (5-mm) within 16 h.

Sample throughput and sample preparation time can largely be improved using special hardware. Sample preparation and sample transfer of 5-mm tubes can be automated using a Tecan Genesis sample-handling robot in conjunction with a Bruker SampleRail system *(33)*. This equipment would allow a cycle time of 2 min for a proton NMR spectrum. Alternatives to conventional tube NMR measurements are parallel NMR measurements based on chemical shift imaging (CSI) *(34,35)* or flow-injection NMR *(36–38)*. In CSI-based parallel NMR the detection volume of the NMR probe is loaded with a bundle of 19 capillaries, each containing a different compound. The 1-mm capillaries can be filled automatically using a Probot robot *(39)*. After measurement of spatially resolved NMR data using the CSI technique, individual spectra of high resolution are obtained as traces extracted from the 3D Fourier-transformed data set. A cycle time of 6 min would be possible for a parallel proton NMR measurement of one bundle, thus 20 s per sample. In flow-injection NMR (**Fig. 6** [*36,37*]), e.g., samples are placed in a 96-well plate and dissolved in deuterated or nondeuterated solvent. An autosampler is required to pick samples from a well and transfer them to the spectrometer. Back mixing between samples is avoided by keeping gaps (e.g., gas bubbles) in between them during the transfer. After acquisition the sample can be sent to a waste receptacle or recovered by transfer back into the starting vial. A cycle time of less than 2 min for a proton NMR spectrum is possible.

1.5. Analysis of Compounds Library 1 Using AutoDROP

Once the ^1H NMR and 2D HSQC spectra have been acquired, the automated calculation steps to interpretate all 96 spectra are performed in less than 5 min. From the 96 compounds analyzed using AutoDROP based on 2D NMR, 68 structures were found true (one false positive), 11 proposed structures were

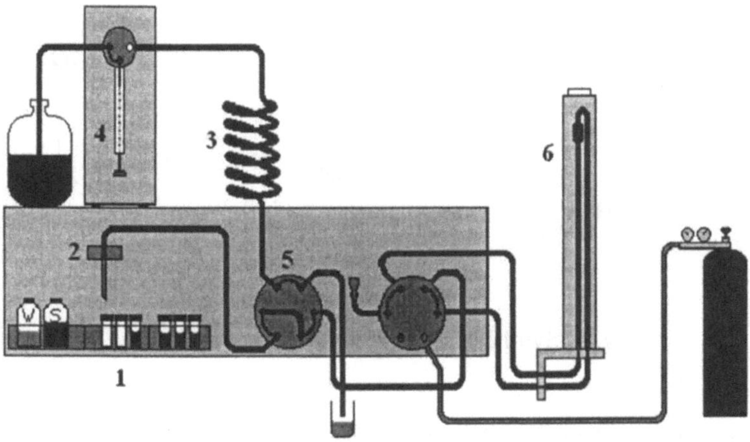

Fig. 6. BEST NMR (Bruker Efficient Sample Transfer) ***(36,37)***. The sample is sucked from a well plate **1** by the autosampler needle **2** and transferred into a sample loop **3**. The sample can be surrounded by gas and liquid buffer regions or gaps to separate it from the transport liquid. A dilutor syringe **4** is then filled with the appropriate volume of transfer liquid, which is used to push the contents of the sample loop **3** after appropriate valve switching **5** into the probehead **6**.

assigned false (one false negative), and 17 proposed structures were found "unclear" (*see* **Fig. 9**). Spectra, structures, and results of analysis can be visualized in a flexible way using a viewer with a direct interface to AutoDROP (**Fig. 7**). Further analysis of the NMR spectra showed that the synthesis of the 4-phenylbenzopyrans led to mixtures of diastereomers in an 1.8:1 (syn:anti) ratio. Two side products **2** and **3** were identified (**Fig. 8**).

1.6. ESIMS Analysis

The MS analysis of a library can be performed by an automated control of the expected molecular ions ***(40,41)***. The highest intensity of each spectrum is set to 100% after subtraction of a noise level and all peaks are scaled accordingly. The mass peaks of the expected molecular ions ($[M+H]^+$, $[M+Na]^+$, and $[M+NH_4]^+$) are then searched in each spectrum. The mass was verified when the sum of the relative intensities is above 30%. For a relative intensity below 10% the mass was not validated. A relative intensity between 10% and 30% lead to the category "unclear." ESIMS analysis of library **1** indicates 73 true structures, 17 wrong, and 6 unclear (*see* **Fig. 9**).

1.7. HPLC Analysis

The retention times of library compounds can be exploited for an automated structure control ***(42)***. The principle relies on the fact that the individual con-

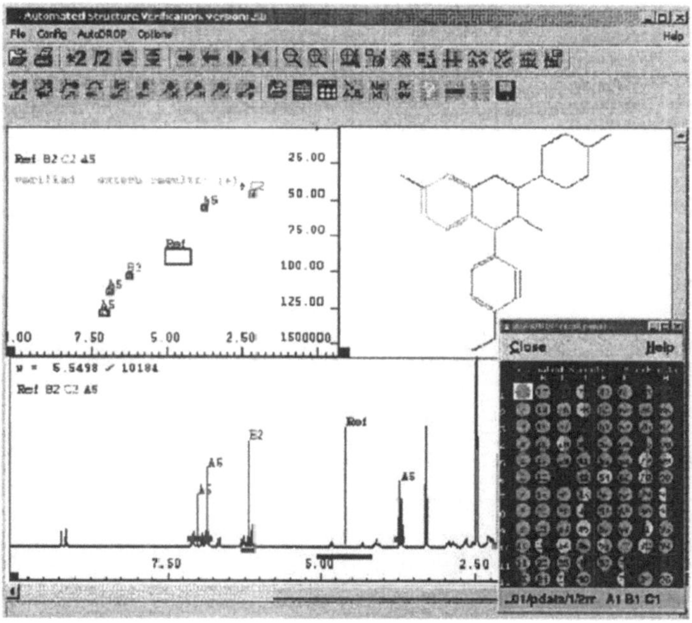

Fig. 7. Graphical results from an AutoDROP calculation shown in a data panel from which spectra, structures, and graphical patterns can be selected. The calculation of NMR and MS methods is visualized in the data panel (NMR results: left hemisphere, MS results: right hemisphere).

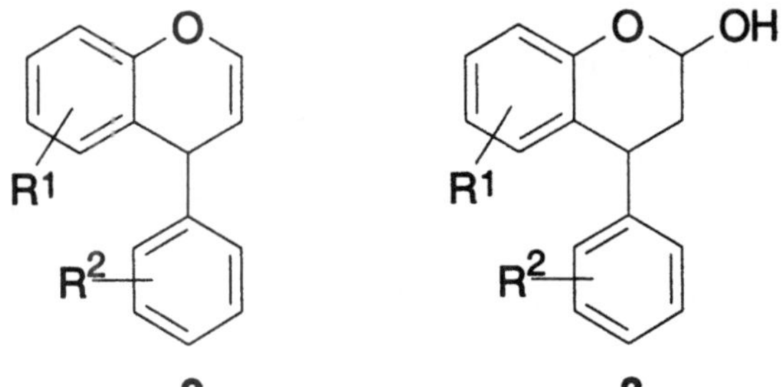

Fig. 8. Identified side-products **2** and **3**.

tribution to the retention time of a substructure does not strongly depend on the total composition of the compound. Thus, relative retention time contributions can be assigned to each individual substructure. These were obtained from the differences in retention between compounds (**Table 2**). The retention time of a molecule is then calculated as the sum of the relative retention time contributions.

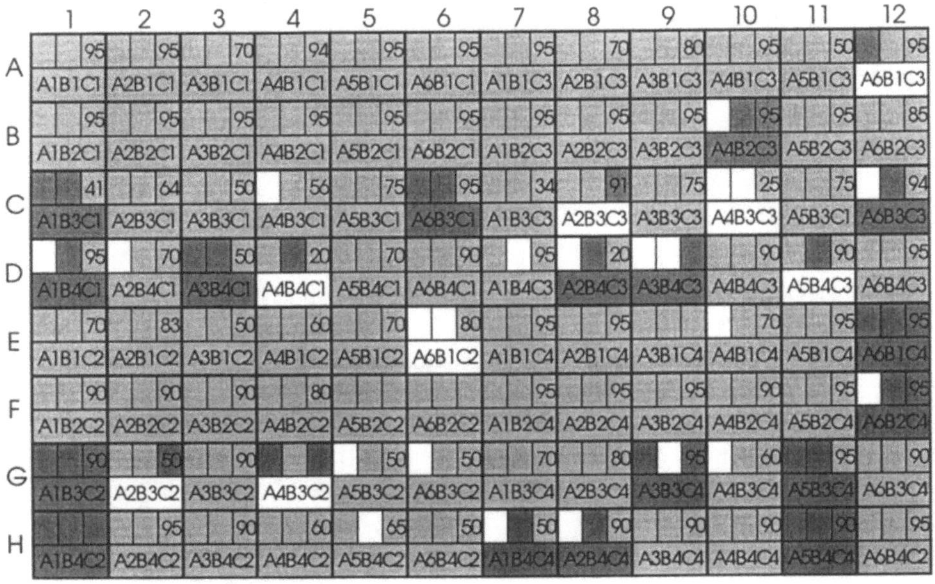

Fig. 9. Results of the automated NMR, ESIMS, and HPLC analysis. Each cell contains the expected structure code, the final assignment, and the data for NMR (top left), ESIMS (top middle), and HPLC (top right). Light grey coloration means that the proposed structure is true in NMR, gives the expected molecular ion in ESIMS, and shows the expected retention time in HPLC. Dark grey means that the proposed structure is false following NMR, does not give a diagnostic molecular ion in ESIMS, or the retention time differs from the expected one. White is given for unclear results in both NMR and ESIMS. HPLC purity is given in% (top right). Combined results are given in the structure code field (light grey: true, dark grey: false, white: unclear). The classification "true" of the HPLC analysis was not taken into consideration for the final assignment. Contradictory results lead to the final category "unclear." Eighteen compounds were not obtained by the synthetic procedure (B10, C1, C6, C12, D1, D3, D4, D8, D9, E12, F12, G1, G9, G11, H1, H7, H8, H11).

Appropriate reference compounds were selected to determine the relative retention time contributions.

Substructures A_4, B_2, and B_3 showed a significant influence on the retention times. The product is defined "true" when the variance between calculated and experimental retention time is between ± 0.3 min (**Table 3**). The HPLC analysis indicated 84 structures as correct and 12 as wrong (*see* **Fig. 9**). Side product **3** had a similar retention time as the target compound. Therefore, for five samples, false positive assignments were obtained.

Table 2
Relative HPLC Retention Time Increments of Individual Substructures[e]

Ref. compound	Retention time difference	Relative incremental retention time
$A_1B_2C_1$	0[a]	A1: =1[c]
$A_2B_2C_1$	0.06[a]	A2: 1.06
$A_3B_2C_1$	−0.04[a]	A3: 0.96
$A_4B_2C_1$	−0.90[a]	A4: 0.10
$A_5B_2C_1$	0.02[a]	A5: 1.02
$A_6B_2C_1$	0.09[a]	A6: 1.09
$A_5B_1C_1$	0[b]	B1: = 1[c]
$A_5B_2C_1$	−0.57[b]	B2: 0.43
$A_5B_3C_1$	0.83[b]	B3: 1.83
$A_5B_4C_1$	−0.10[b]	B4: 0.90
$A_1B_2C_1$	0[a]	C1: 0.97[d]
$A_1B_2C_2$	0.11[a]	C2: 1.08
$A_1B_2C_3$	0.03[a]	C3: 1.00
$A_1B_2C_4$	0.03[a]	C4: 1.00

[a] Retention time difference relative to $A_1B_2C_1$.
[b] Retention time difference relative to $A_5B_1C_1$.
[c] The increments of A_1 and B_1 were set to 1.
[d] The increment of C_1 was calculated from the retention time using increments of A_1 and B_2.
[e] The increments were calculated relative to A_1, B_1, and C_1 respectively.

Table 3
Calculated and Measured Retention Time[a]

	Retention time		
	Calculated[b]	Found[c]	Result
$A_3B_1C_3$	2.93	2.86/2.94	true
$A_3B_2C_3$	2.39	2.38/2.46	true
$A_3B_3C_3$	3.79	3.90	true
$A_3B_4C_3$	2.86	—[d]	false
$A_3B_1C_2$	3.04	2.85/2.93	true
$A_3B_2C_2$	2.47	2.39/2.47	true
$A_3B_3C_2$	3.87	2.50	false
$A_3B_4C_2$	2.94	2.84	true
$A_4B_1C_3$	2.10	2.04	true
$A_4B_2C_3$	1.53	2.21/2.36	false

[a] In minutes.
[b] Obtained from the sum of the corresponding increments.
[c] Usually two retention time resulting from each diastereomere are found.
[d] Not enough substance.

1.8. Comparison of the Automated 2D NMR, MS, and HPLC Analysis

An automated procedure should give a final result for each sample, take in account contrary results of the different methods, and extract samples that cannot be reliably assigned a result. For 81 compounds out of 96 the results of the automated 2D NMR, ESIMS, and HPLC analysis were in accordance (**Fig. 9**). When only MS and NMR results are compared, 92 results were in accordance. The HPLC analysis was of limited diagnostic value, since 11 substructures did not have a significant influence on the retention time.

The results of the automated 2D NMR, MS, and HPLC analysis were combined to a final assignment. Classification "true" of the HPLC analysis was not taken into consideration. Contradictory results lead to the final category "unclear." Seven samples remained unclassified and in these cases the classification was done by the manual interpretation of the ^{1}H NMR spectra. This analysis showed that 78 compounds were obtained by the synthetic procedure.

2. Materials

All chemicals and solvents are purchased from Fluka or Aldrich.

2.1. NMR

Bruker DRX 400-spectrometer (^{1}H, 400.1 MHz; ^{13}C, 100.6 MHz); *Bruker* 5-mm inverse broadband NMR probe; *Bruker* B-ACS 120 sample changer for acquisition of spectra using standard NMR probes (5-mm). Bruker Efficient Sample Transfer (BEST-NMR) for flow-injection acquisition, 4-mm flow cell, 120 μL cell volume, Gilson 215 Liquid Handler is used for sample transfer. Data processing using Xwinnmr 2.6. AutoDROP is implemented in AMIX software (Bruker). With the exception of Gilson 215 Liquid Handler (Gilson), all items are supplied by Bruker Analytik GmbH, Rheinstetten, Germany.

2.2. MS

PE Sciex API 300. Data processing using MultiView version 1.3. ESIMS spectra were measured in positive mode. Mass range was 210–900 amu. Spectra were centroided and exported for analysis as text.

2.3. HPLC

Standard gradient HPLC analytical and preparative system can be used.

2.4. Preparation of 4-Phenylbenzopyran Library 1

1. Prepare a 0.5 *M* stock solution in ethanol of all reagents.
2. Add a solution of a phenol (400 μL, 0.2 mmol) to the corresponding reactor.
3. Add a solution of the corresponding unsaturated aldehyde (400 μL, 0.2 mmol).

4. Add a solution of the appropriate secondary amine (400 µL, 0.2 mmol).
5. Close the reactors and heat reaction solutions to 70°C for 3 h (*see* **Note 1**).
6. Allow mixtures to reach room temperature.
7. Collect 24 compounds that did precipitate by decantation.
8. Purify the remaining 72 compounds using preparative HPLC (*see* **Note 2**).

3. Methods

3.1. Acquisition of NMR Spectra

1. Dissolve compounds in 600 µL DMSO-d_6 and measure in 5-mm glass tubes at 298 K using the 120-sample changer. When using flow injection, samples are placed in a 96-well plate and diluted in 300 µL DMSO-d_6 or DMSO (*see* **Note 3**).
2. Measure 16 scans for ^{1}H NMR (*see* **Note 4**).
3. Measure two scans per increment and 128 experiments are for 2D HSQC. The relaxation delay is 1 s (*see* **Note 4**).
4. Cycle time to acquire ^{1}H NMR, 2D HSQC and change the sample is 10 min per sample.
5. Perform automated phase correction procedure (*see* **Note 5**):
   a. Peak picking.
   b. Adding those rows (columns) containing peaks.
   c. Phase correction of the resulting 1D spectrum.
   d. Phase correction of the 2D spectrum using the 1D phase correction values.

3.2. Analysis of the Spectra Using AutoDROP

1. Enter into the software (*see* **Note 6**):
   a. List of codes for the possible molecular fragments involved in the combinatorial reaction.
   b. The paths to the recorded spectra and the associated structure codes.
2. Perform calculation step to define the integration boxes for each molecular fragment. Output are boxes assigned to each fragment.
3. Perform calculation step to determine appropriate reference spectra (*see* **Note 7**).
4. Perform calculation step to integrate all spectra. As an output a graphical display in rack format using three colors (red, green, and yellow) to characterize the samples is shown and a textual result list is written on disk (*see* **Note 8**).

3.3. Analytical HPLC

1. Prepare 5 m*M* solution of each compound in acetonitrile (*see* **Note 9**). For example, the solutions can be placed into the wells of a 96-well plate.
2. Place the samples onto the HPLC autosampler (*see* **Note 10**).
3. Inject 10 µL of each solution on a YMC Pack Pro C_{18} column (5 µm, 75 × 4.6 mm).
4. Analyze the compounds at 254 nm using the following method:
   a. Flow rate 2.5 mL/min.
   b. Solvents 0.05% CF_3COOH in H_2O (A) and acetonitrile (B).
   c. Gradient: within 4 min from 20% B to 95% B; 1.5 min at 95% B; in 0.2 min from 95% B to 20% B; 1.3 min at 20% B.

3.4. Preparative HPLC

1. Inject reaction mixtures on a YMC Pack Pro C_{18} column (5 µm, 120Å, 50 × 20 mm)
2. Analyze the compounds at 254 nm using following method:
   a. Flow rate 35 mL/min (*see* **Note 11**).
   b. Solvents H_2O (A), acetonitrile (B).
   c. Gradient: within 3.3 min from 20% B to 95% B, 1.2 min at 95% B, in 0.1 min from 95% B to 20% B, 0.1 min at 20% B; detection at 254 nm.

4. Notes

1. Caution. Use glass reactor with screw caps when heating ethanol solution at 70°C. Internal pressure. Wait until reactors have reached room temperature before opening them.
2. The reaction mixtures are directly injected on the preparative HPLC column.
3. Solvent suppression might be necessary.
4. Adjust the number of scans according to the sample amount. The relaxation delay can be shortened to save time.
5. Instead of applying phase correction, magnitude calculation can be performed.
6. Has been automated by using Unix shell scripts
7. Manual supervision of the selected reference spectra is recommended.
8. Steps 1 to 4 to analyze all 96 spectra are done in less than 5 min.
9. Caution. Acetonitrile is a toxic solvent. Always wear gloves when working with it.
10. Most providers of HPLC hardware do offer autosamplers for 96-well plates.
11. Make sure the HPLC hardware will handle flow rate of 35 mL/min.

Acknowledgments

We thank Mr. W. Meister for ESIMS measurements and Angewandte Chemie for the kind allowance of printing Figures 1, 3, 5, 8, and Table 1 *(22)*. We are grateful to Dr. David Thorpe (Selectide-Aventis) for a critical reading of the manuscript.

References

1. Jung, G. (ed.) (1999) *Combinatorial Chemistry*, Wiley-VCH, Weinheim, and references cited therein.
2. Sepetov, N. and Issakova, O. (1999) Analytical characterization of synthetic organic libraries. *Comb. Chem. Technol.* 169–203.
3. Pretsch, A. (1998) Möglichkeiten und Grenzen der vollautomatischen Spektreninterpretation. *Nachr. Chem. Tech. Lab.* **46(4, Suppl.),** A71–A73.
4. Shapiro, M. J. and Gounarides, J. S. (1999) NMR methods utilized in combinatorial chemistry research. *Prog. NMR Spectrosc.* **35,** 153–200.
5. Abraham, R. J. (1999) A model for the calculation of proton chemical shifts in non-conjugated organic compounds. *Prog. NMR Spectrosc.*, **35,** 85–152.
6. Pretsch, E. and Bürgin Schaller, R. (1994) A computer program for the automatic estimation of ^{1}H NMR chemical shifts. *Anal. Chim. Acta* **312,** 95–105.

7. Pretsch, E. and Fürst, A. (1990) A computer program for the prediction of ^{13}C-NMR chemical shifts of organic compounds. *Anal. Chim. Acta* **229,** 17–25.
8. PredictNMR, CheckNMR and Assemble by Upstream Solutions (www.upstream.ch).
9. Thiele, H., Paape, R., Maier, W., and Grzonka, M. (1995) Ein datenbank- und informationssystem zur Verwaltung und Interpretation von NMR-Spektren, *GIT Fachz. Lab.* **7,** 668–670.
10. WinSpecedit by Bruker (www.bruker.de).
11. Barth, A. (1992) Specinfo—An integrated spectroscopic information system. *J. Chem. Inf. Comput. Sci.* **32,** 291.
12. SpecInfo and SpecSolv by Wiley-VCH (www.chemicalconcepts.com, www. creonlabcontrol.com).
13. Kalchhauser, H. and Robien, W. (1984) CSEARCH: A computer program for identification of organic compounds and fully automated assignments of carbon-13 nuclear magnetic resonance spectra, *J. Chem. Inf. Comput. Sci.* **25(2),** 103–108.
14. Sadtler Suite by Bio Rad Laboratories (www.sadtlersuite.com).
15. HNMR, CNMR, CombiNMR and Structure elucidator by advanced chemistry development (www.acdlabs.com).
16. gNMR by adept scientific (www.adeptscience.co.uk).
17. HyperNMR by hypercube (www.hyper.com).
18. NMRSCAPE and NMR-SAMS by spectrum research (www.specres.com).
19. Bürgin Schaller, R., Munk, M. E., and Pretsch, E. (1996) Spectra estimation for computer-aided structure determination. *J. Chem. Inf. Comput. Sci.* **36,** 239–243.
20. Schriber, H. and Pretsch, E. (1997) Rule-based system to derive automatically good-list and bad-list entries for structure generators from spectra. *J. Chem. Inf. Comput. Sci.* **37,** 884–891.
21. Badertscher, M., Korytko, A., Schulz, K. P., Madison, M., Munk, M. E., Portmann, P., Junghans, M., Fontana, P., and Pretsch E. (2000) Assemble 2.0: A structure generator. *Chemom. Intell. Lab. Syst.* **51,** 73–79.
22. Will, M., Fachinger, W., and Richert, J. R. (1996) Fully automated structure elucidation—a spectroscopist's dream comes true. *J. Chem. Inf. Comput. Sci.* **32(2),** 221–227.
23. Williams, A., Mityushev, D., Shilay, V., and Kvasha, M. (1999) *NMR prediction software and tubeless NMR—an analytical tool for screening of combinatorial libraries*, Somerset, New Jersey, November 14–19, Eastern Analytical Symposium, Presentation.
24. Schröder, H., Neidig, P., and Rossé, G. (2000) High throughput structure verification of a substituted 4-phenylbenzopyran library using 2D NMR techniques. *Angew. Chem., Intl. Ed.* **39,** 3816–3819.
25. Schröder, H., Neidig, P., and Rossé, G. (2000) *AUTODROP, a novel method for the automated structure verification in combinatorial chemistry*, Pacifichem, International Chemical Congress of Pacific Basin Societies, Honolulu, Hawaii, December 14–19, Poster presentation.

26. Schröder, H. and Neidig, P. (1998) *Method of verifying the synthesis of organic molecules in combinatorial chemistry*. DE-19849231–C2 and US sn 09/422,639.
27. Schröder, H., Rossé, G., and Neidig, P. (1999) *Automated structure verification of combinatorial library members using 2D NMR techniques*, 37th IUPAC Congress/27th GDCh General Meeting, Berlin, Germany, August 14–19, Poster presentation.
28. Schröder, H. and Neidig P. (1999) AutoDROP, a new method of automated structure verification in combinatorial chemistry. *Bruker Report* **147,** 18–21.
29. Fischer, C., Neidig, P., and Schröder, H. (2000) New AutoDROP development: structure verification in combinatorial chemistry based on 1D NMR spectra. *Bruker Report* **148,** 27.
30. Jurd, L. (1991) Synthesis of 4-phenyl-2*H*-1-benzopyranes. *J. Heterocycl. Chem.*, **28,** 983–986.
31. Ugi, I., Dömling, A., and Ebert, B. (1999) in *Combinatorial Chemistry* (Jung G., ed.), Wiley-VCH, Weinheim, 125–165.
32. Weber, L., Illgen, K., and Almstetter, M. (1999) Discovery of new multi component reactions with combinatorial methods. *Synlett* **3,** 366–374.
33. Ross, A. and Senn, H. (2001) Automation of measurements and data evaluation in biomolecular NMR screening, *DDT* **6(11),** 583–593.
34. Ross, A., Schlotterbeck, G., and Senn, H. (2000) *Spectroscopic measurement method using NMR*, EPA 00810338.4.
35. Ross, A., Schlotterbeck, G., Senn, H., and von Kienlin, M. (2001) *Application of chemical shift imaging for simultaneous and fast acquisition of NMR spectra on multiple samples*, ENC, Orlando, Florida 11–16 March, Poster presentation.
36. Spraul, M., Hofmann, M., Ackermann, M., Nicholls, A. W., Damment, S. J. P., Haselden, N. J., Shockcor, J. P., Nicholson, J. K., and Lindon, J. C. (1997) Flow injection proton nuclear magnetic resonance spectroscopy combined with pattern recognition methods: Implications for rapid structural studies and high throughput biochemical screening. *Anal. Commun.* **34,** 339–341.
37. Spraul, M., Hofmann, M., and Neidig, P. (1999) High-throughput flow-injection NMR and its applications. *Bruker Report* **147,** 14–17.
38. Keifer, P. A., Smallcombe, S. H., Williams, E. H., Salomon, K. E., Mendez, G., Belletire, J. L., and Moore, C. D. (2000) Direct-injection NMR (DI-NMR): A flow NMR technique for the analysis of combinatorial chemistry libraries. *J. Comb. Chem.* **2,** 151–171.
39. Schlotterbeck, G., Ross, A., Senn, H., Hochstrasser, R., Tschirky, H., Seydoux, R., Marek, D., Kühn, T., Schett, O., and Warden, M. (2001) *High resolution NMR in capillary tubes a new miniaturized 1mm TXI Probe*, ENC, Orlando, Florida 11–16 March, Poster presentation.
40. Hegy, G., Görlach, E., Richmond, R., and Bitsch, F. (1996) High throughput electrospray mass spectrometry of combinatorial chemistry racks with automated contamination surveillance and results reporting. *Rapid Commun. Mass Spectrom.* **9,** 1894–1900.

41. Görlach, E., Richmond, R., and Lewis, I. (1998) High-throughput flow injection analysis mass spectroscopy with networked delivery of color-rendered results. 2. Three-dimensional spectral mapping of 96–well combinatorial chemistry racks. *Anal. Chem.* **70,** 3227–3234.
42. Graf von Roedern, E. (1998) A new method for the characterization of chemical libraries—solely by HPLC retention times. *Mol. Diversity* **3,** 253–256.

10

Rapid Liquid-Phase Combinatorial Synthesis of Heterocyclic Libraries

Chung-Ming Sun

1. Introduction

Over the past few years, combinatorial chemistry has emerged as an exciting new approach to discover novel drug candidates. It contains many tactics and processes for the rapid synthesis of large, organized collections of low-molecular-weight compounds without tedious and time-consuming purification. In combination with high-throughput screening, combinatorial organic synthesis of pharmaceutically interesting molecules may revolutionize the drug discovery program. Combinatorial chemistry has become increasingly popular in academic institutes and industries. It has been reported *(1)* that in the early years of 1992–1998, 63% contributors of combinatorial chemistry were from industry and 37% from academic laboratories. In the past two years, both academic institutes and pharmaceutical companies have developed novel combinatorial techniques for library synthesis equally. The majority of academic publications showed new synthetic methodologies; however, pharmaceutical companies used this technique for rapid screening and synthesizing of biologically active compounds.

Solid-phase organic synthesis (SPOS) is the core technology of combinatorial chemistry and has received much attention for its application to generate diverse screening libraries *(2–4)*. It offers benefits in easy, fast purification to separate excess reagents and side products from the desired compounds attached to the insoluble carrier. However, it also has several disadvantages, such as heterogeneous reaction conditions, reduced rate of reactions, solvation of the bound species, and mass transport of reagents. The range of chemical procedures that can be applied on the solid phase is also limited. Furthermore,

From: *Methods in Molecular Biology, Combinatorial Library Methods and Protocols*
Edited by: L. B. English © Humana Press Inc., Totowa, NJ

it is difficult to monitor reaction progress on solid-phase reaction. Although solid-phase synthesis of drug-like molecules offers significant advantages over many conventional solution-phase methods, solid-phase synthesis still requires extensive development time and research effort. Several laboratories *(5–7)* have focused on liquid-phase combinatorial synthesis (LPCS) of libraries using soluble polymer supports such as polyethylene glycol monomethyl ether (MeO-PEG-OH). Unlike an insoluble matrix, the polymer PEG support is soluble in many organic solvents and tends to precipitate in diethyl ether or ethanol. When reactions are complete, the products remain covalently bound to the support. After precipitation, products are purified simply by filtering and washing away the unwanted material. Liquid-phase combinatorial synthesis offers several unique advantages. For example, reactions can be carried out in homogeneous phase, and convenient product purification just like that of the solid-phase method is achieved by simple filtration and washing. The large excess of reagents typically used in solid-support synthesis is normally not required in liquid-phase synthesis. This method combines the strategic features of classical solution-phase chemistry and solid-phase synthesis. Furthermore, this nondestructive method allows routine analytical methodologies (e.g., ^{1}H-NMR, ^{13}C-NMR, IR, TLC) to monitor progress of the reaction transformations and to determine the structures of compounds directly attached to the polymer support. Because liquid-phase reactions can be carried out in homogeneous solution, the wealth of known chemical transformations available to solution-phase chemistry may be applied directly, and reaction optimization time should be reduced as well. It is also of note that biological assays can be directly evaluated on the PEG-linked small-molecule libraries, which is a particularly attractive feature of PEG-supported library synthesis *(8)*.

However, generating libraries by liquid-phase methods has its disadvantages. To fulfill all the requirements of a stepwise synthesis without intermediate purification, PEG should be stable during all stages of synthesis. Nevertheless, there is still a need to attach a linker to PEG and sever target compounds from the support at the end of the synthesis. Complete separation of the desired products from the support is physically difficult, whether using precipitation or size exclusion chromatography, because some PEG-supported products may not recrystalize easily and coprecipitation of other reagents might occur.

In an effort to develop new synthetic methodologies for liquid-phase synthesis of small organic molecules, we have explored the scope of liquid-phase synthesis for generating heterocyclic molecules using MeO-PEG-OH. For our study we chose MeO-PEG-OH as a carrier because it is inexpensive, commercially available with a wide range of molecular weights, and easily functionalized with different spacers and linkers.

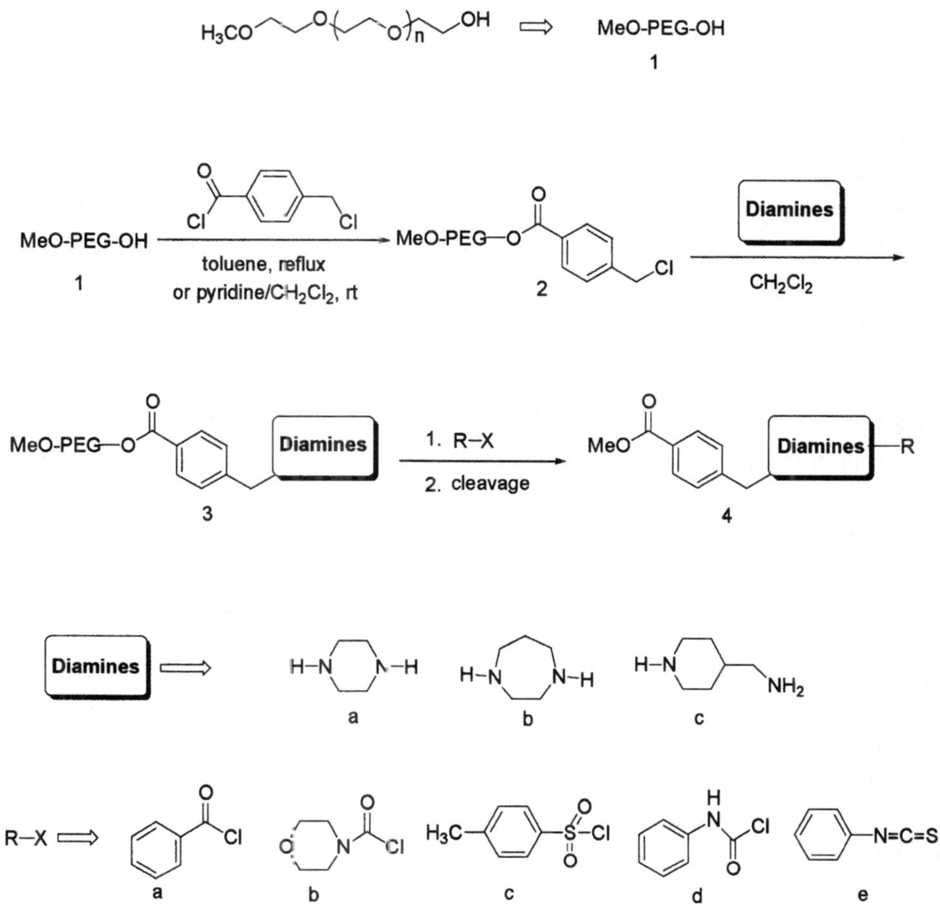

Fig. 1. Liquid-phase synthesis of benzyl arylpiperazine and piperidine derivatives.

Our first application of combinatorial liquid-phase synthesis is the preparation of biologically active aryl benzyl piperazine derivatives as shown in **Fig. 1**. PEG-bound benzyl chloride **2** was prepared from MeO-PEG-OH **1** (MW: 5000) and 4-chloromethyl benzoyl chloride with pyridine via ester linkage. The reaction also proceeded smoothly in toluene reflux and delivered PEG-bound benzyl chloride **2** in excellent yield. The conversion of the linker attachment estimated by ^{1}H NMR was greater than 95%. After the reaction was complete, the excess 4-chloromethyl benzoyl chloride was removed from the reaction mixture by precipitating the polymer with diethyl ether. The resulting polymer-supported benzyl chloride **2** was then reacted with various diamine templates, including piperidine, piperazine, and homopiperazine to give corresponding

PEG-supported organic moieties **3**. The PEG-bound amines **3** were then treated with various electrophiles such as benzoyl chloride, morpholine carbonyl-chloride, phenyl isocyanate, and sulfonyl chloride at ambient temperature to give corresponding acylated products. Treatment of the acylated products with 1% KCN/methanol resulted in efficient cleavage from the polymer support to give corresponding products **4** in 99% yield with 84–94% HPLC crude purity *(9)*.

Progress of the reactions was easily monitored by conventional ^{1}H NMR as shown in **Fig. 2**. Spectrum **A** is the initial stage of the reaction mixtures recorded after stirring 30 min at room temperature. Proton NMR spectrums **B**, **C**, **D** were recorded after 3 h intervals. We observed that benzylic protons ($ClCOC_6H_4\underline{CH_2}Cl$, δ 4.64 ppm) of 4-chloromethyl-benzoyl chloride (marked as ●) were shifted slightly upfield at δ 4.61 ppm (marked as ▼) after attaching to the support. Similarly, the two bunches of aromatic protons of 4-chloromethyl-benzoyl chloride (δ 7.5 and 8.1 ppm, marked as ★ and ■) were shifted upfield at δ 7.4 and δ 8.0 ppm. Spectrum **B** and **C** showed the gradually decreasing concentration of 4-chloromethyl-benzoyl chloride, which was reacted with PEG. In spectrum **D**, aromatic absorption of 4-chloromethyl-benzoyl chloride disappeared completely after the reaction was done. After confirmation of the loaded PEG structure, we proceeded with further reactions.

Nucleophilic substitution reactions of PEG-bound benzylic chloride were easily monitored by ^{1}H NMR spectroscopy. For example, **Fig. 3** shows a typical change in chemical shift during the nucleophilic substitution reaction with piperidine. After the coupling reaction was completed, the benzylic protons (δ 4.61 ppm, marked as ▼) of immobilized benzyl chloride **2** completely disappeared and were shifted upfield into the PEG protons absorption area (~δ 3.4–3.8 ppm), but no change was observed in the aromatic region. This evidence showed complete transformation of **2** to a PEG bound amine. Similar results were also obtained by employing other nucleophiles such as piperazine, 4-(aminomethyl)piperidine, and homopiperazine.

We next studied the cleavage condition and checked the reaction progress by regular proton NMR. The NMR spectra **F** of PEG-bound piperidine showed that the absorption of α-methylene protons (marked as ★) at the PEG attached site ($-OCH_2\underline{CH_2}OCO-$, δ 4.46 ppm) slowly disappeared during transesterification and was shifted to the upfield site at δ 3.6 ppm as shown in **Fig. 4**. After work up of the reaction, the CH_2 absorption of PEG fragment completely disappeared and a new singlet of three protons $\underline{CH_3}OCO-$appeared at δ 3.9 ppm (marked as ◆).

Taking advantage of PEG-bound diamine synthons **3**, we expanded our methodology to increase the guanidine functional group into the diamine libraries. It has been well demonstrated that the guanidine-containing compounds are very important in medicinal chemistry, as they possess a broad range of biological

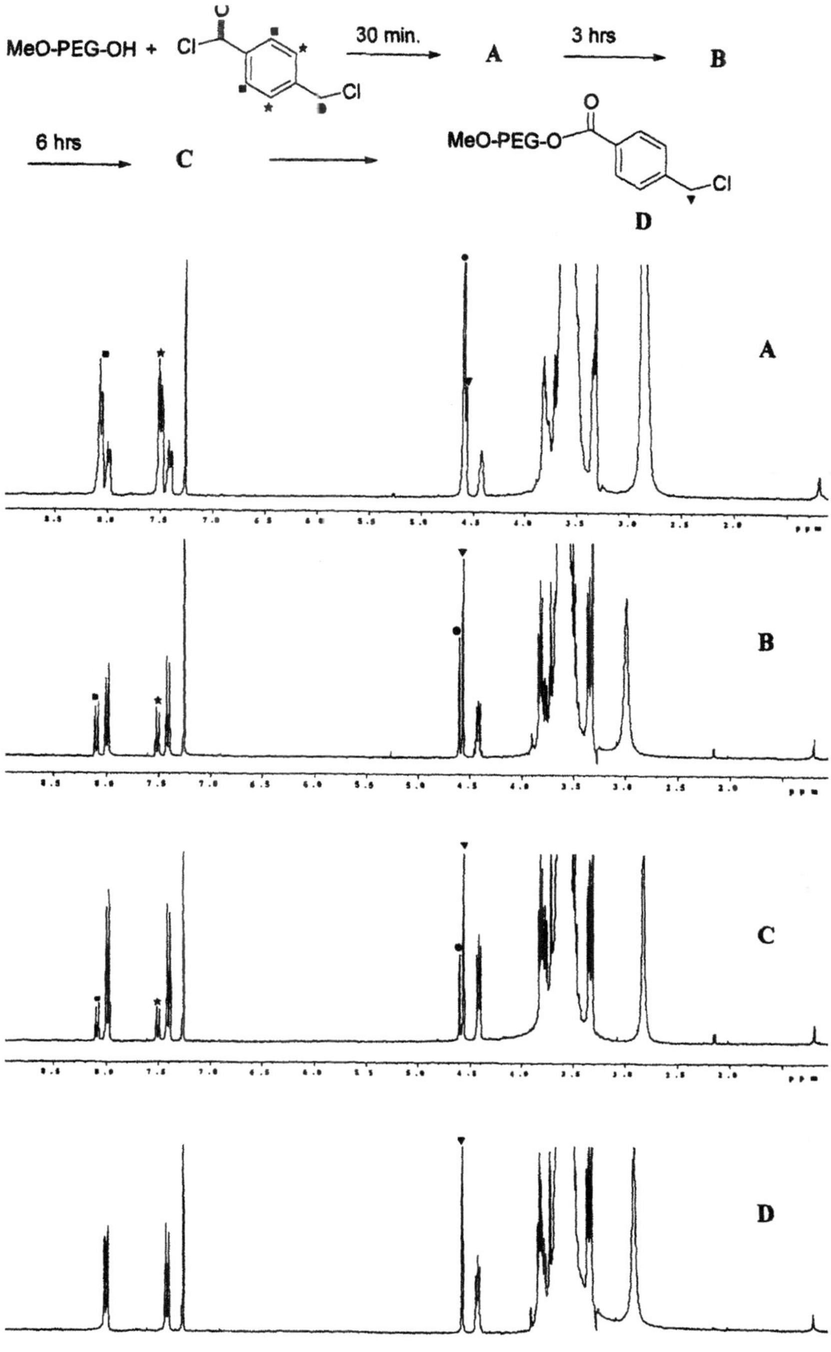

Fig. 2. ^{1}H-NMR spectrums recorded during the progress of loading linker on MeO-PEG-OH.

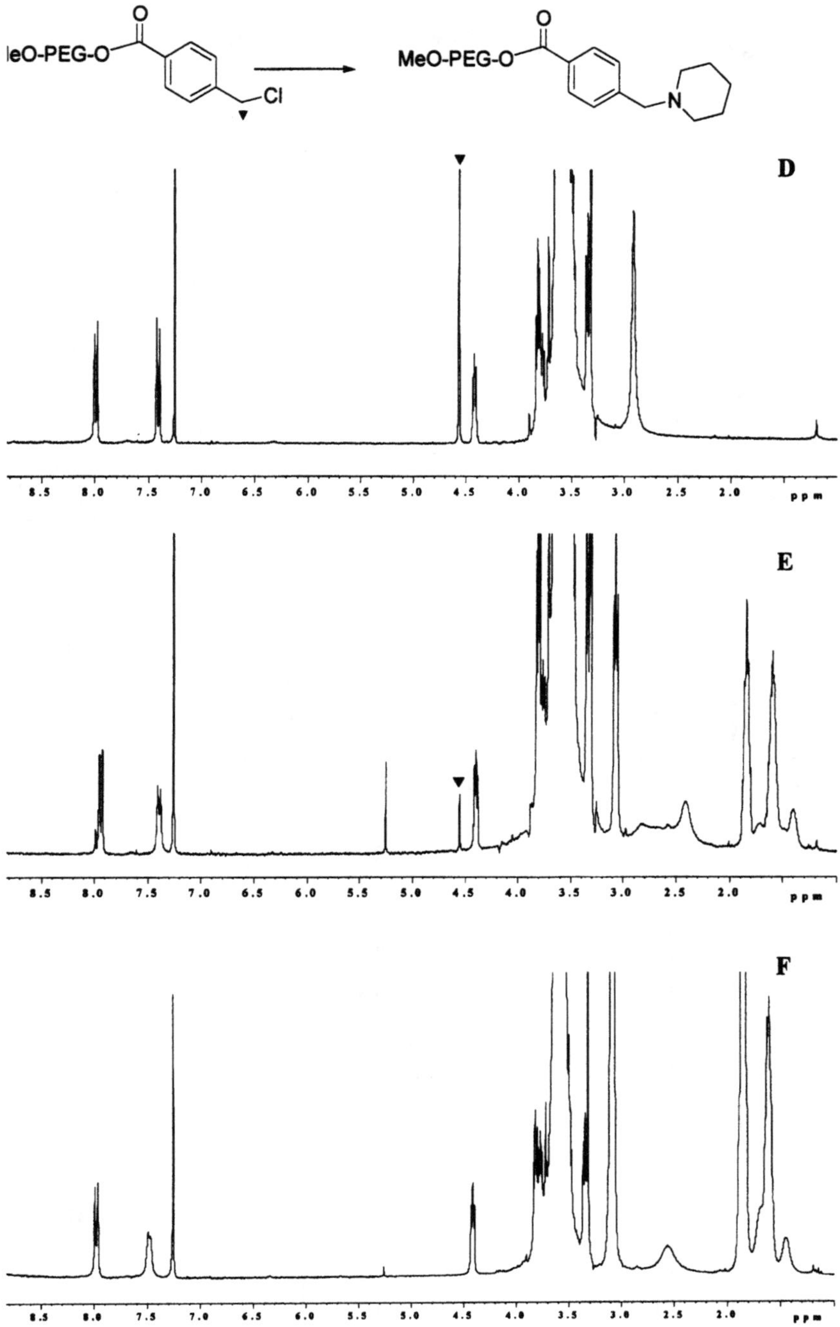

Fig. 3. ^{1}H-NMR spectrums recorded during nucleophilic substitution reaction (spectrum **D**: pure compound **2**; spectrum **E**, recorded after 2 h reaction; spectrum **F**, pure compound **3**).

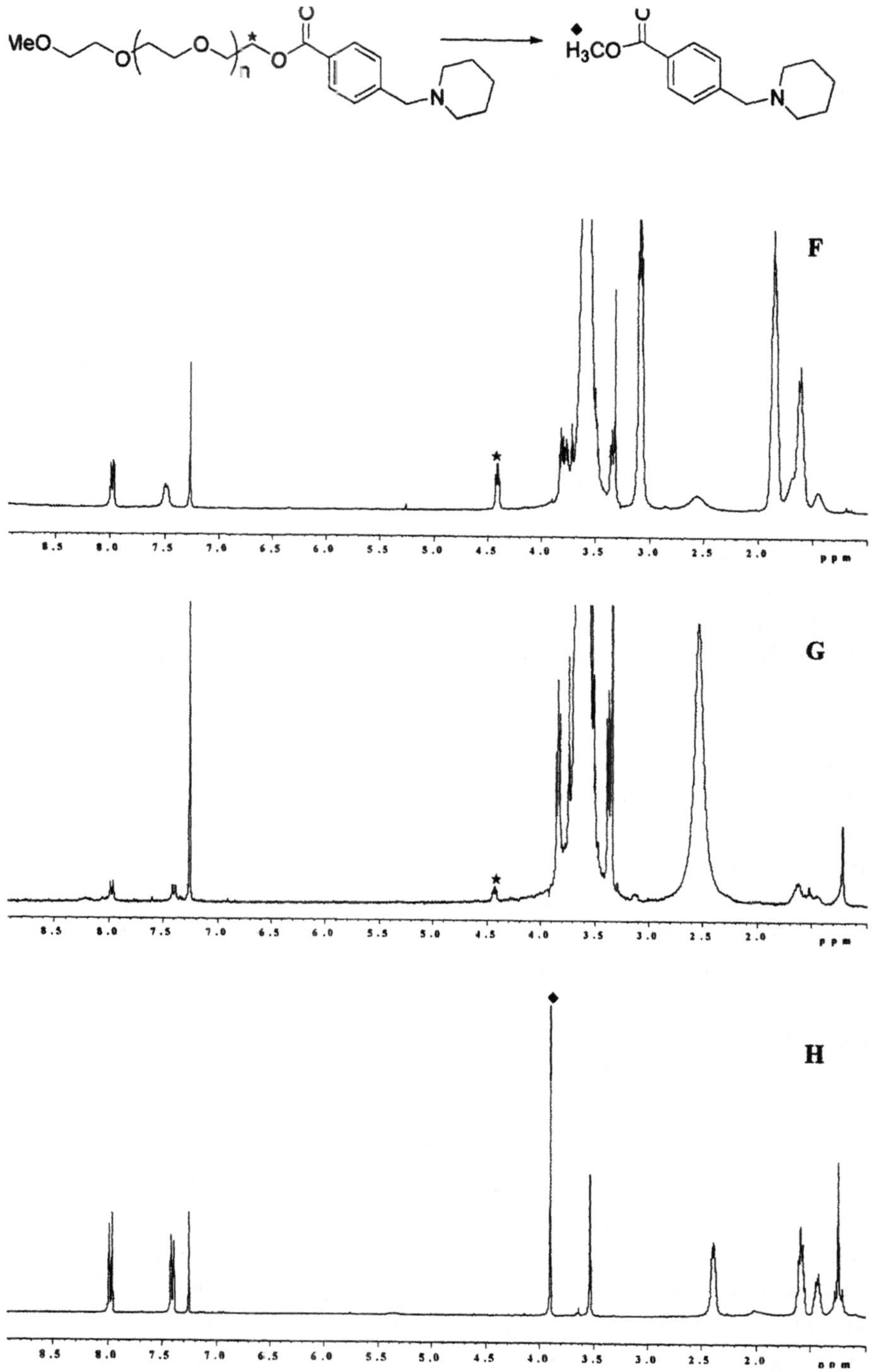

Fig. 4. ^{1}H NMR spectrums recorded during the cleavage of PEG support.

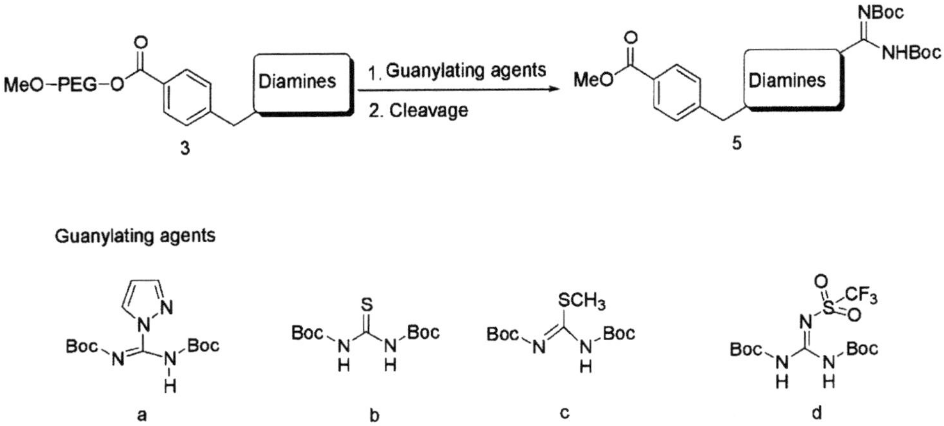

Fig. 5. Generation of *N,N′*-di(Boc)-protected guanidines.

activities, including hypertensive and neurological effects. In our laboratory, we have recently developed a novel method for synthesizing guanidine- containing piperazine and a piperidine moiety **5** using a soluble polymer support ***(10,11)***. As shown in **Fig. 5**, the PEG bound diamines **3** were reacted with various guanylating agents such as *N,N′*-bis-Boc-1-guanylpyrazole (a), *N,N′*-bis-Boc-thiourea (b), N,N'-di-(tert-butoxyl-carbonyl)-S-methyl isothiourea (c), and N,N'-di-boc-N"-triflyguanidine (d) at room temperature in dichloromethane to give corresponding polymer-bound products in good yield. We observed that guanidine formation with *N,N′*-bis-Boc-1-guanylpyrazole and **3c** proceeded smoothly within several hours without any activation since 1-pyrazolyl is a good leaving group. However, the PEG-bound secondary amines **3a** and **3b** were less reactive, and reaction with the guanylating agent (a) was unsuccessful even at elevated temperatures (e.g., refluxing toluene). It is possible that sterically crowded amines (**3a** and **3b**) may hinder the approach of the center amidine carbon. Further deprotection of di-Boc-guanidines **5c** was carried out in trifluoroacetic acid/dichloromethane solution (50%) in quantitative yield.

With a more reactive reagent such as *N,N′*-di-boc-*N″*-triflyguanidine (d), guanidine formation with **3a–3c** proceeded efficiently at 25°C with triethylamine. The guanidinylation reactions with gulanylating agents (b) and (c) were very sluggish, and proper activation using mercury chloride or diisopropylcarbodimide (DICDI) was necessary to drive the reactions to completion. Finally, guanidine-containing piperazine and piperidine libraries were obtained in 80–98% yield with high purity by detaching the MeO-PEG-OH support using 1% KCN-MeOH.

After achieving guanidine-containing piperazine libraries, we were interested in the increasing chemical diversity of the molecules that could be achieved by reacting PEG-bound guanidine with various amines (**Fig. 6**).

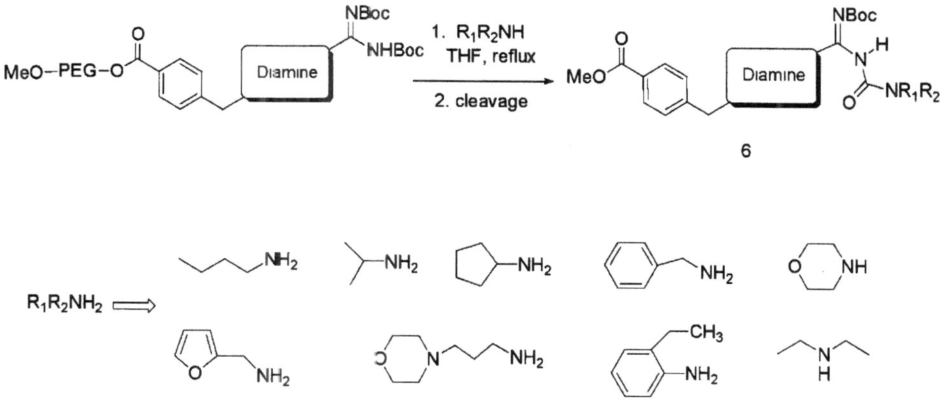

Fig. 6. Synthetic procedure for the Boc-substituted amidinoureas library.

Although the cleaved product was expected from the polymer support by transamination, no products were actually liberated from the PEG-supported guanidines in refluxing tetrahydrofuran (THF). Instead of the desired aminolysis product, the amidinourea **6** was recovered. The scope of the coupling method was assessed using a variety of aromatic and aliphatic amines, including aniline. Because aminolysis of many carbamates is well understood, the Boc-protecting group is known to be particularly stable under basic conditions and is also strongly resistant to various nucleophilic reagents. Boc-substituted amidinoureas **6** were obtained in 80–95% yield, with 82–85% purity after cleavage *(12)*.

As shown in **Fig. 7**, we also synthesized piperazine-containing ureas. The stable scaffolds **7** were achieved by reacting PEG-immobilized piperazine and homopiperazine with triphosgene in the presence of triethylamine *(13)*. The resulting PEG-bound trichloromethyl carbamates **7** were then reacted with various primary and secondary amines at room temperature in dichloromethane to give corresponding ureas **8**. The desired compounds **9** were liberated from the support using NaOMe in 85–98% yields. Purity assessment of each library member by HPLC was usually good, ranging from 76–94%. Complete cleavage of PEG was verified by the observation of the downfield shift of α-methylene protons at the polymer attached site from δ 4.4 ppm to δ 3.6 ppm in ^{1}H NMR.

We next explored the applications of liquid-phase synthesis of benzofused heterocycles from a common building block PEG-bound nitro-activated aryl fluoride (**Fig. 8**). In these synthetic strategies, we created multiple accesses to the structurally diverse core molecules based on a fundamental scaffold **10**. We focused on the construction of a variety of benzimidazole libraries starting from a versatile synthon 4-fluoro-3-nitrobenzoic acid via liquid-phase nucleo-

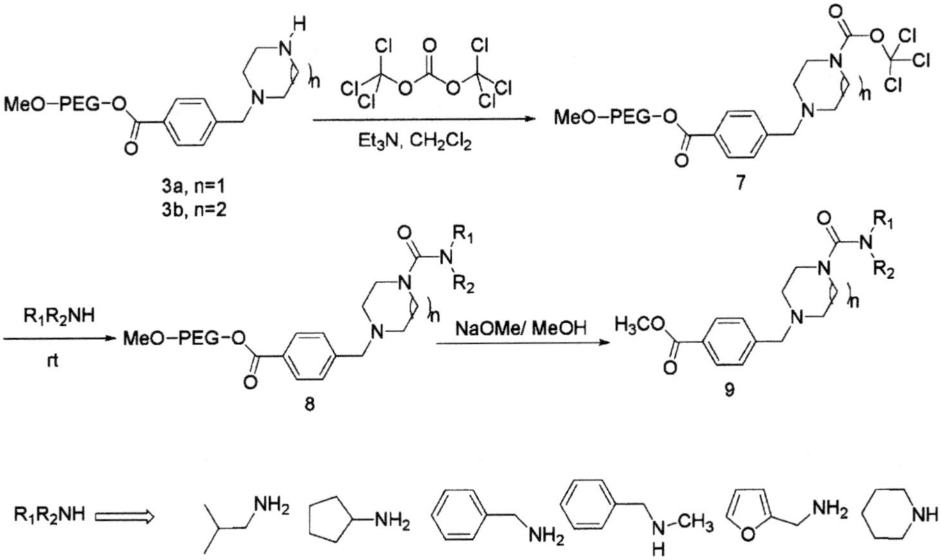

Fig. 7. Synthesis of piperazine containing ureas.

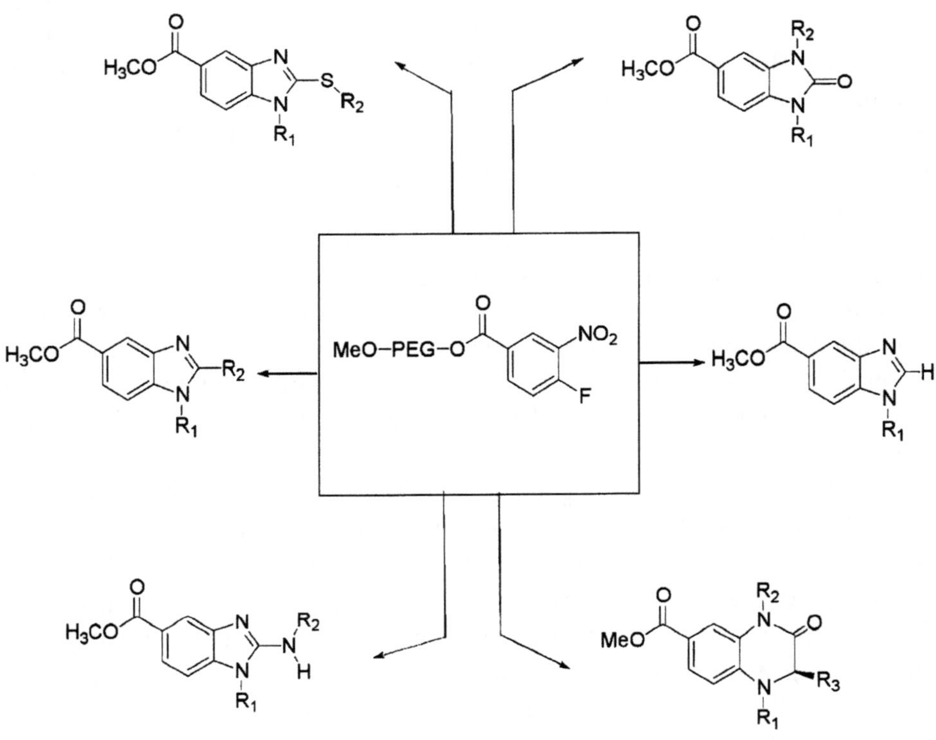

Fig. 8. General approach towards the synthesis of benzofused heterocycles.

philic aromatic substitution (S_NAr) reactions with nitrogen necleophiles. These types of heterocycles possess a broad spectrum of biological activity. Numbers of compounds from this class have been used clinically as anti-ulcer, anti-cancer, anti-viral, and pesticide agents.

The synthetic route described in **Fig. 9** was used to prepare a representative library. In the first step of the reaction sequence, 4-fluoro-3-nitro-benzoic acid was attached to MeO-PEG-OH **1** by the DCC/DMAP coupling in dichloromethane at room temperature to afford **10**. As illustrated in **Fig. 9**, *o*-nitroaniline **11** was produced by reacting **10** with various primary amines at ambient temperature through nucleophilic aromatic substitution (S_NAr). Once the reaction was complete, excess reagents were washed away by precipitation and filtration. Reduction of the aromatic nitro group was first attempted with aqueous 2 *M* $SnCl_2$ as reported by Pavia *(14)*. After a work-up reaction, we obtained an amorphous solid and the filtrate contained a trace amount of tin(II) chloride that was very difficult to separate from the product. It was found that both Zn/NH_4Cl and Pd-C/NH_4COOH reagents easily reduced PEG-bound nitroanilines **11** to PEG-bound diamines **12** at ambient temperature. After reduction, the inorganic material was removed first by fritted paper and PEG-bound diamines were isolated by precipitation. Treatment of the resulting diamine with triphosgene or thiophosgene in the presence of triethylamine at room temperature yielded corresponding cyclized benzimidazolones **13** and benzimidazole-2-thiones **16**, respectively. Benzimidazole formation with disuccinimidocarbonate (DSC) did not give satisfactory results because of the low solubility of DSC in dichloromethane. N-alkylation of benzimidazolone **13** was carried out by deprotection of N–H with NaH followed by quenching with various electrophiles. S-alkylation of benzimidazole-2-thione **16** was carried out by reaction with various alkyl halides in triethylamine at 25°C. Immobilized N- and S-alkylated products were conveniently separated from the PEG support using 1% KCN/MeOH to liberate the desired benzimedazolone **15** and benzimidazole **18** derivatives *(15–18)*.

During the construction of the benzimidzolone core structure, the formation of immobilized intermediates was easily monitored by ^1H NMR spectroscopy. The aromatic region of the proton NMR absorption of ***11–14*** (δ 6-9 ppm region) is shown in **Fig. 10**. Regular ^1H NMR studies clearly show the changes in the chemical shift in the aromatic region during each step of the transformations from **11** to **14** (R_1 = isobutyl group, R_2 = CH_3).

In order to investigate the further application of immobilized 4-fluoro-3-nitro benzoic acid synthon, we expanded our methodology to prepare 1,2-disubstituted benzimidazoles ***23–25***. **Figure 11** shows our detailed efforts toward the preparation of a benzimidazole library. To achieve this goal, one pot cyclization of PEG-bound diamine **12** with trimethylorthoformate in TFA/

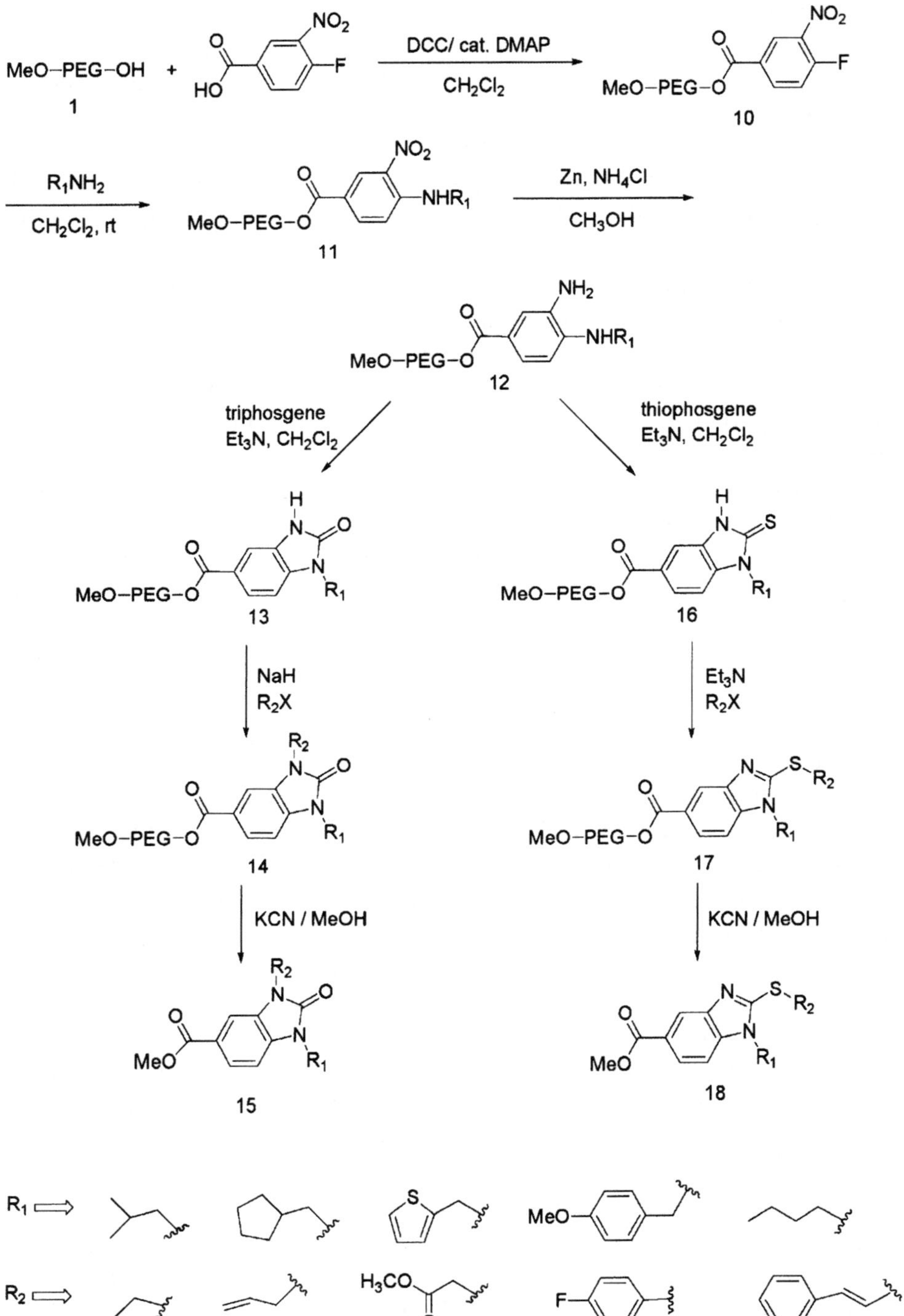

Fig. 9. Preparation of benzimidazolone and benzimidazole library.

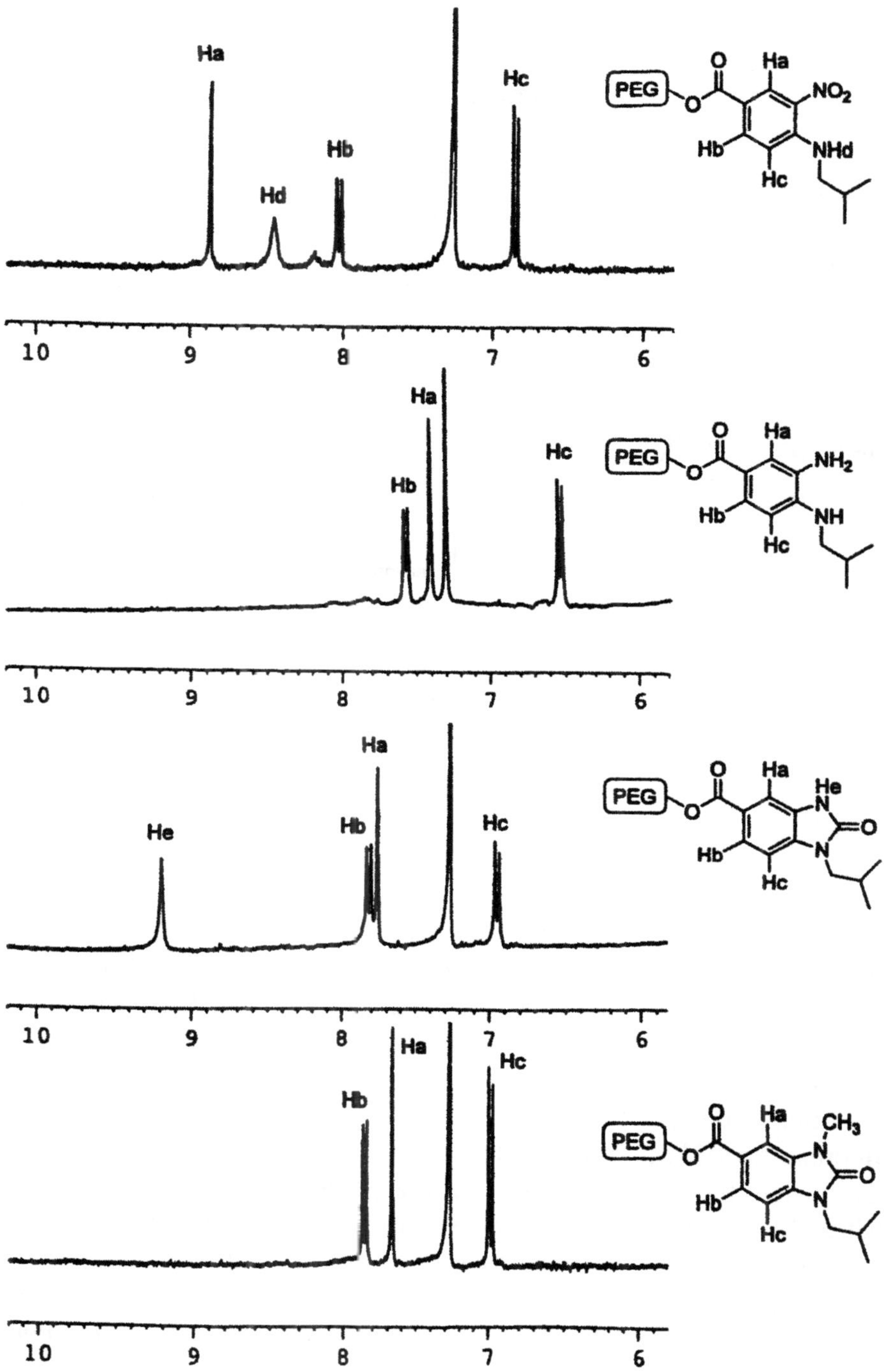

Fig. 10. Stepwise transformation of MeO-PEG-bound nitro aniline **11** to MeO-PEG-bound benzimidazolone **14** (top to bottom).

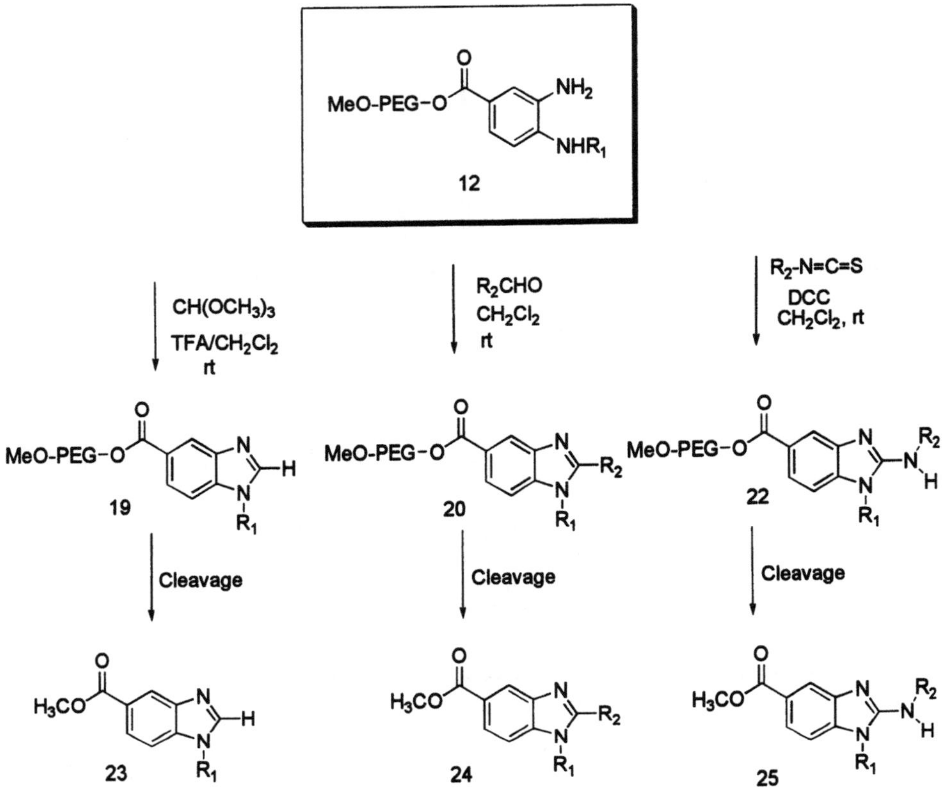

Fig. 11. Synthesis of substituted benzimidazoles.

CH_2Cl_2 mixture was carried out at room temperature to afford quantitatively the compound benzimidazoles **19**. The proposed mechanism of this TFA-catalyzed ring closure reaction of immobilized diamine **12** was initiated by the reaction of trimethylorthoformate exclusively with aniline funcationality (**Fig. 12**). The resulting intermediate-imine was reacted with a secondary amine with further elimination of methanol to afford the desired benzimidazole **19**. The possible trifluoroacetic acid catalyzed mechanism is shown in **Fig. 12**.

Following ether and ethanol washes after precipitation, PEG bound cyclized products **19** were subjected to cleavage from the support with NaOMe in methanol to provide the desired compounds **23** in overall yields ranging from 71 to 94% with 73–94% purity *(19)*. As shown in **Fig. 11**, substituted benzimidazoles **20** were also synthesized at room temperature by reacting PEG-bound diamine **12** with various alkyl or aryl aldehyes in dichloromethane. A similar condensation was performed by refluxing reactants in nitrobenzene at 150°C for 8–12 h between tetraaminobiphenyl and aromatic aldehyde *(20)*. The published

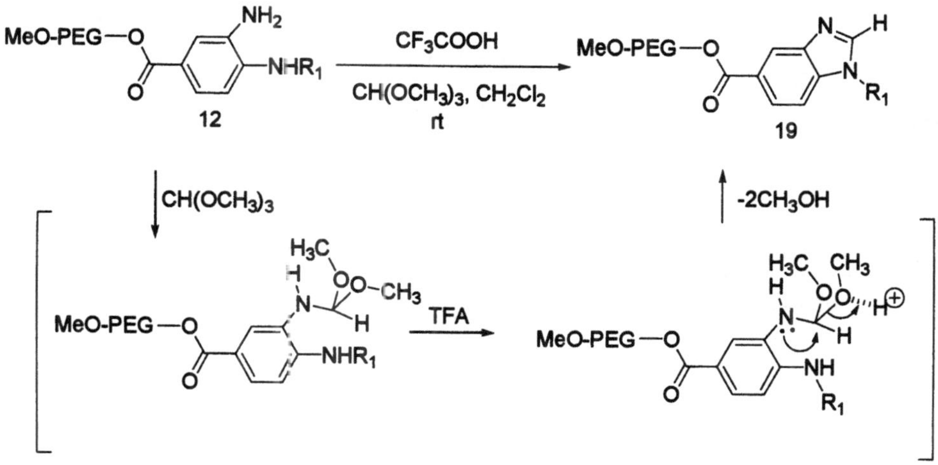

Fig. 12. TFA-catalyzed mechanism in benzimidazole formation.

one-pot procedure of Mayer was required with the addition of 2 equivalents of dichlorodicyanobenzoquinone (DDQ) and aldehydes to *o*-phenylenediamine in *N,N*-dimethylformamide for solid-phase application *(21)*.

Treatment of PEG-supported *o*-phenylenediamine **12** with isothiocyanate in dichloromethane provided *N*-(2-aminophenyl)-*N'*-substituted thioureas **21** (**Fig. 13**). In order to check regioselectivity at this stage, compounds **21** were liberated from the support to confirm that isothiocyanate was only attached to the more nucleophilic, secondary aniline nitrogen. Treatment of immobilized thiourea **21** with diisopropylcarbodimide (DICDI) resulted in complete cyclization in 5 h at room temperature.

In the present study, one-pot cyclodesulfurization provided a more efficient route to 2-arylbenzimidazole **22**. We found PEG-bound *o*-phenylenediamine reacted well with DICDI and isothiocyanates in one pot at room temperature. A two-step protocol was not necessary for the cyclodesulfurization. By employing the desired reaction sequence, we were able to subsequently introduce two diverse substitutions that had a large number of building blocks readily available. In both cases, the desired benzimidazoles were obtained in high yield with excellent HPLC purity after cleavage.

In order to demonstrate the versatility of the immobilized 4-fluoro-3-nitro benzoic acid scaffold **10**, we also synthesized benzopiperazinones derivatives. The building block used for the synthesis of benzopiperazinones is shown in **Fig. 14**. Complete acylation of **12** at the secondary aniline site was achieved by a reaction with chloroacetyl chloride in the presence of triethyl amine. Finally, one pot intramolecular cyclization of **25** and cleavage was done with NaOMe/ MeOH at room temperature. According to the thin layer chromatography analy-

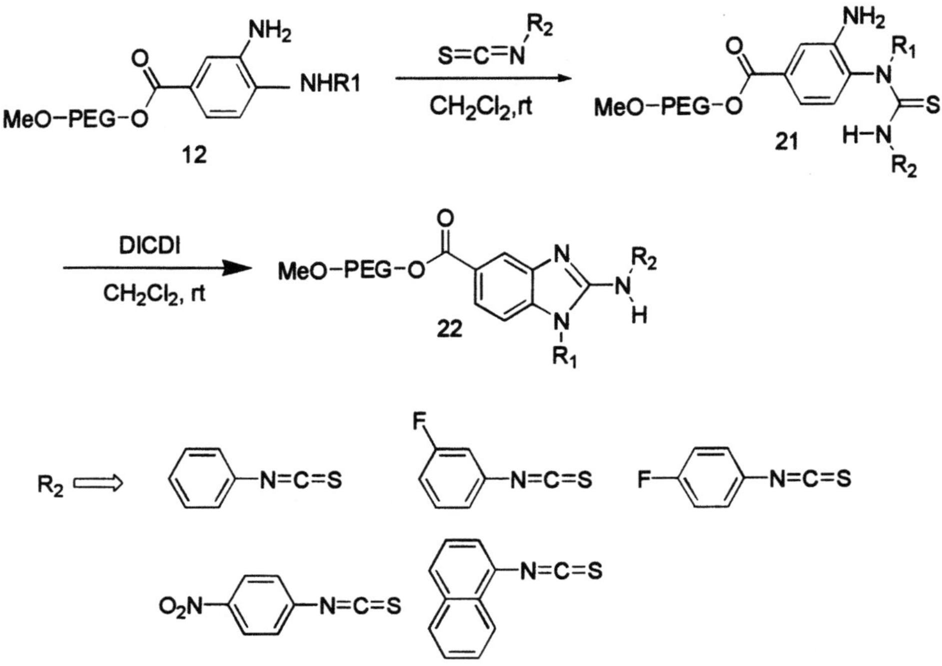

Fig. 13. Synthesis of *N*-(2-aminophenyl)-*N′*-substituted thioureas **21** and converted to **22** on the PEG.

sis (TLC), it seemed that intramolecular cyclization was performed first and then the desired products were cleaved from the support. The benzopiperazinones library **26** was obtained in good yield with excellent HPLC purity.

In summary, liquid-phase high-throughput synthesis is a versatile and efficient method for the fast generation of chemical libraries. We demonstrated that many biologically active molecules could be prepared on a soluble polymer support from common building blocks such as 4-chloromethyl benzoyl chloride and 4-fluoro-3-nitro benzoic acid. Liquid-phase methodology is applicable to not only single-step transformations, but also to multistep synthesis. It provides an alternative route for library generation by combining the positive aspects of traditional solution-phase synthesis and solid-phase synthesis. The main advantage in all cases is that reactions could be carried out in a homogeneous phase, and PEG-bound intermediates could be purified by the precipitation method followed by simple filtration. Selective precipitation has been developed to isolate pure compounds and eliminate the time-consuming purification common to classical solution-phase chemical synthesis. It is worthy to note that formation of intermediates in each step of library synthesis can be monitored by routine proton NMR spectroscopy and TLC. Desired crude products were obtained in high yields with high HPLC purity just by simple pre-

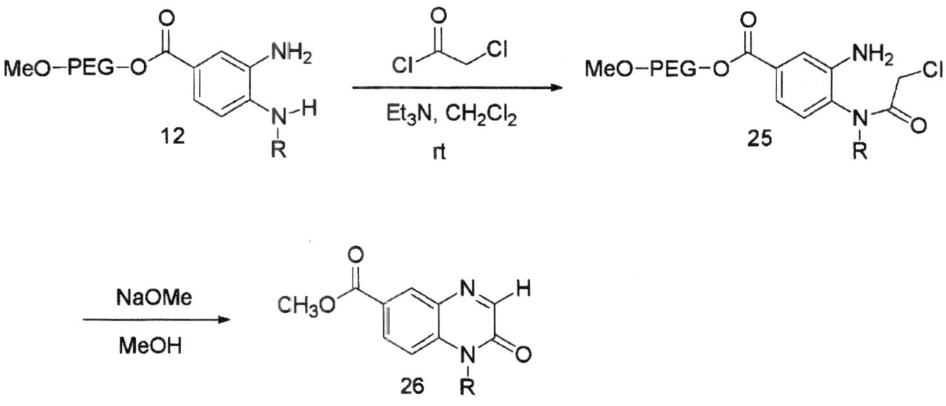

Fig. 14. Benzopiperazinones library synthesis.

cipitation and filtration. We are hopeful that combinatorial liquid-phase methodology as a valuable tool has just started to direct a new avenue in the search for more potent therapeutic lead compounds.

2. Materials

1. All glassware should be dried in an oven (120°C) and after assembly allowed to cool under an atmosphere of dry nitrogen.
2. Nitrogen should be dried by means of a Sicapent (E. Merck) drying tube.
3. Polyethylene glycol monomethyl ether (average MW 5000) (Fluka Chemie GmbH CH-9471 Buchs, Switzerland) should be melted at 70–80°C under high vacuum to remove any trace of moisture before use and stored at room temperature over P_2O_5 under vacuum.
4. Required solvents like dichloromethane, methanol, toluene, and tetrahedronfuran (THF) should be dried and distilled prior to use according to standard procedures.
5. 1,3-Dicyclohexylcarbodimide (DCC) (Aldrich Chemical Company, Inc, Milwaukee, WI) can be used without purification.
6. 4-(Dimethylamino) pyridine (DMAP) (Aldrich) can be used without purification.
7. Zinc dust (Aldrich) can be used without purification.
8. Granular ammonium chloride (J.T Baker, Inc, Phillipsburg, NJ) can be used without purification.
9. Triethylamine (Aldrich) should be freshly distilled.
10. Triphosgene (Aldrich).
11. Thiophosgene (Aldrich).
12. 1% KCN-methanol solution should be prepared in a hood.
13. All PEG-bound compounds are stable at room temperature for several months and should be stored over P_2O_5 under vacuum desiccator.
14. Vigorous stirring is required as large quantities of precipitate form during the addition of diethyl ether.
15. More heating causes most of the polymer bound intermediates to liquefy.

3. Methods

3.1. Polyethylene Glycol Monomethyl Ether Loading onto the Linker-4-Chloromethyl Benzoyl Chloride (2)

1. Equip a 50 mL one-necked round bottom flask with a magnetic stirrer bar and an efficient water condenser fitted with rubber septum through which a needle-tipped nitrogen gas is inserted for 10 min (*see* **Note 1**).
2. Add 1 g MeO-PEG-OH, 0.0456 g (1.2 eq) 4-chloromethyl benzoyl chloride, 10 mL dry toluene, and 0.091 mL (5 eq) pyridine to the flask (*see* **Note 2**).
3. Immerse the reaction flask in a preheated oil bath (120°C) and reflux the contents vigorously overnight.
4. After completion of reaction, remove excess toluene from the reaction mixture with a rotary evaporator and precipitate the polymer-supported benzylic halide by slow addition of cold diethyl ether (30 mL).
5. For completion of the precipitation, leave the suspension of the reaction mixture at 0°C for another 30 min.
6. Collect the precipitate on a glass funnel and thoroughly wash with diethyl ether (20 mL, three times) (*see* **Note 3**).
7. Dry the resulting polymer *in vacuo* to yield PEG-supported benzylic halide **2** quantitatively (*see* **Note 4**).

3.2. Preparation of Acylated Benzyl Arylpiperazines (4)

1. Equip an oven-dried 50 mL round bottom flask with a Teflon-coated stirring bar and nitrogen inlet rubber septum.
2. To this flask, add PEG-bound benzylic halide **2** (1 g), 50 mg (3 eq) piperazine, and 15 mL dichloromethane, and stir the reaction mixture at room temperature (*see* **Note 5**).
3. After completion of reaction, remove excess dichloromethane from the reaction mixture with a rotary evaporator and precipitate the polymer-supported benzylic halide by slow addition of cold diethyl ether (30 mL).
4. Collect the precipitated polymer on a glass funnel and thoroughly wash with diethyl ether (20 mL, three times) to remove unreacted piperazine.
5. Dry the resulting polymer *in vacuo* to yield PEG-supported piperazine **3** quantitatively (*see* **Note 6**).
6. Transfer the derivatized polymer **3** (510 mg) to another 30 mL round bottom flask equipped with Teflon-coated stirring bar and nitrogen inlet rubber septum.
7. Add (by syringe) 5 mL dichloromethane and 35 mg (1.2 eq) phenyl isocyanate to this flask.
8. Stir the resulting mixture at room temperature for 12 h.
9. Check for completion of reaction with proton NMR.
10. Remove excess dichloromethane with a rotary evaporator.
11. Add 20 mL diethyl ether to the concentrated solution and collect the precipitated PEG-bound acylated product on a glass funnel.
12. Dissolve the polymer-bound acylated product (522 mg) in 5 mL of 1% KCN-

methanol solution, and stir the reaction mixture overnight at room temperature (*see* **Note 7**).
13. Precipitate the MeO-PEG-OH from the homogeneous solution by adding diethyl ether (20 mL).
14. Filter the detached MeO-PEG-OH through a glass funnel and wash the polymer thoroughly with diethyl ether (15 mL, three times).
15. Concentrate the combined filtrate and washings to get the corresponding acylated piperazine **4e** in 91% yield.

3.3. Synthesis of N,N'-di(Boc)-Protected Guanidines (5)

The required starting material (PEG-supported piperazine **3a**) for the synthesis of *N,N'*-di(Boc)-protected guanidines was prepared by using the above procedure (*see* **Subheadings 3.1.** and **3.2.**).

1. Equip a 30 mL one-necked round bottom flask with a magnetic stirrer bar and fitted with rubber septum through which a needle-tipped nitrogen gas is inserted and keep the flask under nitrogen atmosphere for 10 min.
2. Add 520 mg PEG-supported piperazine, 5 mL dichloromethane, 1.2 eq diisopropylcarbodimide (DICDI), and *N,N'*-bis(Boc)thiourea (1.4 eq) to the flask.
3. Stir the reaction mixture for 8 h at room temperature under nitrogen gas.
4. After completion of reaction, slowly add cold *tert*-butyl methyl ether (20 mL) to the reaction mixture.
5. Collect the precipitated PEG-bound *N,N'*-di(Boc)-protected guanidine on a glass funnel and thoroughly wash with cold *tert*-butyl methyl ether (20 mL, twice).
6. Dry the resulting PEG-bound *N,N'*-di(Boc)-protected guanidine *in vacuo* to get the desired product in excellent yield.
7. Dissolved the PEG-bound *N,N'*-di(Boc)-protected guanidine (500 mg) in 5 mL 1% KCN-methanol solution and stir the reaction mixture overnight at room temperature.
8. Precipitate the polymer from the homogenous solution by adding *tert*-butyl methyl ether (20 mL).
9. Filter the detached MeO-PEG-OH through a glass funnel and wash the polymer thoroughly with *tert*-butyl methyl ether.
10. Concentrate the combined filtrate and washings to get the *N,N'*-di(Boc)-protected guanidine.

3.4. Synthesis of Boc Substituted Amidinoureas (6)

1. Equip a 30 mL one-necked round bottom flask with a magnetic stirrer bar and an efficient water condenser fitted with rubber septum through which a needle-tipped nitrogen gas is inserted, and keep the flask under nitrogen atmosphere for 15 min.
2. Add 500 mg PEG-bound *N,N'*-di(Boc)-protected guanidine, 5 mL THF and amine (5 eq) to the flask.
3. Immerse the reaction flask in a preheated oil bath (70°C) and reflux the contents vigorously for 8 h.

4. After completion of reaction, remove excess THF and unreacted amine from the reaction mixture with a rotary evaporator and precipitate the PEG-bound *N,(N'-tert*-butoxycarbonylamidino)urea by slow addition of cold *tert*-butyl methyl ether (25 mL).
5. For completion of the precipitation, leave the suspension of the reaction mixture at 0°C for another 30 min.
6. Collect the precipitate on a glass funnel and thoroughly wash with *tert*-butyl methyl ether (10 mL, three times).
7. Dry the resulting polymer *in vacuo* to yield PEG-bound *N,(N'-tert*-butoxy-carbonyl- amidino)urea in 98% yield.
8. Dissolve the PEG-bound *N,(N'-tert*-butoxycarbonylamidino)urea (500 mg) in 5 mL of 1% KCN-methanol solution and stir the reaction mixture overnight at room temperature.
9. Check for completion of reaction with ^{1}H NMR.
10. Precipitate the polymer from the homogenous solution by adding *tert*-butyl methyl ether (20 mL).
11. Filter the detached MeO-PEG-OH through a glass funnel and wash polymer thoroughly with *tert*-butyl methyl ether (10 mL, three times).
12. Concentrate the combined filtrate and washings to yield bright yellow solid in excellent yield.

3.5. Synthesis of PEG-Bound Aryl Fluoride (10)

1. Equip an oven dry, 50 mL round bottom flask with a Teflon-covered magnetic stirrer bar, fitted with rubber septum through which a needle-tipped nitrogen gas is inserted, and keep the flask under nitrogen atmosphere for 10 min.
2. To this flask, add 1 g MeO-PEG-OH, 468 mg (1.2 eq) 4-fluoro-3-nitro benzoic acid, 494 mg (1.2 eq) DCC, 47 mg (0.05 eq) DMAP, and 10 mL dichloromethane, stir the resulting mixture at room temperature for 24 h under nitrogen gas.
3. After 24 h of stirring, slowly add diethyl ether (30 mL) to the reaction mixture (*see* **Note 8**).
4. For completion of the precipitation, leave the suspension of the reaction mixture 0°C for another 30 min.
5. Collect the precipitate on a glass funnel and thoroughly wash with diethyl ether several times to remove unreacted DCC, DMAP, and 4-fluoro-3-nitro benzoic acid.
6. Dry the resulting polymer *in vacuo* to yield PEG-supported aryl fluoride **10** quantitatively.

3.6. Synthesis of Piperazine-Containing Ureas (9)

1. Equip an oven dried 50 mL round bottom flaked with a Teflon-coated stirrer bar fitted with rubber septum through which a needle-tipped nitrogen gas is inserted, and keep the flask under nitrogen atmosphere for 10 min.
2. To this flask, add PEG-supported benzylic halide 2 (1 g), 50 mg (3 eq) piperazine, and 15 mL dichloromethane; stir the reaction mixture at room temperature for 8 h.
3. Check for completion of reaction with proton NMR.

4. Remove excess dichloromethane with a rotary evaporator.
5. Slowly add 30 mL diethyl ether to the concentrated solution and collect the precipitated polymer on a glass funnel and thoroughly washed with diethyl ether (20 mL, three times) to remove unreacted piperazine.
6. Dry the resulting polymer *in vacuo* to yield PEG-bound piperazine **3a** quantitatively.
7. Transfer the derivatized polymer **3** (510 mg) to another 30 mL round bottom flask equipped with a magnetic stirrer bar and nitrogen inlet rubber septum.
8. To this flask add (by syringe) 5 mL dichloromethane, triethyl amine (3 eq), and triphosgene [bis(trichloromethyl)carbonate] (3 eq).
9. Stir the resulting reaction mixture at room temperature for 24 h under nitrogen atmosphere.
10. Remove excess dichloromethane with a rotary evaporator.
11. Add 20 mL diethyl ether to the concentrated solution, and collect the precipitated PEG-bound carbamoly chloride **7** on a glass funnel.
12. Dry the polymer *in vacuo* before continuation of the next reaction step.
13. Dissolve the PEG-bound carbamoly chloride **7** in 5 mL anhydrous dichloromethane in 30 mL round bottom flask.
14. Add 3 eq amine (R_1NH_2,) and continue to stir at room temperature for 12 h.
15. After completion of the reaction, remove excess dichloromethane from the reaction mixture with a rotary evaporator and precipitate the PEG-bound urea by slow addition of 20 mL diethyl ether.
16. Collect the precipitate on a glass funnel and thoroughly wash with diethyl ether (20 mL, twice).
17. Dry the PEG-bound urea *in vacuo* overnight.
18. Dissolve the PEG-bound urea (500 mg) in 5 mL 1% KCN-methanol solution, and stir the reaction mixture overnight at room temperature under nitrogen gas atmosphere.
19. Precipitate the MeO-PEG-OH from the homogeneous solution by adding 25 mL cold diethyl ether.
20. Filter the detached MeO-PEG-OH through a glass funnel and wash the polymer thoroughly with 30 mL diethyl ether.
21. Concentrate the combined filtrate and washings to get the ureas **9** in excellent yield with high HPLC purity.

3.7. General Procedure for the Synthesis of Benzimidazolones (14)

1. Equip an oven-dried 30 mL round bottom flask with a Teflon-coated stirrer bar.
2. To this flask, add PEG-bound aryl fluoride (517 mg), primary amine (R_1NH_2, 2 eq), and 5 mL dichloromethane, stir the reaction mixture at room temperature.
3. Check for completion of the reaction with proton NMR.
4. After completion of the reaction, remove excess dichloromethane from the reaction mixture with a rotary evaporator and precipitate the polymer *o*-nitroaniline by slow addition of diethyl ether (20 mL).
5. For completion of the precipitation, leave the suspension of the reaction mixture at 0°C for another 30 min.
6. Collect the precipitated *o*-nitroaniline derivatives **11** on a glass funnel and thor-

oughly wash with diethyl ether (20 mL, three times).

7. Dry the resulting polymer *in vacuo* to yield immobilized nitro aniline derivative **11** in quantitative yield.
8. Dissolve 500 mg PEG-bound 3-nitro-4-amine **11** in 10 mL methanol, to this solution add zinc dust (20 eq), ammonium chloride (6 eq), and stir the heterogeneous mixture at room temperature for 3 h.
9. Check for completion of the reaction with proton NMR.
10. After completion of the reaction, filter heterogeneous material through fritted paper and wash the residue with 10 mL methanol.
11. Concentrate the combined filtrate and washings with a rotary evaporator and precipitate the PEG-bound diamine **11** by slow addition of diethyl ether (20 mL).
12. Collect the precipitated polymer on a glass funnel.
13. Dry the PEG-bound diamine *in vacuo* before the next reaction step.
14. Transfer 510 mg immobilized diamine to another 30 mL round bottom flask equipped with a magnetic stirrer bar and nitrogen inlet rubber septum.
15. To this flask, add (by syringe) 5 mL dichloromethane, 3 eq triethyl amine, and 3 eq triphosgene (*see* **Note 9**).
16. Stir the resulting mixture at room temperature for 8 h under nitrogen atmosphere.
17. After completion of the reaction, remove excess dichloromethane from the reaction mixture with a rotary evaporator and precipitate the PEG-bound benzimidazolone-2-one by slow addition of cold diethyl ether (20 mL).
18. Collect the precipitate on a glass funnel and thoroughly wash with diethyl ether (20 mL, twice).
19. Dry the resulting polymer *in vacuo* to yield PEG-bound benzimidazolone-2-one **13**.
20. Dissolve the resulting PEG-bound benzimidazolone-2-one **13** in 5 mL dichloromethane, to this add 10 eq sodium hydride and 2 eq alkyl halide, and stir the reaction mixture at room temperature for 10 h under nitrogen gas atmosphere (*see* **Note 10**).
21. Add 20 mL diethyl ether to the reaction mixture and collect the participated polymer bound alkylated product **14** on a glass funnel.
22. Dry the precipitate *in vacuo* to yield PEG-bound N-alkylated product **14** in excellent yield.
23. Dissolved the PEG-bound N-alkylated product **14** (500 mg) in 5 mL of 1% KCN-methanol solution, and stir the reaction mixture overnight at room temperature.
24. Check for the completion of the reaction with proton NMR.
25. After completion of the reaction, slowly add diethyl ether (20 mL) to the mixture.
26. Filter the detached MeO-PEG-OH through a glass funnel and wash the PEG thoroughly with diethyl ether.
27. Concentrate the combined filtrate and washings to get the benzimidazolone **15**.

3.8. Synthesis of PEG-Bound Benzimidazole (17)

1. Equip an oven-dried 30 mL round bottom flask with magnetic stirrer bar and nitrogen inlet rubber septum.

2. To this flask, add PEG-bound diamine 500 mg, 5 mL dichloromethane, triethyl amine (5 eq), and thiophosgene (3 eq); stir the resulting homogeneous solution at room temperature for 8 h under nitrogen gas atmosphere.
3. After 8 h of stirring, slowly add 30 mL diethyl ether to the reaction mixture.
4. Collect the precipitated polymer on a glass funnel and thoroughly wash with diethyl ether to remove unreacted thiophosgene.
5. Dry the PEG-bound benzimidazole-2-thione *in vacuo*.
6. Transfer 506 mg PEG-bound benzimidazole-2-thione to another 30 mL single neck round bottom flask equipped with a magnetic stirrer bar and a septum fitted with a nitrogen inlet.
7. To this flask, add 32.1 mg (0.145 mmol) 2-(bromomethyl)-naphtalene, 0.054 mL (0.39 mmol) triethyl amine, and 5 mL dichloromethane.
8. Stir the homogeneous solution at room temperature for 8 h under nitrogen atmosphere.
9. Check for completion of the reaction with proton NMR.
10. After completion of the reaction, slowly add diethyl ether (20 mL) to the reaction mixture.
11. Collect the precipitated polymer product on a glass funnel and thoroughly wash with diethyl ether to remove excess halide.
12. Dry the polymer *in vacuo* to get the S-alkylated product **17** in excellent yield.
13. Cleavage condition given in **Subsection 3.7.**

3.9. Synthesis of PEG-Bound Substituted Benzimidazole 23

1. Equip an oven dried 30 mL round bottom flask with magnetic stirrer bar and nitrogen inlet rubber septum.
2. To this flask, add PEG-bound diamine 500 mg, trifluoro acetic acid (0.5 eq), 5 mL dichlormethane, and trimethylorthoformate (5 eq), and stir the reaction mixture at room temperature under nitrogen atmosphere.
3. Check for completion of the reaction with proton NMR.
4. After the reaction is completed, slowly add 30 mL diethyl ether to the reaction mixture.
5. Collect the precipitated PEG-bound benzimidazole on a glass funnel and thoroughly wash with diethyl ether (10 mL, three times).
6. Dry the PEG-bound benzimidazole *in vacuo*.

3.10. Synthesis of PEG-Bound Substituted Benzimidazole 25

1. Equip a 30 mL round bottom flask with a magnetic stirrer bar and a septum fitted with a nitrogen inlet.
2. To the flask, add 500 mg PEG-bound diamine, 10 mL dichloromethane, 1.2 eq DIDCI, and 3 eq isothiocyanate, and stir the reaction mixture at room temperature for 12 h under nitrogen atmosphere.
3. Upon completion of reaction (check by proton NMR), remove excess dichloromethane form the reaction mixture with a rotary evaporator and precipitate the PEG-bound benzimidazole by slow addition of diethyl ether (20 mL).

4. Collect the precipitate on a glass funnel and thoroughly wash with diethyl ether (20 mL, twice).
5. Dry the resulting polymer *in vacuo* to yield PEG-bound benzimidazole **22**.
6. Transfer the PEG-bound benzimidazole (500 mg) to another round bottom flask equipped with Teflon-coated stirrer bar and nitrogen inlet rubber septum.
7. To this flask, add 5 mL methanol and 3 eq sodium methoxide.
8. Stir the reaction mixture at room temperature overnight.
9. Precipitate the MeO-PEG-OH from the homogeneous solution by adding cold diethyl ether (20 mL).
10. Filter the detached MeO-PEG-OH through a glass funnel and wash the polymer thoroughly with diethyl ether (10 mL, three times).
11. Concentrate the combined filtrate and washings to get the corresponding benzimidazole.
12. Compounds **20** and **22** were cleaved from the support by the same protocol to give **24** and **25**.

3.11. Synthesis of Benzopiperazinone 26

1. Dissolved the PEG-supported diamine (500 mg) in 5 mL dichloromethane at room temperature in 25 mL round bottom flask.
2. To this homogeneous solution, add (by syringe) 3 eq triethylamine and 3 eq chloroacetyl chloride (use freshly distilled).
3. Stir the reaction mixture at room temperature for 24 h under nitrogen atmosphere.
4. After completion of the reaction, slowly add diethyl ether (30 mL) to the reaction mixture.
5. Collect the precipitated polymer bound acylated product on a glass funnel and thoroughly wash with diethyl ether (30 mL, three times).
6. Dry the acylated polymer *in vacuo* before the next step of the reaction sequence.
7. Dissolve the acylated polymer in 5 mL dry methanol. To this solution add 3 eq sodium methoxide, and stir the reaction mixture at room temperature under nitrogen atmosphere.
8. Check for completion of the reaction with proton NMR.
9. After completion of the reaction, precipitate the MeO-PEG-OH from the homogeneous solution by adding diethyl ether.
10. Filter the detached MeO-PEG-OH through a glass funnel and wash the polymer thoroughly with diethyl ether (15 mL, twice).
11. Concentrate the combined filtrate and washings to get the benzopiperazinone **26**.

4. Notes

1. Every reaction should be carried out in nitrogen gas atmosphere.
2. All reactions should be carried out in dry solvents.
3. In each step of the reaction sequence, the polymer-bound products were precipitated selectively and excess reagents and by-products were removed by filtration methods before the next step of reaction sequence.
4. The polymer-bound products should be dried overnight under high vacuum.

5. Do not expose PEG-bound compounds to the air.
6. The yields of the PEG-bound compounds were determined by weight with the assumption that molecular weight is 5000 Da for the PEG fragment (the molecular weight of MeO-PEG-OH is actually ranged from 4500–5500).
7. Potassium cyanide is highly toxic and should be handled in a well-ventilated hood.
8. In the first step of the reaction sequence, the by-product DCU (dicylcohexylurea) should be filtered first, then precipitate the polymer.
9. Preparation of PEG-bound benzimidazolone-2-one and benzimidazolone-2-thione should be carried out in a fume hood because the vapors of thiophosgene and triphosgene are very toxic and carcinogenic.
10. Quenching should be performed very carefully as a rapid evolution of hydrogen occurs during the initial phase. Flushing with nitrogen throughout the quenching is recommended.

References

1. Dolle, R. E. (2000) Comprehensive survey of combinatorial library synthesis: 1999. *J. Comb. Chem.* **2(5),** 383–433.
2. Jung, G. (ed.) (1996) *Combinatorial peptide and non-peptide libraries.* VCH, Weinheim, Germany
3. Wilson, S. R. and Czarnik, A. W. (ed.) (1997) *Combinatorial chemistry synthesis and application.* John Wiley & Sons Inc, NY, USA.
4. Jung, G. (ed.) (1999) Combinatorial chemistry synthesis, analysis, screening. Wiley & VCH, Weinheim, Germany.
5. Gravert, D. J. and Janda, K. D. (1997) Organic synthesis on soluble polymer support liquid-phase methodologies. *Chem. Rev.* **97,** 489–509.
6. Sun, C. M. (1999) Recent advance in liquid-phase combinatorial chemistry. *Comb. Chem. & high throughput screening* **2,** 299–318.
7. Toy, P. H. and Janda, K. D. (2000) Soluble polymer-supported organic synthesis. *Acc. Chem. Res.* **33,** 546–554.
8. Bonora, G. M., Tocco, G., Zaramella, S., Veronese, F. M., Pliasunova, O., Pokrovsky, A., Ivanova, E., and Zarytova, V. (1998) Antisense activity of an anti-HIV oligonucleotide conjugated to linear and branched high molecular weight polyethylene glycols. *Il Farmaco.* **53,** 634–637.
9. Shey, J. Y. and Sun, C. M. (1999) Liquid-phase combinatorial synthesis of benzylpiperazines. *Bioorg. Med. Chem. Lett;* **9,** 519–522.
10. Shey, J. Y. and Sun, C. M. (1998) Soluble polymer supported synthesis of N,N'-di(Boc)-protected Guanidines. *Synlett,* **12,** 1423–1425.
11. Ho, K. C. and Sun, C. M. (1999) liquid-phase parallel synthesis of Guanidines. *Bioorg. & Med. Chem. Lett.* **9,** 1517–1520.
12. Shey, J. Y. and Sun, C. M. (1999) High-throughput synthesis of Boc-substituted aminoureas by liquid-phase approach. *J. Comb. Chem.* **1(5),** 361–363.
13. Huang, K. T. and Sun, C. M. (2001) Liquid-phase parallel synthesis of ureas. *Bioorg. & Med. Chem. Lett.* **2,** 271–273.

14. Meyers, H. V., Dilley, G. J., Durgin, T. L., Powers, T. S., Winssinger, N. A., Zhu, H., and Pavia, M. R. (1995) Multiple simultaneous synthesis of phenolic libraries. *Molecular Diversity.* **1,** 13–20.
15. Pan, P. C. and Sun, C. M. (1999) High-throughput combinatorial synthesis of substituted benzimidazolones. *Tetrahedron Lett.* **40,** 6443–6446.
16. Yeh, C. M. and Sun, C. M. (1999) Rapid parallel synthesis of benzimidazoles. *Synlett.* **6,** 810–812.
17. Yeh, C. M. and Sun, C. M. (1999) Liquid-phase synthesis of 2-(alkylthio)benzimidazoles. *Tetrahedron Lett.* **40,** 7247–7250.
18. Yeh, C. M., Tung, C. L., and Sun, C. M. (2000) Combinatorial liquid-phase synthesis of structurally diverse benzimidazole libraries. *J. Comb. Chem.* **2,** 341–348.
19. Chi, Y. C. and Sun, C. M. (2000) Soluble polymer supported synthesis of a benzimidazole library. *Synlett.* **5,** 591–594.
20. Mann, J., Baron, A., Opoku-Boahen, Y., Johansson, E., Parkinson, G., Kelland, L. R., and Neidle, S. (2001) A new class of symmetric bisbenzimidazole-based DNA minor groove-binding agents showing antitumor. *J. Med. Chem.* **6,** 851–856.
21. Mayer, J. P., Lewis, G. S., McGee, C., and Bankaitis-Davis, D. (1998) Solid-phase synthesis of benzimidazoles. *Tetrahedron Lett.* **39,** 6655–6658.

11

Soluble Polymer-Supported Methods for Combinatorial and Organic Synthesis

Paul Wentworth, Jr. and Carsten Spanka

1. Introduction

Cross-linked polymer supports are now ubiquitous throughout the fields of combinatorial chemistry, organic synthesis, and catalysis *(1,2)*. However, problems associated with the heterogeneous nature of the ensuing chemistry and with "on-bead" spectroscopic characterization have meant that soluble polymers are developing as alternative matrices for combinatorial library production and organic synthesis *(3,4)*. Synthetic approaches that utilize soluble polymers, termed "liquid-phase" chemistry or soluble polymer-supported chemistry, couple the advantages of homogeneous solution chemistry—high reactivity, lack of diffusion phenomena, and ease of analysis—with those of solid phase methods—use of excess reagents and easy isolation and purification of products. Separation of the functionalized matrix can be achieved by a number of processes. Methods include solvent or heat precipitation, membrane filtration, and size-exclusion chromatography. This chapter will detail both the different areas of synthetic and combinatorial chemistry in which soluble polymer supports are used and the experimental techniques exploited.

2. Materials

Unless otherwise stated the polymers, solvents, reagents, and enzymes are commercially available from Aldrich Chemical Co.

2.1. Polymers

1. Poly(ethylene) glycol (PEG) of various molecular weights: typically 3400, 5000.
2. Monomethoxy poly(ethylene) glycol (MPEG).
3. 2% Divinylbenzene crosslinked polystyrene beads.

From: *Methods in Molecular Biology, Combinatorial Library Methods and Protocols*
Edited by: L. B. English © Humana Press Inc., Totowa, NJ

4. Diamino-functionalized poly(ethylene glycol) 6000 MW (Rapp-Polymere, Tübingen, Germany).

2.2. Solvents

1. Toluene.
2. Methylene chloride (DCM).
3. Diethyl ether.
4. Tetrahydrofuran (THF).
5. Methanol.
6. Ethyl acetate.
7. Hexanes.
8. Isopropylalcohol.
9. Acetonitrile.
10. Dimethylformamide (DMF).
11. Heptane.
12. Dimethylacetamide (DMA).
13. Benzene.
14. *tert*-Butanol.
15. Dimethyl sulfoxide.

2.3. Reagents

1. TMSCl.
2. Bis(cyclopentadienyl)zirconium chloride hydride.
3. 1-Hexyne.
4. $Li_2CuCNMe_2$.
5. MeLi.
6. Celite.
7. $MgSO_4$.
8. Trifluoromethanesulfonic anhydride.
9. 2-Butyn-1-ol.
10. 2,6-Di-*tert*-butylpyridine.
11. Ariethylamine (Et_3N).
12. Ammonium chloride.
13. Dichlorodi-isopropylsilane.
14. 1,8-Diaza-bicyclo[5.4.0]undec-7-ene (DBU).
15. Tetrabutylammonium fluoride.
16. CaH_2.
17. Triton CF-54.
18. Benzophenone.
19. 10-Camphorsulfonic acid.
20. L-selectride (1 *M* in THF).
21. Sodium hydroxide.
22. Hydrogen peroxide (30%).
23. Vinyl acetate.

24. Hydrofluoric acid.
25. Cuprous iodide.
26. Cesium carbonate.
27. 2-Iodobenzyl bromide.
28. 4 Å molecular sieves.
29. $Pd(PPh_3)_2Cl_2$.
30. 1-Pentyne.
31. Manganese(II) chloride.
32. Phenylboronic acid.
33. *N*-isopropylacrylamide.
34. *N*-(acryloxy)-succinimide.
35. 2,2'-azobisisobutyronitrile.
36. Bis(acetonitrile)dichloropalladium(II).
37. Ceramide (Funakoshi, Japan).

2.4. Enzymes and Buffers

1. CloneZyme™ library of thermophilic enzymes (ESL-001-01 to -07) (Sigma).
2. Novozym™ 435 (immobilized *Candida antarctica* lipase) (Novozorsk).
3. CRL (*C. rugosa* lipase) (Sigma).
4. α-(2→3)-Sialyltransferase (from rat liver, Sigma).
5. Bovine serum albumine (BSA, Sigma).
6. Calf intestinal alkaline phosphatase (CIAP, Sigma).
7. Sodium cacodylate (Sigma).
8. Leech ceramide glycanase (Sigma).
9. Penicillin G acylase (penicillin amidase from *Escherichia coli*, EC 3.5.1.11, in 0.1 *M* phosphate buffer pH 7.5, approx 30 U/mg, approx 70 mg/mL, approx 2100 U/mL, Fluka).
10. Sodium citrate buffer (Sigma).

3. Methods

3.1. Syntheses Utilizing Soluble Polymer Supports

3.1.1. Syntheses of Prostanoids on a Non-Crosslinked Soluble Polystyrene Support

Overview: The utility of combinatorial chemistry within drug discovery is ultimately linked to the ability to rapidly construct complex molecules on polymer supports. With this in mind, a polymer-supported approach to the prostaglandin core was seen as an important benchmark in the progress of this chemistry. First a two-step "liquid-phase" version of Noyori's *(5)* three-component coupling strategy was realized. It allowed the successful synthesis of PGE_2 methyl ester **1a** *(6)* and $PGF_{2\alpha}$ *(7)* **1b** (**Fig. 1**).

The synthetic strategy hinged upon the choice of a soluble polymer support that could withstand extreme reaction and work-up conditions. While PEG is

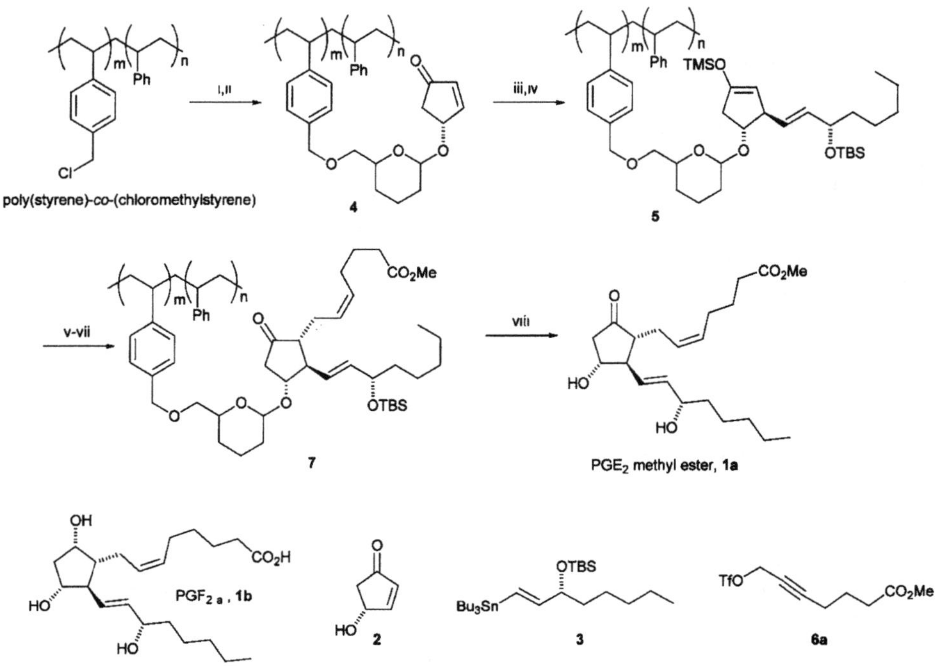

Fig. 1. *Reagents and conditions.* (i) 6-(hydroxymethyl)-3,4-dihydro-2*H*-pyran (3 eq), NaH (3.3 eq), DMA, room temperature, 24 h; (ii) **2** (3.0 eq), PPTS (0.5 eq), CH_2Cl_2, 40°C, 16 h; (iii) **3** (4.2 eq), $Li_2CuCNMe_2$ (3.9 eq), THF, –78°C, 15 min; (iv) chlorotrimethylsilane (15 eq), –78°C, 30 min; triethylamine (TEA, 30 eq), 0°C, 15 min; (v) MeLi (3 eq), THF, –23°C, 30 min; (vi) **6** (6 eq), –78°C, 10 min; –23°C, 30 min; (vii) H_2, 5% Pd-$BaSO_4$, quinoline, benzene/cyclohexane (1:1), room temperature, 48 h; (viii) 48% aqueous HF/THF (3:20, v/v), 45°C, 6 h.

ostensibly the polymer of choice for most facets of liquid-phase chemistry, its use in this case was contraindicated for two reasons: insolubility in THF at low temperatures and its solubility in water, which precluded aqueous extraction/removal of organometallic by-product.

Methods: The various preparation methods follow.

3.1.1.1. Preparation of Prostanoid 1a

1. Prepare a non-crosslinked copolymer of styrene and chloromethylstyrene (3 mol %) *(8)* with optimal loading of 0.3 mmol/g loading as the polymer matrix (*see* **Note 1**).
2. Attach cyclopentanoid alcohol **2** to the soluble *co*-polymer via Ellman's tetrahydropyran linker *(9)*.
3. Add vinylstannane ω-chain **3** to **4** in the presence of $Li_2CuCNMe_2$ in THF at –78°C. Isolate the stable polymer-bound silyl enol ether **5** following reaction of the intermediary enolate with TMSCl.

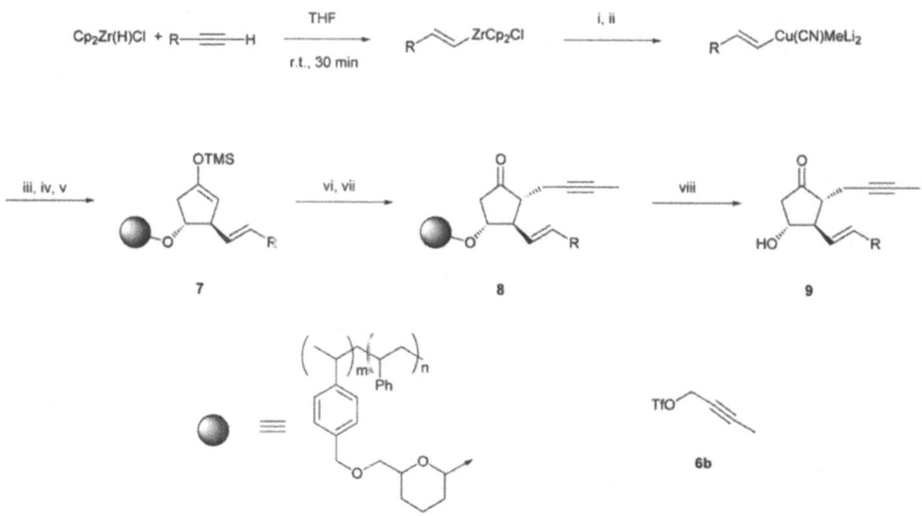

Fig. 2. *Reagents and conditions.* (i) MeLi (10 eq), –50°C, 15 min; (ii) CuCN (5 eq), –50°C, 15 min, MeLi (5 eq), –50°C, 15 min; (iii) **4** (1.0 eq) in THF, –50°C, 40 min; (iv) TMSCl (25 eq), –50°C, 50 min; (v) NEt_3 (50 eq), –50°C → 0°C, 15 min; (vi) MeLi (4.5 eq), –25°C, 1 h, then –70°C; (vii) *in situ* prepared **6b** (20 eq), –70°C, 10 min, –25°C, 50 min; (viii) 48% aqueous HF/THF (1:8, v/v), 45°C, 6 h.

4. Incorporate the α-chain, as its respective triflate **6**, by trapping of the intermediate enolate formed following addition of MeLi to **5** in THF (–23°C).
5. Following partial reduction of the α-chain alkyne, the polymer-bound *Z* alkene **7** is then cleaved from the support, with accompanying deprotection of the silyl ether protecting group to give **1a** in an overall yield of 37% (*see* **Note 2**).

In **Subheadings 3.1.1.2–3.1.1.3.**, a recently developed and improved version of the protocol described above, which allows the preparation of novel prostanoids with yields not otherwise obtainable by the conventional solution phase method, is described (**Fig. 2**) ***(10)***.

3.1.1.2. Preparation of Poly(styrene)-Supported Silylenolether 7

1. To a flame-dried 10 mL round bottom flask equipped with a stir bar and septa, add bis(cyclopentadienyl)zirconium chloride hydride (122 mg, 0.45 mmol) followed by dry THF (1.8 mL) and 1-hexyne (53 µL, 0.45 mmol). Protect the suspension from light with aluminum foil and stir for 30 min at room temperature.
2. Cool the now clear, homogeneous solution to –50°C, treat with methyl lithium (1.4 *M* solution in diethyl ether, 0.64 mL, 0.9 mmol) and stir for 15 min at –50°C.
3. Transfer the reaction mixture via cannula to a precooled (–50°C) 25 mL round bottom flask containing cuprous cyanide (40 mg, 0.45 mmol) and stir for 15 min

at –50°C after which a second portion of methyl lithium solution (0.32 mL, 0.45 mmol) should be added.

4. After 15 min, add a solution of polymer-supported enone **4** (loading: 0.3 mmol/g THP-linker, 300 mg, 0.09 mmol) in THF (3.6 mL) over a period of 20 min. Stir the resulting thick mixture for 40 min at –50°C.
5. Add chlorotrimethylsilane (245 mg, 2.25 mmol) dropwise to afford a clear solution and stir for 50 min and then add triethylamine (455 mg, 4.50 mmol). Allow the mixture to warm to 0°C and pour into a 1:1 mixture of deionized water and ethyl acetate, filter, and dry over magnesium sulfate.
6. Filter and concentrate to give a colorless oil. Redissolve the oil in THF (2 mL) and precipitate the polymer from cold methanol (–30°C). Remove the polymer by filtration and wash several times with cold methanol. Dry the polymer under high vacuum to give polymer-bound silylenolether **7** as a colorless powder (340 mg, 98% polymer recovery).

3.1.1.3. Preparation of Polystyrene-Supported Prostanoid 8 (*see* Note 3)

1. Cool a 25 mL round bottom flask, equipped with an efficient stir bar and septum, to –25°C. Add trifluoromethanesulfonic anhydride (633 mg, 2.24 mmol) followed by dropwise addition of a mixture of 2-butyn-1-ol (143 mg, 2.04 mmol) and 2,6-di-*tert*-butylpyridine (468 mg, 2.46 mmol) in dichloromethane (1.5 mL) over 3 min. Continue stirring at –25°C for 1 h. Cool a 25 mL round bottom flask, containing a solution of polymer **7** (340 mg, 0.102 mmol) in THF (8 mL), to –25°C and add methyl lithium (1.4 *M* solution in diethyl ether, 0.33 mL, 0.46 mmol) in one portion. Continue stirring at –25°C for 1 h. Add hexanes (5 mL) dropwise and then cool to –70°C with vigorous stirring. Filter the thick colorless suspension through a pad of anhydrous magnesium sulfate under argon into a precooled (–70°C) pear-shaped 25 mL flask and wash the filter cake with cold hexanes (1 mL). The filtrate is then rapidly concentrated (T < 0°C), then cooled to –70°C, and THF (1 mL) is added to the flask (*see* **Note 4**).
2. The freshly prepared triflate **6a** is then added via cannula to the cold (–70°C) yellow anion solution of polymer **7**. Stir the resulting solution for 10 min at –70°C, then for 50 min at –25°C, quench with saturated ammonium chloride solution (5 mL) and allow to warm to room temperature.
3. Pour the mixture into saturated ammonium chloride solution (50 mL) and extract with ethyl acetate (3 × 50 mL). Wash the combined organic extracts with brine, filter through Celite, dry over magnesium sulfate, and concentrate.
4. Isolate the polymer-supported prostanoid **8** as a colorless powder (313 mg, 93% polymer recovery) by the standard precipitation procedure described in **Subheading 3.1.1.2.**

3.1.1.4. Release of Hydroxyprostanoid 9 from Polymeric Support 8.

1. Add aqueous hydrofluoric acid (48%, 0.2 mL) to a solution of polymer **8** (150 mg, 0.04 mmol) in THF (1.6 mL) in a small polypropylene vial and stir at 45°C for 6 h. Neutralize the reaction with saturated sodium bicarbonate solution and wash the aqueous layer with ethyl acetate. Combine the organic fractions and wash them

with saturated sodium bicarbonate solution, brine, dry over magnesium sulfate and then concentrate.

2. Add a small amount of THF and methanol (10 mL) to the residue and cool to –30°C with vigorous stirring. This precipitates the polymer. Remove the polymer by filtration. Concentrate the filtrate to give a crude oil (15 mg) that can be purified by flash chromatography (ethyl acetate/hexanes 1:3) to afford **9** as a colorless oil (3.3 mg, 0.012 mmol, 31%).

3.1.2. Synthesis of Glycosteroids on a Poly(ethylene glycol) Support

Overview: Poly(ethylene glycol)s have been used for the polymer-supported preparation of oligosaccharides ***(11–13)***. Recently Kirschning and coworkers ***(14)*** reported a method for the construction of glycoconjugates by using polyethylene glycol monomethyl ether (MPEG) supported 6-deoxyglycals as acceptors for the aglycon building blocks (**Fig. 3**). The glycals were attached to the polymeric support via a bis-*O*,*O*-silyl linker. The most stable linker moiety dichlorodiisopropylsilane was identified. The linker was installed by *very slow* addition of MPEG (**10**) to a slight excess of dichlorodiisopropylsilane (1.1 eq) in the presence of 1,8-diaza-bicyclo[5.4.0]undec-7-ene (DBU) as base. The chlorosilane intermediate **11** formed was not isolated but rather coupled directly to glycal **12** in the presence of DBU. The glycosidation reaction of polymer-bound glycal **13** with digitoxigenin **14** was carried out in the presence of a catalytic amount of 10-camphorsulfonic acid. The resultant mixture of polymer-bound glucoconjugate anomers **15** formed was released from the support by treatment with ammonium-fluoride-buffered tetrabutylammonium fluoride (3:1). In this way the TBS protecting group was removed as well and post-cleavage modifications of product **16** (cleavage of the glycosidic bond, rearrangements) could be suppressed. After purification by flash chromatography both anomers of **16** could be separated and isolated in 39% total yield (based on the loading capacity of the MPEG used: 0.2 mmol/g).

The results obtained with soluble MPEG support **10** were compared with those from classical solution-phase synthesis and a solid-phase variant using 2% divinylbenzene crosslinked polystyrene beads in combination with succinate linker **17**. Both polymer-utilizing methodologies did not give improved yields in comparison to the solution-phase synthesis of **16**. Interestingly, significantly improved selectivities in the glycosylation step were observed (MPEG-based method: **16** α/β = 7:1, total yield: 39%; solution phase synthesis: **16** α/β/Ferrier products = 2.8:1.7:1, total yield: 77%). However, the MPEG-based synthesis of **16** exhibited some problems that are intrinsic to the linker and support used (acid lability of the silyl linker and insufficient polymer recovery due to incomplete precipitation caused by the presence of very lipophilic groups on the polymeric support).

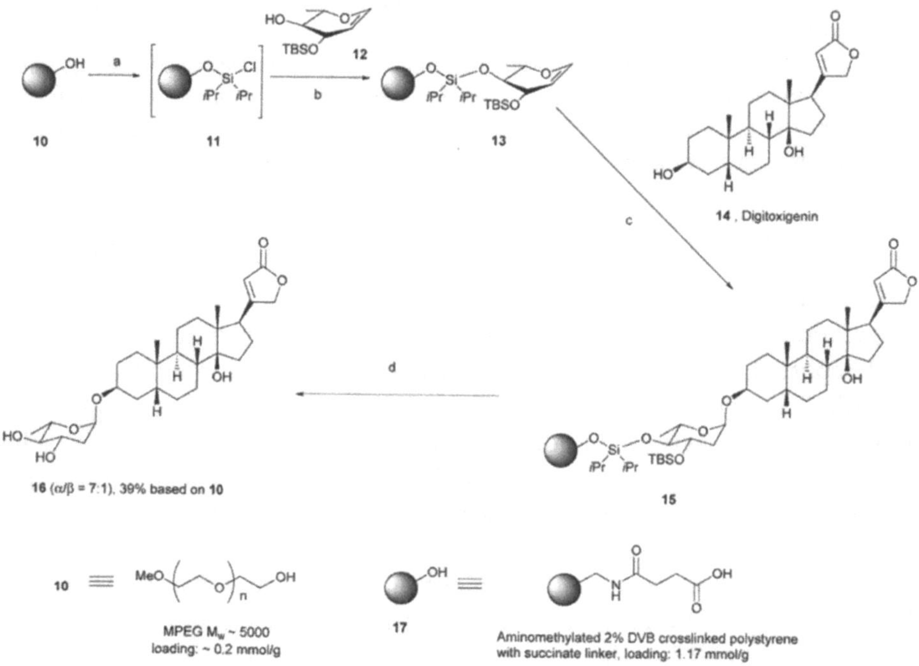

Fig. 3. *Reagents and conditions.* (a) DBU (1.5 eq), dichlorodiisopropylsilane (1.1 eq), DCM, very slow addition of **10** (1.0 eq) within 48 h, then 45°C, 12 h; (b) glycal **12** (1.1 eq), DBU (1.5 eq), DCM, 50°C, 24 h; (c) **13** (1.0 eq), digitoxigenin (**14**) (1.33 eq), 10-camphorsulfonic acid (0.16 eq), molecular sieves 4 Å, DCM, room temperature 24 h, then 10-camphorsulfonic acid (0.08 eq), room temperature, 24 h; (d) Bu_4NF (1 eq), NH_4F (3 eq), room temperature, 48 h.

Methods: The MPEG polymer (polyethylene glycol monomethyl ether, Mw approx 5000, free OH approx 0.2 mmol/g) was purified by precipitation of a DCM solution into diethyl ether. Then, the freshly precipitated polymer was dissolved in dry DCM/MeOH (50:1) and was dried by gel filtration under argon over a mixture of silica gel, magnesium sulfate, and Celite® (1:0.5:1) followed by concentration under reduced pressure under argon (*see* **Note 5**).

3.1.2.1. Preparation of Poly(ethylene glycol)Methyl Ether Supported Glycal 13

1. *Slowly* (0.02 mL/min, 2 d) add a solution of polymer **1** (5.0 g, 1 mmol) and DBU (0.22 mL, 1.5 mmol) in DCM (5 mL) by syringe pump to a solution of DBU (0.22 mL, 1.5 mmol, 1.5 eq) and dichlorodiisopropylsilane (204 mg, 1.1 mmol, 1.1 eq) in dry DCM (5 mL) under argon. Stir the solution for 12 h at 45°C, by which time the color of the solution should change to blue-green.

2. Add a solution of glycal **3** (268 mg, 1.1 mmol) and DBU (0.22 mL, 1.5 mmol) DCM (10 mL) and continue stirring for 24 h at 50°C. Precipitate the polymer-supported product by pouring the reaction mixture into vigorously stirred ice cold Et_2O (300 mL). Filter the precipitate under argon, wash with dry Et_2O (2 × 100 mL), and dry *in vacuo* to give polymer supported glycal **13** polymer as a colorless powder (3.68 g, 70% polymer recovery).

3.1.2.2. Synthesis of Polymer-Bound Glycosteroid 15

1. Add molecular sieves (4 Å, 1.5 g) and 10-camphorsulfonic acid (37 mg, 0.16 mmol) to a solution of polymer **4** (5.037 g, ≤ 1 mmol glycal content) and digitoxigenin (**14**) (0.5 g, 1.33 mmol) in dry DCM (30 mL). Stir the reaction mixture for 24 h at room temperature, then add an additional portion of 10-camphorsulfonic acid (19 mg, 0.08 mmol) and stir for a further 24 h.
2. Remove the molecular sieves by filtration and precipitate the polymer-bound glycoconjugate **6** by pouring the filtrate into vigorously stirred ice cold dry ether (150 mL) under argon. Wash the precipitate with dry cold ether (3 × 100 mL) and dry *in vacuo* to afford **15** as a colorless powder.

3.2. Methods Involving Enzymatic Transformations

3.2.1. Soluble Polymer-Supported Lipase-Mediated Acylation

Overview: The dihydropyranyl-linker functionalized poly(ethylene) glycol monomethyl ether (average molecular weight 5000) **II** was successfully used for a novel synthetic approach to the C_{21}–C_{27} fragment **I** of the bryostatins that incorporates a stereoselective enzymatic transformation (**Fig. 4**). This provided an approach to a composite fragment of the bryostatin family of natural products *(15)*. Selective *syn* reduction of precursor ketone *rac*-**18** using L-selectride, generated a mixture of racemic alcohols (*syn*-**19** and *syn*-**20**) in almost quantitative yield, and with complete diastereoselectivity (determined by ^{1}H-NMR). This then set-up the prospect for a potential stereoselective enzyme-catalyzed resolution of *syn*-**19** and *syn*-**20** with a suitable lipase. A panel of commercially available esterase/lipase enzymes were examined; including the CloneZyme™ library of recombinant thermophilic enzymes (ESL-001-01 to -07), Novozym™ 435 (immobilized *C. antarctica* lipase) and CRL (*C. rugosa* lipase) (*see* **Note 6**). Optimization of the reaction conditions revealed that the Novozym™ 435 catalyzed acylation of *syn*-**19** (*R*-alcohol) to acetate **21** can be carried to almost complete conversion with excellent stereoselectivity in anhydrous toluene as the solvent: 70°C, 4 h, 40% conversion: > 99% ee; 70°C, 6 h, 50% conversion: 92% ee (determination of ee by chiral capillary GC of the product **22** after its cleavage from the PEG support).

The enzymatic resolution allowed ready separation of alcohol **22** and diol **23**, by silica gel chromatography, after cleavage. Both compounds **22** and **23** can be used for the preparation of orthogonally protected building blocks that offer an entry point into diastereomeric libraries of bryostatin analogs.

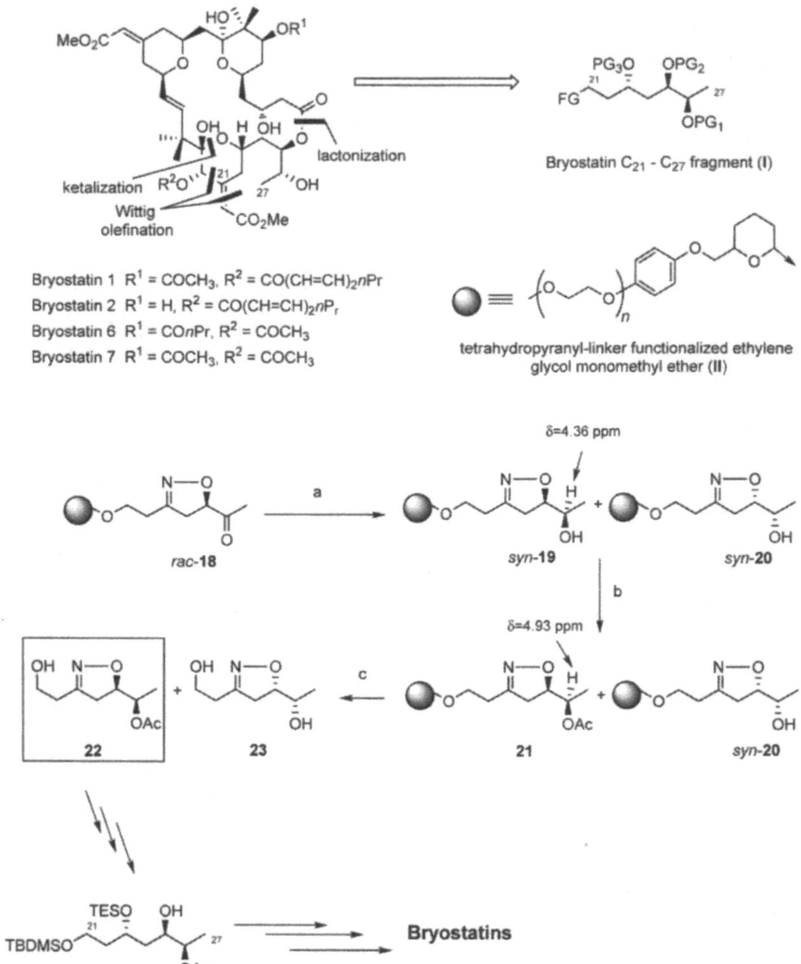

Fig. 4. *Reagents and conditions.* (a) L-Selectride (2.5 eq), DCM, –78°C, 6 h (90% PR); (b) Novozym 453, vinyl acetate (20 eq), toluene, 70°C, 4 h (40% conversion, >99% ee, 96% PR); (c) HF(aq.), acetonitrile, 40°C, 8 h (54%).

Methods: The following subheadings give some reparation methods.

3.2.1.1. Preparation of the Racemic Mixture *syn*-19/*syn*-20

1. Dissolve polymer-supported ketone **18** (average molecular weight 5348, 5.05 g, 0.91 mmol) in dry methylene chloride (25 mL) and cool to –78°C and add L-Selectride (2.28 mL, 1 *M* in THF) in one portion.
2. After 6 h, slowly add sodium hydroxide (1 *N*, 2 mL) and hydrogen peroxide (30%, 2 mL) and allow the mixture to warm to room temperature. Separate and

isolate the organic layer and dry over $MgSO_4$. Precipitate into isopropylalcohol (200 mL, 0°C, plus 3 × 30 mL washings) and recover the polymer by filtration to yield a racemic mixture of polymer-supported alcohols *syn*-**19** and *syn*-**20** (4.56 g, 90% mass recovery).

3.2.1.2. Enzymatic Resolution of Polymer-Bound *syn*-19/*syn*-20

1. Dry the polymer obtained in **Subheading 3.2.1.1.** (average molecular weight 5350, 4.49 g, 0.84 mmol) by azeotropic removal of water with toluene.
2. Dissolve the polymer in anhydrous toluene (25 mL) and add vinyl acetate (1.54 mL, 16.76 mmol) and Novozym™ 435 (800 mg). Gently stir the mixture at 70°C (*see* **Note 6**).
3. After approx 40% conversion cool the solution and remove the enzyme by filtration. Following the standard precipitation into ether (*see* **Note 7**) the mixture of polymer-supported acetate **21** and unconverted alcohol *syn*-**20** are isolated as a yellowish solid (4.33 g, 96% mass recovery).

3.2.1.3. Cleavage of Acetate 22 from the Polymer Support

1. Dissolve the mixture of polymer-supported acetate **21** and *syn*-**20** (4.33 g) in acetonitrile (20 mL) and place in a plastic vial. Add hydrofluoric acid (45% in water, 2.5 mL) and stir the mixture at 40°C for 8 h.
2. Add a saturated solution of sodium bicarbonate to neutralize. Pour the mixture into brine and extract with methylene chloride (3 × 50 mL). Combine the organic fractions and dry over magnesium sulfate. Remove the dessicant and concentrate to approx 5 mL. The polymer was then removed by the standard precipitation method (*see* **Note 7**).
3. Evaporate the filtrate to yield a mixture of alcohol **22** and diol **23** as a yellow oil.
4. Alcohol **22** can then be isolated by column chromatography of the yellow oil using a mixture of hexanes and ethyl acetate (1:1) as eluent (*see* **Note 8**). Alcohol **22** is isolated as a colorless oil. The expected yield should be 27% from *syn*-**20**, assuming 100% conversion in all the steps carried out with polymer-bound compounds).

3.2.2. Regioselective Enzymatic Sialylation of a Poly(acrylamide) Supported Substrate

Overview: Enzyme-assisted strategies for the synthesis of oligosaccharides are recognized as promising alternatives to chemical synthesis because of high regio- and stereoselective reactions without the need for protecting groups. An efficient methodology for the construction of carbohydrates, including oligosaccharide and sphingoglycolipids, has been developed by Nishimura and Yamada ***(16,17)***. They synthesized a vinylic oligosaccharide monomer **24** that, when treated with acrylamide under radical polymerization conditions, formed a water-soluble copolymer **25** (**Fig. 5**). This water-soluble conjugate **25** was then used as a primer for a regioselective sialylation reaction catalyzed by rat liver βGal1→3/4GlcNAc α-2,3-silalyltransferase, in the presence of

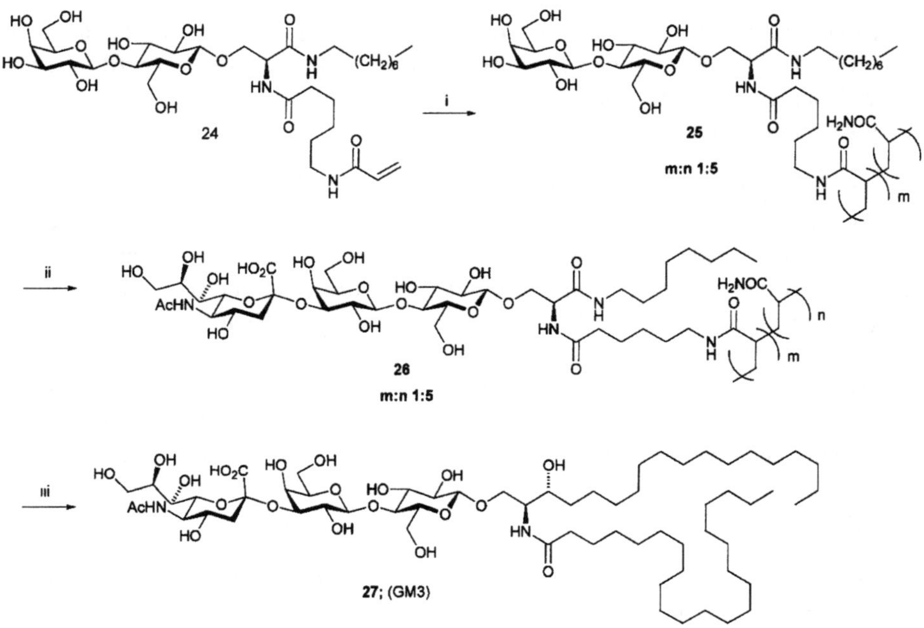

Fig. 5. *Reagents and conditions.* (i) acrylamide (4.0 eq), TMEDA (0.4 eq), ammonium peroxodisulfate (APS) (0.2 eq), 50°C, 1 d (92%); (ii) CMP-NeuAc (1.2 eq), α-2,3-sialyltransferase (0.3 unit), BSA, $MnCl_2$, calf intestinal alkaline phosphatase (20 unit), sodium cacodylate buffer (50 m*M*, pH 7.4), 37°C, 3 d (> 99%); (iii) ceramide (4.85 eq), ceramide glycanase (0.01 unit), Triton CF-54 (1 drop), sodium citrate buffer (50 m*M*, pH 6.0), 37°C, 17 h (61%).

CMP-NeuAc *(18)*, to generate the soluble polymer-supported trisaccharide **26** in quantitative yield.

Purification of **26** from the enzyme and excess sugar building block by gel-filtration chromatography was facilitated by the presence of the poly(acrylamide) support. Subsequent treatment of **26** with leech ceramide glycanase in the presence of an excess of ceramide as an acceptor generated the transglycosidated product GM3 **27** in 61% yield. Thus, this polymer-supported enzyme approach afforded the glycolipid GM3 in 56% yield from the readily available precursor/monomer **24**, a remarkable improvement in both the ease of synthesis and overall yield when compared to that of chemical synthesis *(19)*.

Methods: Various preparation methods follow.

3.2.2.1. Preparation of Co-polymer 25

1. Add a solution of acrylamide (60.3 mg, 0.84 mmol, 4 eq) in water (2 mL) to a solution of glycomonomer **24** (150 mg, 0.21 mmol) in dimethyl sulfoxide (2 mL). Degas this solution by applying a waterpump vacuum with stirring.

2. Add *N,N,N',N'*-tetramethylethylene diamine (12.7 µL, 84 µmol) and ammonium persulfate (7.7 mg, 34 µmol). Stir the clear solution under nitrogen at 50°C for 24 h.
3. Purify the reaction mixture by gel-filtration chromatography on a Sephadex G-25 column (30 mm diameter × 400 mm) and elute with deionized water.
4. Concentrate polymer-containing fractions to a small volume and then lyophilize to give co-polymer **25** as an amorphous powder (193 mg, 92%) (*see* **Note 9**).

3.2.2.2. Regioselective Sialylation of Polymer 25

1. Incubate a solution of polymer **25** (22 mg, approx 20 µmol of substrate), cytidine-5'-monophospho-N-acetyl neuraminic acid (CMP-NeuAc, Sigma C 8271) (15 mg, 24.4 µmol), α-(2→3)-sialyltransferase (from rat liver, Sigma S 2769) (0.3 unit), bovine serum albumin (BSA, 4 mg), and calf intestinal alkaline phosphatase (CIAP, Sigma P 7923) (20 unit) in 50 m*M* sodium cacodylate buffer (pH 7.4, 2.0 mL) containing manganese(II) chloride (0.62 mg) and Triton CF-54 (10 µL) at 37°C for 72 h.
2. Purify the reaction mixture directly by chromatography on a Sephadex G-25 column (30 mm diameter × 400 mm) and elute with deionized water. The polymer-containing fractions are collected and lyophilized to afford the GM3 trisaccharide glycopolymer **26** (22.0 mg) (*see* **Note 10**).

3.2.2.3. Transglycolsylation by Ceramide Glycanase

1. Add Triton CF-54 (1 drop) to a mixture of glycopolymer **26** (22 mg, approx 16.2 µmol of sialyllactose residue) and ceramide (Funakoshi, Japan) (50 mg, 78.7 µmol) in 50 m*M* sodium citrate buffer (pH 6.0, 1 mL) and sonicate for 1 min.
2. Add leech ceramide glycanase (Sigma C 2557) (0.01 unit) and incubate the solution at 37°C for 17 h.
3. The polymeric support is removed by chromatography (Sephadex LH-20 column, eluent: 60:30:4.4 chloroform-methanol-water) to give GM3 **27** (12 mg, 63%) (*see* **Note 11**).

3.2.3. Soluble Polymer Synthesis Incorporating Enzyme-Labile Linkers

Overview: Waldmann and his group recently reported the development of a penicillin G acylase sensitive "safety catch linker" for the attachment of alcohols and anilines to a polymeric support (**Fig. 6**) *(20)*. The release of the synthesized target molecule from the support is initiated by the hydrolysis of the phenylacetamide moiety contained in the linker **28** under extremely mild conditions (pH 7.0, room temperature or 37°C). The liberated benzylamine **29** intermediate cyclizes instantaneously to release the desired target molecule with the tetrahydroisoquinolinone **30** formed in the course of this process remaining on the polymer.

Linker **31** was attached via a carbamate-ester to an amino-functionalized polymeric support (**Fig. 7**). Treatment with lithium hydroxide liberated linker acid **32**. First attempts to use this enzyme cleavable linker on crosslinked

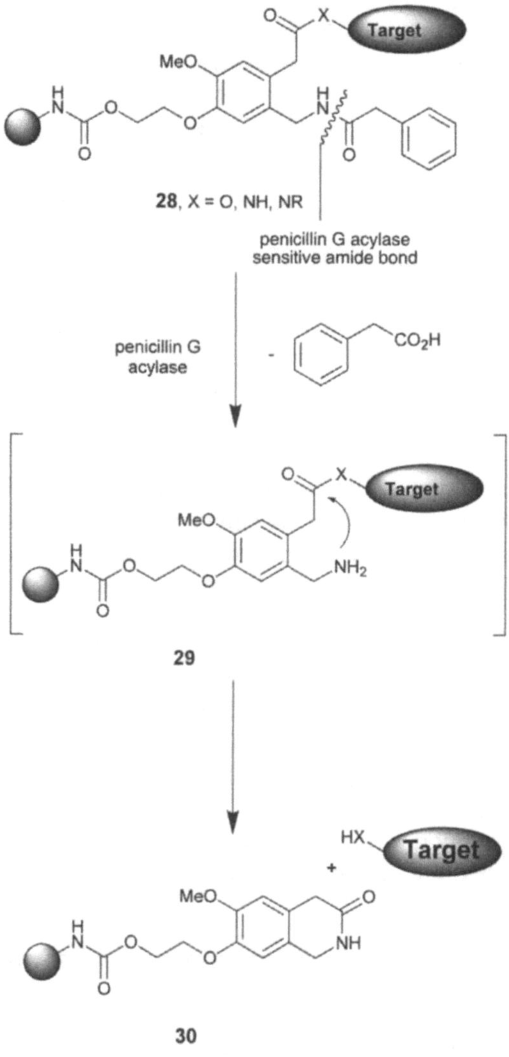

Fig. 6. An enzyme labile phenylacetamide linker. Penicillin acylase cleaves the acetamide group of polymer-supported linker **28**. The benzylamine derivative **29** then cyclizes to release the target compound.

supports (TentaGel), controlled pore glass (CPG) beads, and PEGA (crosslinked 2-acrylamidoprop-1-yl[2-aminoprop-1-yl]-polyethylene glycol) were unsuccessful most likely due to insufficient penetration of the three-dimensional polymeric matrix by the enzyme. However, this approach worked very efficiently when diamino-functionalized polyethylene glycol was used as the sup-

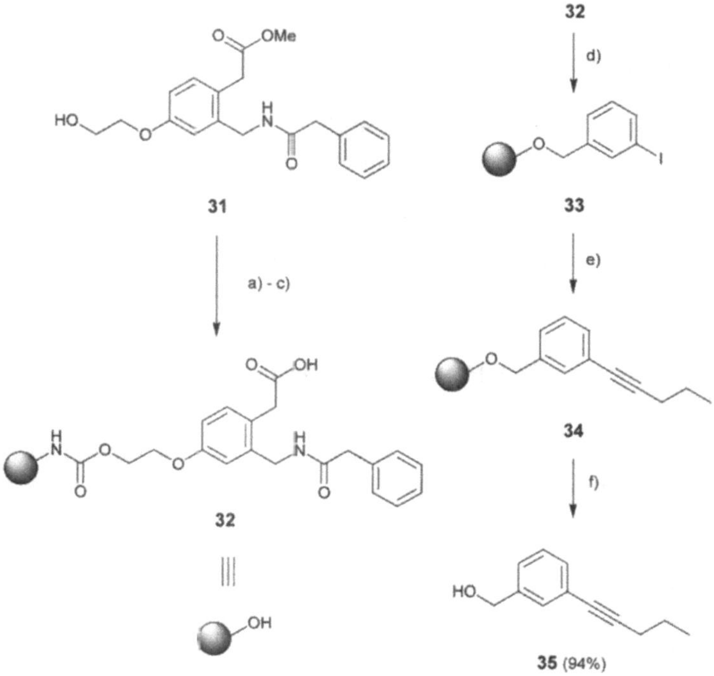

Fig. 7. *Reagents and conditions*. (a) phosgene (1.93M in toluene) (6 eq), DCM, room temperature, >98%; (b) diamino-functionalized poly(ethylene glycol) (MW 6000), DMAP (11 eq), HOBt (3 eq), DCM, room temperature, 93%; (c) LiOH 0.25 *M* in THF/H_2O 1:2, room temperature, 99%; (d) 2-iodobenzylbromide, Cs_2CO_3, (5 eq), molecular sieves 4 Å, DMF, 50°C, 24 h, 95%; (e) 1-pentyne (10 eq), [$Pd(PPh_3)_2Cl_2$] (0.1 eq), CuI (0.25 eq), dioxane/NEt_3 2:1, 20°C, 24 h, 97%; (f) penicillin G acylase, pH 7.0, MeOH/0.2 *M* sodium phosphate buffer pH 7.0, 37°C, 48 h.

port: the desired target compound could be obtained in high yields and purities. The linker tolerates a large number of reaction conditions and temperatures up to 80°C. A diverse set of reactions was successfully performed by using this linker on polyethylene glycol as support: Pd(0)-catalyzed Heck-, Suzuki-, and Sonogashira reactions as well as the Mitsunobu reaction and Diels-Alder- and [2+3]-cycloadditions.

Methods: This section below details how the author performed a Sonogashira coupling using this system.

3.2.3.1. Preparation of 2-Iodobenzylester Functionalized PEG **33**

1. Stir a mixture of PEG_{6000} bound carboxylic acid **32** (234 mg, 69 µmol), cesium carbonate (58 mg, 178 µmol, 2.5 eq), 2-iodobenzyl bromide (104 mg, 249 µmol, 3.6 eq), and 4 Å molecular sieves (500 mg) in DMF (50 mL) under argon at 50°C.

2. Remove the molecular sieves by filtration and evaporate the filtrate to dryness.
3. Dissolve the residue in DCM (15 mL) and wash with saturated ammonium chloride solution (10 mL). Extract the aqueous layer with DCM (2 × 15 mL) and combine the organic extracts and dry them over $MgSO_4$.
4. Concentrate the obtained organic solution to approx 2 mL and precipitate the polymer by slow addition into *vigorously* stirred dry ice cold ether (100 mL) under argon. Isolate the precipitate by filtration and wash it with cold ether (25 mL) and ethanol (25 mL). The polymer obtained was dissolved in fresh DCM (2 mL) and the precipitation procedure described in **Note 7** was repeated to give the polymer as a colorless powder (234 mg, 64 μmol, 95%, M_n = 7231 Da) (*see* **Note 12**).

3.2.3.2. Preparation of 3-(1-Pentynyl)Phenyl]Methyl Ester-Functionalized PEG 34

1. Degas a solution of polymer bound iodide **33** (153 mg, 42 μmol) and 1-pentyne (41 mg, 423 μL, 10 eq) in Et_3N/dioxane 1:2 (12 mL) for 15 min by ultrasonication.
2. Add cuprous iodide (2 mg, 11 μmol, 0.25 eq) and $[Pd(PPh_3)_2Cl_2]$ (3 mg, 4 μmol, 0.1 eq) and stir for 24 h at room temperature under argon.
3. Add $CHCl_3$ (15 mL) and wash the mixture with saturated ammonium chloride solution (10 mL). Isolation of the polymer by the same method as described in **Subheading 3.2.3.1.** affords the desired polymer **34** as a colorless powder (147 mg, 41 μmol, 97%, M_n = 7111 Da).

3.2.3.3. Penicillin G Acylase-Catalyzed Release of [3-(1-Pentynyl)Phenyl]Methanol 35

1. Dissolve polymer bound acetylene **34** (19 mg, 5.3 μmol) in a mixture of methanol (1 mL) and 0.2 *M* aqueous sodium phosphate buffer (pH 7, 0.9 mL).
2. Add penicillin G acylase (2 μL, 4.2 U, suspension in 0.1 *M* aqueous phosphate buffer pH 7.5, 2100 U/mL) and incubate the reaction mixture for 48 h at 37°C with shaking.
3. Add further aliquots (2 μL) of the enzyme suspension after 12, 24, and 48 h.
4. Extract the reaction mixture with diethyl ether (6 × 20 mL), combine the organic extracts, and dry them over $MgSO_4$. Evaporation of the solvent affords alcohol **35** as a colorless oil (0.9 mg, 5.0 μmol, 94%); TLC: R_f = 0.20 (ethyl acetate/hexanes 1:6 v/v) (*see* **Note 13**).

3.3. Thermomorphic Supports

Bergbreiter and coworkers ***(21)*** recently described the development of a new polymeric support with interesting physical properties: poly(*N*-isopropylacrylamide) (PNIPAM). Several co- and ter-polymers containing PNIPAM have been successfully prepared. These materials exhibit an inverse temperature solubility profile in water. If heated above their lower critical solution temperature (LCST), they precipitate quantitatively from solution. Unlike a protein, PNIPAM derivatives do not denature. The LCST of the PNIPAMs can

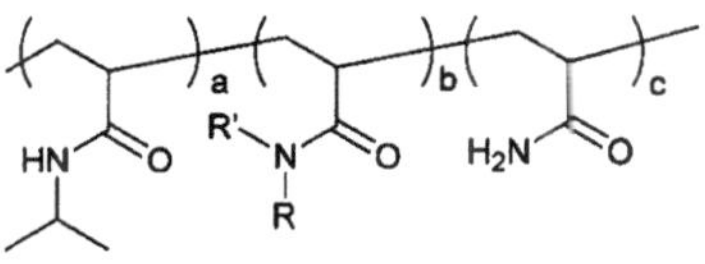

PNIPAM co-polymers

36:	R = -C_6H_4OH	R' = H	LCST = 27¡C	a:b:c = 20.1:0
37:	R = -$C_6H_4CH_2OH$	R' = H	LCST = 38¡C	a:b:c = 29:1:0
38:	R = -$C_6H_4CO_2H$	R' = H	LCST = 29¡C	a:b:c = 20:1:0
39:	R = -$C(Me_2)CH_2SO_3H$	R' = H	LCST = 44¡C	a:b:c = 20.1:0
40:	R = -$(CH_2)_3PPh_2$	R' = H	LCST = 24¡C	a:b:c = 50.2:3
41:	R = -CH2CH2PPh2	R' = R	LCST = 18¡C	a:b.c = 100:1:9

Fig. 8. Thermomorphic PNIPAM co-polymers.

be tuned over a wide range of temperatures (room temperature to above 100°C). Thus, PNIPAMs can be used as supports for the preparation of "smart" and easily recoverable substrates and catalysts. Several functionalized derivatives were successfully prepared and used either as carriers for substrates (**36–39**) or the preparation of PNIPAM supported catalysts for hydrogenation [co-ordination of Rh(I) to phosphines **40** and **41**] or carbon–carbon bond formation [co-ordination of Pd(0) to phosphine **40**] (**Fig. 8**). The latter catalyst could also be used for the preparation of allylamines by coupling of secondary amines to cinnamyl acetate. Both **40** and **41** derived catalysts are compatible with a wide variety of solvents (water, alcohols, THF, heptane, as well as mixtures thereof). However, both polymer-supported phosphine ligands were still very oxygen sensitive *(22)*. In the experimental section, the application of a robust polymer-supported version of a tridentate pincer-type SCS palladium(II) complex **42** *(23)* in Heck- and Suzuki-reactions under "thermomorphic" condition is described. This methodology involves a biphasic catalyst recovery protocol *(24)*.

The system consists of solvent mixture that is biphasic at room temperature and becomes homogeneous when heated (e.g., heptane and 90% DMA/water become miscible in all proportions above 65°C). After completion of the reaction and cooling to room temperature the reaction products are staying in the nonpolar phase and can be isolated by simple phase separation. The polar phase that contains the polymer bound catalyst can be reused in further runs by adding fresh substrate solution in heptane.

Polymer supported catalysts **43** and **44** containing a 5-amido-SCS-ligand proved to be excellent catalysts for the conduction of Heck and Suzuki reactions in DMF and triethylamine as base at high temperatures (95°C) even in an air atmosphere (**Fig. 9**). They could be recycled several times without loss in activity observed after up to five cycles: Quantitative yields were obtained in many cases beyond the second cycle when the polar phase containing the catalyst became saturated by the product formed *(24)*. Catalyst **45** containing a 5-oxo-SCS-ligand slowly decomposed under the reaction conditions (deposition of palladium metal) and was not suitable for recycling *(23)*.

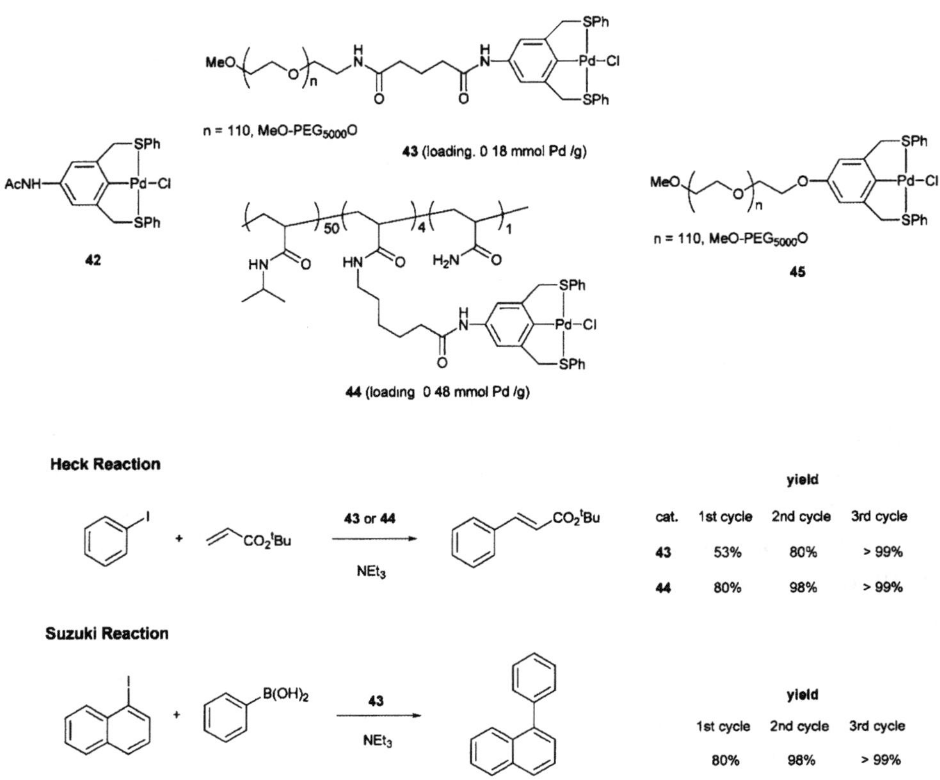

Fig. 9. Polymer supported SCS Pd complexes **43-45** were compared with solution phase catalyst **42** in the Heck and Suzuki reactions.

*3.3.1. General Procedure for Thermomorphic Catalysis with Polymer Bound SCS-Pd(II)-Complexes **43, 44** (24)*

1. Prepare a solution of either **43** or **44** (0.01 mmol of Pd; corresponding to 56 mg of **43** or 21 mg of **44**) in 90% aqueous acetamide (5 mL) in a 30 mL screw cap vial containing a Teflon stirrer bar (*see* **Note 14**).
2. Add a solution of the desired aryl iodide (5 mmol), acceptor (6 mmol), and triethylamine (1.05 mL, 7.5 mmol) in heptane (10 mL) and seal the tube with the screw cap (*see* **Note 15**).
3. Heat the tube to 95°C in an oil bath and monitor the reaction by TLC (silica gel, heptane).
4. After completion of the reaction, allow the system to reach room temperature, remove the upper phase by means of a pipet, and evaporate the solvent under reduced pressure. Dry the remaining residue under vacuum to give the pure coupling products.

5. Catalyst recycling can be affected by adding a fresh heptane solution of the reactants (plus a new portion of phenylboronic acid to the DMA layer for the Suzuki reactions), resealing the system, and heating to 95°C.

4. Notes

1. This copolymer is soluble in THF, dichloromethane (DCM), and ethyl acetate even at low temperatures, but is insoluble in methanol and water, so that purification can involve both aqueous extraction and precipitation techniques.
2. The main features to note are that the polymer recovery mass balance can be > 97% and only one polymer-bound species should be detected by NMR analysis for each step of the synthesis.
3. For reproducible results, it is essential that the preparation of propargyl triflate **6a** and the preparation of the lithium enolate of polymer **7** be conducted simultaneously in separate flame dried equipment.
4. This solution should be purged with argon and stored at –70°C until needed.
5. For best results the PEG support should be further dried by lyophilization for 24 h. During all steps in the precipitation process, moisture has to be excluded thoroughly otherwise the precipitate becomes sticky and/or waxy and is difficult to filter efficiently.
6. The enzyme-catalyzed reaction was routinely monitored by ^{1}H-NMR following removal of aliquots from the reaction mixture. The downfield chemical shift of the resonance of the methine proton on C_{26} from 4.36 ($C_{26}H$-OH, **19**) to 4.93 ($C_{26}H$-OAc, **21**) ppm was a clear indication of the acylation progress.
7. Standard precipitation procedure for the PEG supported-derivatives: A dry solution of the PEG polymer in DCM (approx 5 wt%) was dripped slowly into a 10-fold volume of vigorously stirred ice cold dry isopropanol or dry diethyl ether. All manipulations were performed under a blanket of argon to exclude moisture and in order to obtain a finely and easy to filter precipitate. The filter cake was washed three times with fresh solvent used in the precipitation step and dried *in vacuo* to afford the polymer as fine powder.
8. Chiral GC of alcohol **22** [1 m*M* solutions in methylene chloride, γ-cyclodextrin trifluoroacetyl column (30 m × 0.25 mm, head column pressure = 60 psi, oven at 90°C)] is routinely used to determine the % ee of the enzymatic resolution.
9. Typically the molecular weight of the resultant polymer can be determined by gel-permeation chromatography (Asahipak GS-510 column; standards: pullulans 5.8, 12.2, 23.7, 48.0, 100, 186, 380 kD, Shodex standard P-82) and is usually > 380,000.
10. The degree of sialylation can be estimated from integration data of the ^{1}H-NMR spectrum. It is usually quantitative.
11. A small amount of sialyllactose, formed by hydrolysis, is also isolated as a byproduct.
12. The reaction progress can be determined by 500 MHz ^{1}H-NMR spectrum (in $CDCl_3$).The polymer support can be efficiently isolated from the aqueous layer by DCM extraction (4 × 20 mL). The extracts are then combined, dried over

$MgSO_4$, and evaporated to dryness. Drying the residue under reduced pressure gives the polymer as a colorless powder.

13. Water (10% by volume) has to be added to the DMA phase in order to suppress the miscibility-increasing property of the poly(ethylene glycol) based catalyst **45**. In the case of the PNIPAM-bound catalyst **45** the water content in the DMA phase prevented the precipitation of the catalyst while the reaction mixture is hot [increase of the lower critical solution temperature (LCST) above the reaction temperature of 95°C].
14. For Suzuki couplings, phenylboronic acid is first dissolved in the 90% DMA phase due to its insufficient solubility in heptane.

References

1. Fruchtel, J. S. and Jung, G. (1996) Organic chemistry on solid supports. *Angew. Chem., Int. Ed. Engl.* **35,** 17–42.
2. Thompson, L. A. and Ellman, J. A. (1996) Synthesis and applications of small molecule libraries. *Chem. Rev.* **96,** 555–600.
3. Wentworth Jr., P. and Janda, K. D. (1999) Liquid-phase chemistry: recent advances in soluble polymer-supported catalysts, reagents and synthesis. *Chem. Commun. (Cambridge)*, 1917–1924.
4. Wentworth Jr., P. (1999) Recent developments and applications of liquid-phase strategies in organic synthesis. *TIBTECH* **17,** 448–452.
5. Noyori, R. and Suzuki, M. (1984) New synthetic methods (49). Prostaglandin syntheses by three-component coupling. *Angew. Chem., Int. Ed. Engl.* **23,** 847–849.
6. Chen, S. and Janda, K. D. (1997) Synthesis of prostaglandin E2 methyl ester on a soluble-polymer support for the construction of prostanoid libraries. *J. Am. Chem. Soc.* **119,** 8724–8725.
7. Chen, S. and Janda, K. D. (1998) Total synthesis of naturally occurring prostaglandin F2a on a non-cross-linked polystyrene support. *Tetrahedron Lett.* **39,** 3943–3946.
8. Narita, M. (1978) Liquid Phase Peptide Synthesis by the Fragment Condensation on Soluble Polymer Support. I. Efficient Coupling and Relative Reactivity of a Peptide Fragment with Various Coupling Reagents. *Bull. Chem. Soc. Jpn.* **51,** 1477–1480.
9. Ellman, J. A. and Thompson, L. A. (1994) Straightforward and general method for coupling alcohols to solid supports. *Tetrahedron Lett.* **35,** 9333–93336.
10. Manzotti, R., Tang, S., and Janda, K. D. (2000) Improved synthesis of prostanoids on a non-crosslinked polystyrene soluble support. *Tetrahedron* **56,** 7885–7892.
11. Douglas, S., Whitfield, D., and Krepinsky, J. (1991) Polymer-supported solution synthesis of oligosaccharides. *J. Am. Chem. Soc.* **113,** 5095–5097.
12. Krepinsky, J. J. (1996) *Modern Methods of Carbohydrate Synthesis.* (Khan, S. H. and O'Neill, R. A., eds.), Harwood Academic Publishers.
13. Douglas, S., Whitfield, D., and Krepinsky, J. (1995) Polymer-supported solution synthesis of oligosaccharides using a novel versatile linker for the synthesis of

D-mannopentanose, a structural unit of D-mannans of pathogenic yeasts. *J. Am. Chem. Soc.* **117,** 2116–2117.

14. Jesberger, M., Jaunzems, J., Jung, A., Jas, G., Schönberger, A., and Kirschning, A. (2000) Soluble versus insoluble polymers as supports for the preparation of new glycosteroids—a practicability study. *Synlett* 1289–1293.
15. Lopéz-Pelegrín, J. A., Wentworth Jr., P., Sieber, F., Metz, W. A., and Janda, K. D. (2000) Soluble polymer-supported chemoenzymatic synthesis of the C_{21}-C_{27} fragment of the bryostatins. *J. Org. Chem.* **65,** 8527–8531.
16. Nishimura, S.-I. and Yamada, K. (1997) Transfer of ganglioside GM3 oligosaccharide from a water soluble polymer to ceramide by ceramide glycanase. A novel approach for the chemical-enzymatic synthesis of glycosphingolipids. *J. Am. Chem. Soc.* **119,** 10555–10556.
17. Yamada, K., Fujita, E., and Nishimura, S.-I. (1998) High performance polymer supports for enzyme-assisted synthesis of glycoconjugates. *Carbohydr. Res.* **305,** 443–461.
18. Sabesan, S. and Paulson, J. C. (1986) Combined chemical and enzymatic synthesis of sialyloligosaccharides and characterization by 500-MHz ^{1}H and ^{13}C NMR spectroscopy. *J. Am. Chem. Soc.* **108,** 2068–2080.
19. Ito, Y. and Paulson, J. C. (1993) A novel strategy for synthesis of ganglioside GM3 using an enzymatically produced sialoside glycosyl donor. *J. Am. Chem. Soc.* **115,** 1603–1605.
20. Grether, U. and Waldmann, H. (2001) An enzyme-labile safety catch linker for synthesis on a soluble polymeric support. *Chem. Eur. J.* **7,** 959–971.
21. Bergbreiter, D. E., Case, B. L., Liu, Y.-S., and Caraway, J. W. (1998) Poly(*N*-isopropylacrylamide) soluble polymer supports in catalysis and synthesis. *Macromolecules* **31,** 6053–6062.
22. Bergbreiter, D. E., Liu, Y.-S., and Osburn, P. L. (1998) Thermomorphic rhodium(I) and palladium(0) catalysts. *J. Am. Chem. Soc.* **120,** 4250–4251.
23. Bergbreiter, D. E., Osburn, P. L., and Liu, Y. S. (1999) Tridentate PCS palladium(II) complexes: New, highly stable, recyclable catalysts for the Heck reaction. *J. Am. Chem. Soc.* **121,** 9531–9538.
24. Bergbreiter, D. E., Osburn, P. L., Wilson, A., and Sink, E. M. (2000) Palladium-catalyzed C-C coupling under thermophilic conditions. *J. Am. Chem. Soc.* **122,** 9058–9064.

12

Analytical Methods for Optimization and Quality Control of Combinatorial Synthesis

Bing Yan, Liling Fang, Mark M. Irving, Jiang Zhao, Diana Liu, Gianine M. Figliozzi, Frank Woolard, and Clinton A. Krueger

1. Introduction

Rapid lead discovery in the drug development process relies on high-throughput screening of diverse compounds against protein targets. These compounds come from traditional organic synthesis, natural products isolation, and combinatorial synthesis *(1)*. For compounds from combinatorial synthesis, the hit rate highly depends on the quality of the compounds. We employ parallel solid-phase *(2)* and solution-phase *(3)* synthesis methods to make lead discovery libraries containing approx 6000 compounds in each library.

To convert a multistep synthesis method initially developed to produce only a few compounds to a general synthesis protocol to produce approx 6000 compounds is an involved process. To synthesize a final library with a generally high purity is particularly challenging. High-throughput purification is still a very expensive and time-consuming process. Synthesis purity libraries made directly for screening are still an option if compounds are made in sufficient purity. It is a misconception that libraries made for purification may not need substantial optimization. In cases when a reaction was not well optimized, we found that only trace amounts of compounds could be recovered after purification owing to either low synthesis yield (with possible high purity) or/and the presence of impurities. Therefore, the chemistry used for the synthesis of libraries for purification must be optimized to the same level as the chemistry for libraries screened directly. We optimize the library synthesis protocol by thorough characterization of synthetic products on solid support or in solution at every stage of synthesis development. In this chapter, we discuss some

From: *Methods in Molecular Biology, Combinatorial Library Methods and Protocols*
Edited by: L. B. English © Humana Press Inc., Totowa, NJ

analytical methodologies we apply at optimization and production stages of combinatorial synthesis.

2. Materials

2.1. Library Standard Compounds and Quality Control (QC) Compounds

Because purity determined by liquid chromatography (LC) or LC–mass spectrometry (LC–MS) with ultraviolet (UV) or evaporative light scattering detector (ELSD) detection represents only relative purity, the presence of impurities such as inorganics, materials with poor chromophores and/or good evaporative properties, and materials that tend to be retained by a guard column cannot be detected. Therefore, we measure the absolute purity of six representative compounds in each library at both optimization and production stages.

1. QC compounds are six compounds with scaffolds and side chains representing the library diversity.
2. Standard compounds have the same structures as QC compounds and are synthesized separately and purified by reversed-phase and normal-phase high-performance liquid chromatography (HPLC). After purification, standard compounds are rigorously characterized by ^{1}H- and ^{13}C-nuclear magnetic resonance (NMR) and LC–MS and their purity established as > 99.6% by combustion elemental analysis.
3. The structures of six standards (and QC compounds) of a 1,2,5-trisubstituted benzimidazole library and a 4,5,6-trisubstituted pyrimidine library are shown in **Figs. 1** and **2**.

2.2. Quantitative NMR Standard

1. The ^{1}H-NMR standard, 4,4-dimethoxybenzhydrol, was purchased from Aldrich (Milwaukee, WI) and was purified to 99.6% by normal-phase HPLC as determined by combustion CHN analysis.
2. The ^{19}F-NMR standard α,α,α-trifluorotoluene, Aldrich (Milwaukee, WI), was used as received (>99%).
3. Either deuterated methanol or acetonitrile, Cambridge Isotope Laboratories, Inc. (Andover, MA), was used as solvent.
4. All samples were analyzed in Kontes (Vineland, NJ) 5-mm × 7-inch ultraprecision NMR tubes.

2.3. Validation and Qualification Libraries

To develop a general synthesis protocol for the final production library, it is necessary to evaluate diverse building blocks under the reaction conditions to be used. This evaluation is performed in parallel synthesis fashion with procedures that reflect as closely as possible those that will be used to produce the entire library.

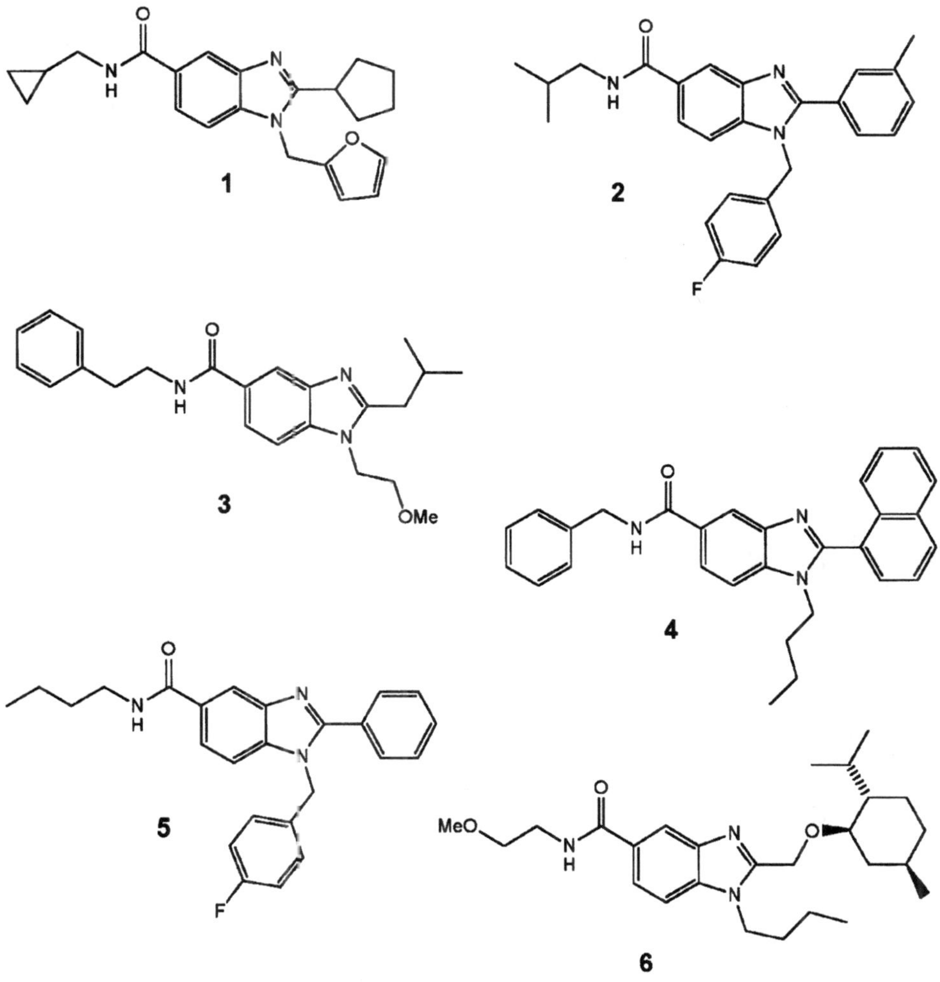

Fig. 1. Structure formulae for compounds 1–6.

1. Validation library. There are two formats for validation libraries. In parallel validation libraries, one site of diversity is varied, and the remaining sites of diversity are held constant. In matrix validation libraries, all sites of diversity are varied. Whereas a parallel library helps determine how one set of precursors performs with a specific substrate, a matrix library provides information of cross-compatibility of all sites of diversity. The average size of a validation library is 400 compounds. About 30 mg of final product per well is expected as the result of synthesis. A daughter plate of 0.3 mg/mL concentration in methanol or acetonitrile is made for LC–UV–ELSD–MS analysis.
2. Qualification library. A qualification library usually is the same size as the validation library and incorporates a selected set of validated building blocks. It

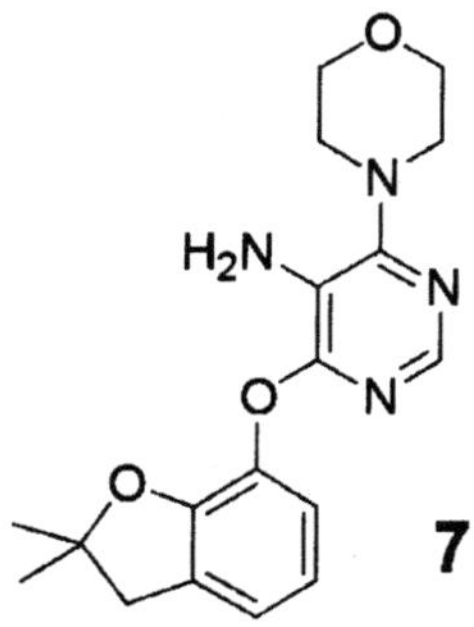

7

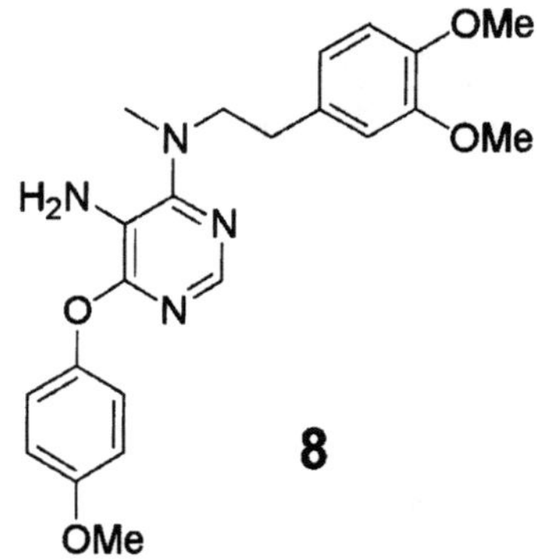

8

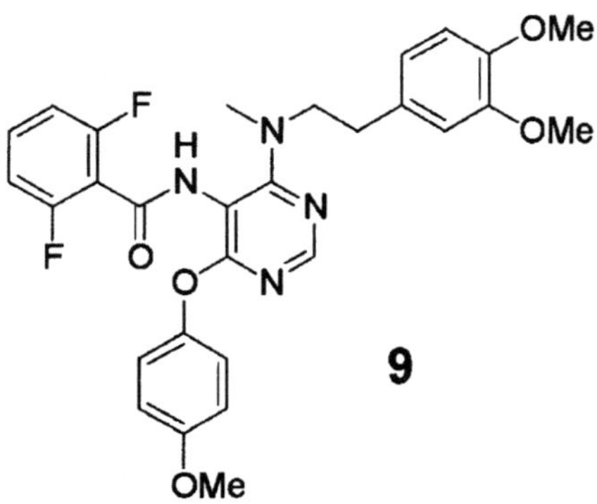

9

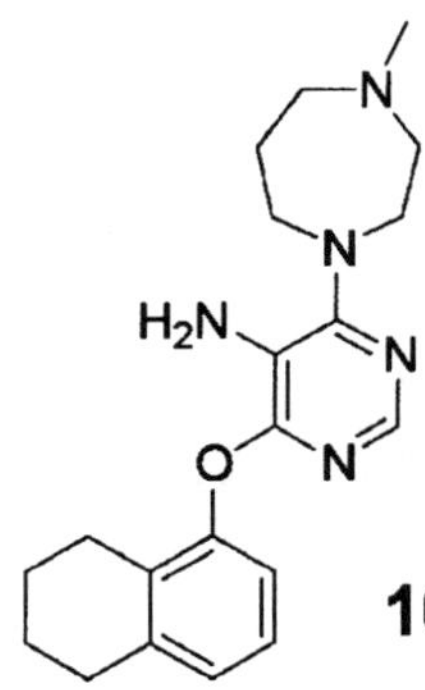

10

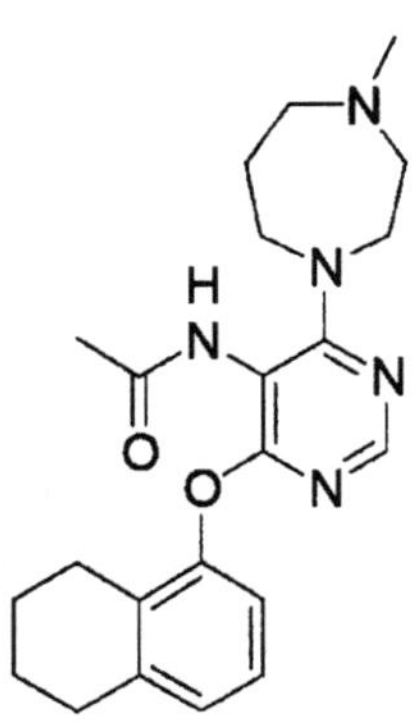

11

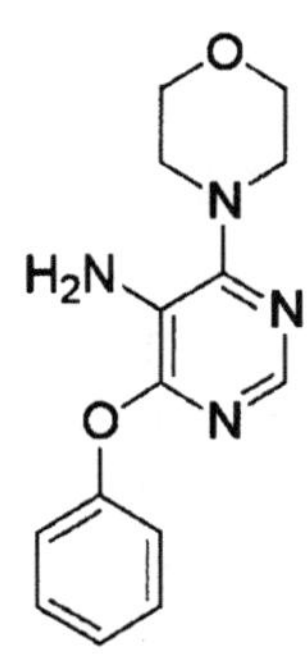

12

Fig. 2. Structure formulae for compounds 7–12.

serves as a pilot run to ensure that the synthesis protocol is effectively transferred in the scale-up synthesis of the final library.

2.4. Production Library

The production library contains all validated building blocks in all combinations. A 1,2,5-trisubstituted benzimidazole library and a 4,5,6-trisubstituted pyrimidine library were synthesized in parallel in 96-deep well plates by solid-phase and solution-phase methods, respectively. They were made at a 50 mg per well scale. Both libraries contained 6336 compounds.

3. Methods

3.1. Calibration Curves for Quantitative Purity Measurement

1. Perform HPLC separation on a HP1100 system (Agilent, Palo Alto, CA), consisting of a vacuum degasser, binary pump, autosampler, column compartment, and diode array detector.
2. Process data with HP Chemstation software.
3. Carry out reversed-phase HPLC on a C18 column (3.0 × 100 mm, 5 μm, 100A) from Phenomenex (Torrance, CA, USA) at 40°C with a flow rate of 1.0 mL/min.
4. Employ two mobile phases (mobile phase A: 99% water, 1% acetonitrile, 0.05% trifluoroacetic acid [TFA]; mobile phase B: 1% water, 99% acetonitrile, 0.05% TFA) to run a gradient condition from 0% B to 100% B in 6.0 min, 100% B for 2.0 min and reequilibrate at 0% B for 2.0 min.
5. Use an injection volume of 10 μL.
6. Weigh all standards to the nearest 0.02 mg on an AT261 DeltaRange analytical balance by Mettler (Toledo, Columbus, OH).
7. Prepare stock solutions of 1.00 mg/mL (for making calibration curves) for each standard using either methanol or acetonitrile as solvent.
8. Dilute the stock solution to make a series of calibration solutions with concentrations of 0, 25, 50, 75, 100, 150, and 300 μg/mL.
9. Use a solution from one of the six standards at 100 μg/mL as an external standard (ES) to compensate for instrumental fluctuation and other systematic errors.
10. Analyze samples from each standard on an HPLC and monitor by UV at 214, 220, and 254 nm.
11. Divide the peak area at each concentration by the peak area of the ES to give the peak ratio (peak ratio = $[\text{peak area}]/[\text{peak area}]_{ES}$).
12. Plot the peak area ratio vs concentration to get calibration curves. The correlation coefficient (R^2) was > 0.99 for all calibration curves.
13. Prepare a 100 μg/mL solution of each standard and analyze in triplicate by HPLC with UV detection at 214, 220, and 254 nm to validate standard calibration curves.
14. Make sure the determined concentrations from the corresponding calibration curves are accurate to ±5%. A flow chart describing this analysis process is shown in **Fig. 3**.

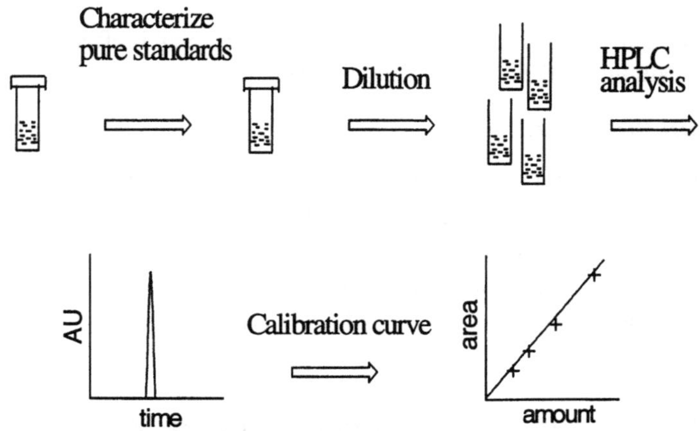

Fig. 3. Flow chart showing the procedure for generating the calibration curve in quantitative purity measurement.

3.2. Determination of the Quantitative Purity of QC Compounds

1. Synthesize the final library, which contained the six QC compounds, in 96-well plates.
2. Dissolve each QC compound in the well with 2 ml of acetonitrile, and transfer the solution to a preweighed vial. Then rinse the well with three 0.5-mL aliquots of acetonitrile, and transfer the rinse solutions to the vial. Remove solvent on a rotary evaporator, freeze the vial, and lyophilize for 14 h to remove trace amounts of moisture before weighing. To ensure complete moisture removal, lyophilize the vial for 2 additional hours and weigh until the weight change is < 0.10 mg.
3. Assuming the compounds were pure, make a solution of approx 200 µg/mL using either methanol or acetonitrile.
4. Analyze the samples by HPLC–UV in triplicate (*see* **Note 1**).
5. Determine the concentration of QC compounds from the peak area ratio relative to ES and individual standard calibration curves. The measured concentration should be within 5% from all three runs.
6. Determine the product quantity by the actual concentration multiplied by the sample volume.
7. Determine the quantitative purity (*see* **Note 2**) from the ratio of determined quantity to the total sample weight. A flow chart describing this analysis process is shown in **Fig. 4**. The quantitative purity data for QC compounds in the 1,2,5-trisubstituted benzimidazole and 4,5,6-trisubstituted pyrimidine libraries are listed in **Tables 1** and **2**. The quantity and quantitative purity of these QC compounds were also determined by qNMR.

3.3. Quantitative NMR

1. Acquire proton *(4)* and fluorine *(5)* NMR spectra on a JEOL (Peabody, MA) Eclipse 270 FT spectrometer with tunable probe and Stackman™ automatic sample changer.

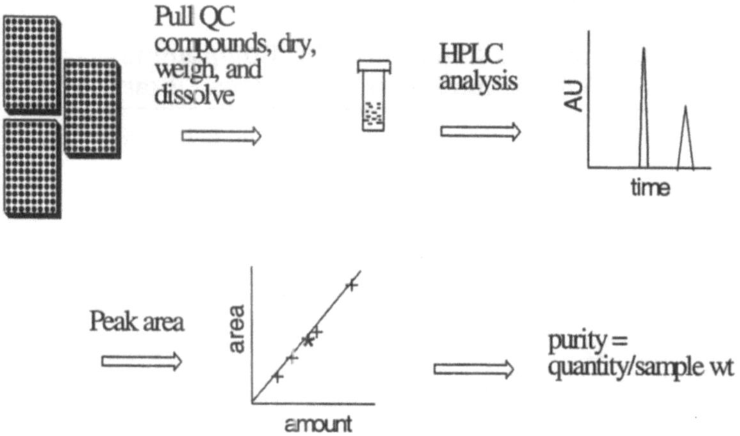

Fig. 4. Flow chart showing the procedure for determining the quantitative purity of QC compounds.

Table 1
Quantitative Analyses of QC Compounds in 1,2,3-Trisubstituted Benzimidazole Library

QC compd	QL 1 quant. purity (%)	qNMR quant. purity (%)	QL 1 TFA (%) by ^{19}F-NMR	QL 2 quant. purity (%)	PL quant. purity (%)
1	24.9	27.3	20.5	57.3	13
2	13.1	12.4	20.7	61.9	86.5
3	18.8	23.5	17.1	59.2	65.6
4	15.3	24.7	24.1	62.8	70.7
5	2.4	10.0	28	34.6	50.5
6	13.3	22.2	54.3	8.6	49.7

2. Use Delta Version 3.1 software to control the instrument and for data processing.
3. For the ^{1}H-NMR experiments, (*see* **Note 3**) tune the probe to a frequency of 270.17 MHz with acquisition parameters as follows: receiver gain 22, pulse width 10.4 µs (π/2), spectral width 4053 Hz, offset of 5 ppm, digital resolution 0.25 Hz, data acquisition time 4.04 s, relaxation time 25 s, and total number of scans 32 (*see* **Note 4**).
4. For the ^{19}F-NMR experiments, tune the probe to 254.18 MHz with the following acquisition parameters: receiver gain 18, pulse width 10.0 µs (π/2), spectral width 50.8 kHz, offset of –100 ppm , digital resolution 3.10 Hz, data acquisition time 0.32 s, relaxation delay 5 s, and total number of scans 32.

Table 2
Quantitative Analyses of QC Compounds in 4,5,6-Trisubstituted Pyrimidines Library

QC compd	QL quant. purity (%)
7	79.6
8	76.1
9	60.9
10	86.9
11	63.4
12	84.9

5. Use the following equation for quantification:

$$C_s n_s / A_s M_s = C_a n_a / A_a M_a$$

where *C* represents the concentration, *n* is the number of nuclei responsible for a given peak, *A* is the area under the peak, and *M* is the weight. The subscript "s" represents the values for the standard peaks, and the subscript "a" represents the values for the analyte peaks.
6. Weigh QC compounds accurately (± 0.02 mg), spiked with a known amount (4.41 mg) of 4,4-dimethoxybenzhydrol and diluted to 1.00 mL with either methanol-d_4 or acetonitrile-d_3.
7. From ^{1}H-NMR data, calculate concentration values by comparing the area of the standard resonance peaks to those of the sample (*see* **Note 5**).
8. Another issue in the development of the library synthesis protocol is the amount of TFA in the final product. Because TFA is used for cleavage of product from resin support and it is difficult to remove by vacuum pumping and lyophilization, there is always a certain amount of TFA carried through to the end. The ^{19}F-qNMR experiment was therefore performed on the same samples used for the proton qNMR work.
9. Prepare a set of calibration solutions of fluorine standard α,α,α–trifluorotoluene and make an external calibration curve from α,α,α–trifluorotoluene in acetonitrile-d_3. This compound gives a single fluorine resonance at –63.1 ppm (relative to Freon $CFCl_3$).
10. Determine the concentration of compounds by comparing the fluorine resonance area of the sample to the calibration curve of the standard. Fluorine NMR spectra of all the synthetic samples revealed one major fluorine resonance for TFA at –77.7 ppm. From ^{19}F-qNMR data, it is clear that a significant amount of TFA is present in the 1,2,5-trisubstituted benzimidazole library. Make a second qualification library by incorporating a more thorough base washing procedure that reduces the amount of TFA carryover.
11. **Table 1** summarizes results obtained from the qNMR analysis for the qualification library (QL 1) and production library (PL) of 1,2,5-trisubstituted benzimidazoles. The percent purity is the average value from two to three distinct proton peaks.

3.4. Analysis of Validation and Qualification Libraries by LC–MS–UV–ELSD

1. Perform analysis of validation and qualification of libraries using an API 150 EX instrument from PE Sciex (Concord, Ontario, Canada).
2. Use an HPLC system consisting of a Gilson 215 liquid handler equipped with an 819 injection valve (Middleton, WI), a HP1100 vacuum degasser, binary pump, column compartment, and a diode array detector (Agilent, Palo Alto, CA).
3. Split eluent from the HPLC system 1:5 to the mass spectrometer and a Sedex 55 (S.E.D.E.R.E., Alfortville Cedex, France) evaporative light scattering detector.
4. Operate the mass spectrometer in positive ion mode. Use the following turbo ion spray conditions: temperature at 400°C, ion spray voltage at 5000 V, curtain gas at 12, nebulizer gas at 10. Use a full scan range from 150 to 800 amu in 1.5 s to acquire MS data.
5. Set ELSD drift tube temperature at 40°C, set gain at 10, and set the nitrogen flow rate at 3.3 L/min.
6. Collect both the signal from UV_{214} and the ELSD through a PE Nelson 900 series interface to a Macintosh computer using MassChrom 1.1 at a rate of 50 data points per second.
7. Process all peak areas and their qualitative peak purity for TIC, UV_{214}, and ELSD signals with a customized PurityScript in MultiView 1.4.
8. Carry out reversed-phase HPLC on a Luna C18 column (2.0 × 30 mm, 5 µm, 100A) from Phenomenex (Torrance, CA) at 40°C with a flow rate of 3.0 mL/min.
9. Employ two mobile phases (mobile phase A: 99% water, 1% acetonitrile, 0.1% acetic acid; mobile phase B: 1% water, 99% acetonitrile, 0.1% acetic acid) to run a gradient condition from 10% B to 100% B in 3.0 min, stay at 100% B for 0.5 min, and reequilibrate for additional 0.5 min.
10. Use the MS signal of MH^+ to identify product peak, use the UV signal to assess product purity, and use the ELSD signal to estimate product quantity ***(6,7)***. Based on validation library analysis, a diversity element would be included in the library production only when its product purity was > 85% by UV_{214} detection. **Figure 5** shows that LC–MS–UV–ELSD purity analysis of two qualification libraries. The purity is expressed as a distribution bar chart. The same results can also be presented as a plateview showing the purity in each well (not shown). ELSD data were also used to estimate the yield of the desired product in each well and can be presented as both a distribution bar chart and plateview (not shown).

3.5. Production Library Analysis

1. Analyze each production library on an eight-way parallel LC/MS/UV system.
2. Generate a binary gradient flow at 18.8 mL/min from Gilson (Littleton, WI) 306 HPLC pumps with 25 SC pump heads equally split in eight ways, and introduce samples with a Gilson 215 autosampler with an eight-probe injector.
3. Equip each channel with a MetaChem (Torrance, CA) HPLC column Polaris 5µ C18-A 50 × 4.6 mm with a precolumn frit and a Gilson UV detector set at 214 nm with an analytical flow cell.

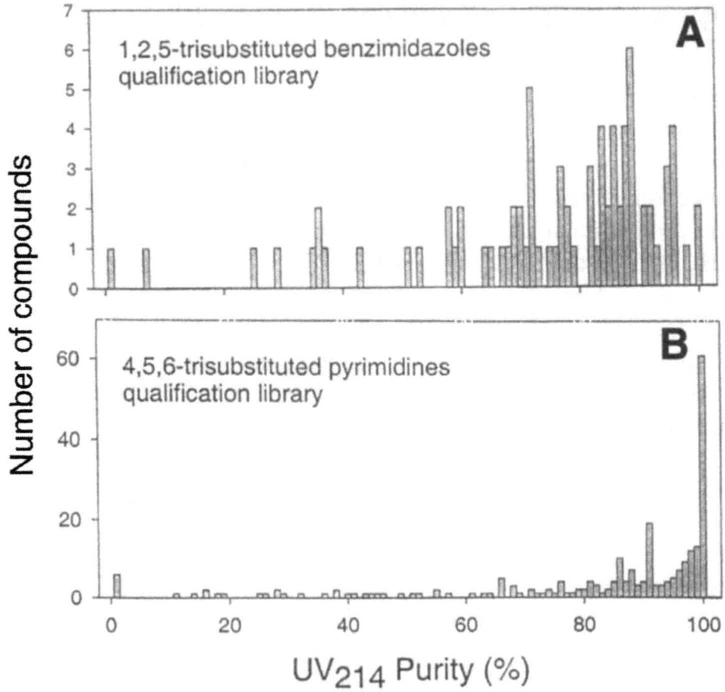

Fig. 5. The purity distribution charts for two qualification libraries.

4. Split the flow 46:1 after the UV detector, and a flow of 50 µL/min from each channel converges on a Micromass (Manchester, UK) LCT equipped with an eight way MUX electrospray ion source.
5. Use an HPLC gradient in this application with 0.05% (v/v) TFA in water as mobile phase A, and 0.05% TFA in acetonitrile as B.
6. Increase B from 0 to 100% in 4 min and hold at 100% for 0.5 min, then reduce to 0% in 0.1 min, and regenerate column for 0.5 min.
7. Use samples loops of 20 µL each, and an injection volume of 50 µL.
8. Drive all eight channels with a single HPLC pump system, and make sure each channel has a similar backpressure so the results will be reproducible. Plumb for each channel identically and use HPLC columns from the same batch.
9. Introduce a quality control injection of Fmoc-ASP(OtBu)-OH solution (0.2 mg/mL) into all eight channels after every 24 runs. The TIC and UV traces of these samples were checked to ensure the system was operating within our specifications.
10. Set the retention time at 3.13 ± 0.2 min, so that the difference between any two channels does not exceed 0.2 min.
11. Include a blank injection of pure methanol in every 24 runs to detect any carry over.

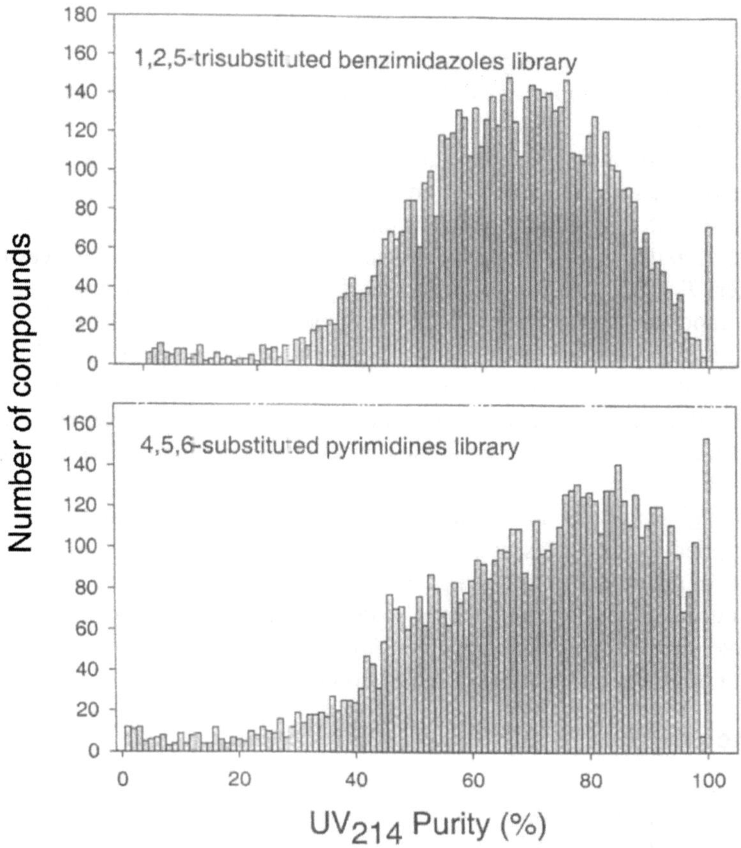

Fig. 6. The purity distribution charts for two production libraries.

12. Process data using Openlynx software. The target ion $(M+H)^+$ was searched within a 1 amu window. The relative peak area of the product as detected by UV_{214} is the relative purity of the target compound.
13. Purity analyses of a 1,2,5-trisubstituted benzimidazoles (6336 compounds) and a 4,5,6-trisubstituted pyrimidines (6336 compounds) are shown in **Fig. 6**. In both libraries, more than 93% of compounds synthesized had purity higher than 40%.

4. Notes

1. Each standard at about 0.1 mg/mL was analyzed first to assess solvent effect and retention time. Solvent effect could cause peak splitting for standards eluted early in the gradient. Therefore, these standards should not be dissolved in pure methanol or acetonitrile. They should be dissolved in solvents with higher water composition such as 50–80% water. If solubility was a problem, < 1% acid or base was added. Standard compounds at about 0.1 mg/mL were also used to optimize ion optics settings and to maximize ion signals in the mass spectrometer.

2. Relative purity measured by LC–UV or LC–ELSD is higher than quantitative purity determined by weight percentage of the compound. This suggests that there are undetectable impurities in the sample. These may include inorganics, TFA, plastic extracts, solvents.
3. The key to obtaining accurate quantitative values using qNMR is to know the values of the spin-lattice relaxation times (T_1) of each sample. This requires doing some sort of T_1 inversion recovery experiment on the sample. The general rule, for a $\pi/2$ pulse, is that the time between pulses should be at least fives times the longest T_1. This time can be shortened if a pulse $< \pi/2$ (Ernst angle) is used.
4. It is important that the phase adjustments are correctly applied to obtain a pure absorption spectrum. The upper half of each integrated peak should be defined by at least four or five points to obtain reliable area values. Since the integration limits of a given peak determine the percent area of a given resonance it is important that these limits be applied in a consistent manner. To include 99% of a NMR peak area, the integrated region must extend 31.5 times the peak width at half-height, on each side of an integrated peak. Other limits and procedures can be applied to obtain lesser or greater degrees of accuracy. To obtain measurements in the 99% certainty range S/N should be on the order of 250:1.
5. The purity values determined by qNMR are in general slightly higher than quantitative standard analysis suggesting there could be hidden (unresolved) peaks from impurities under some of the proton resonances.

References

1. Fenniri, H. (2000) *Combinatorial chemistry. A Practical Approach.* Oxford University Press, New York.
2. Seneci, P. (2000) *Solid-Phase Synthesis and Combinatorial Technologies.* John Wiley & Sons, New York.
3. Baldino, C. M. (2000) Perspective articles on the utility and application of solution-phase combinatorial chemistry. *J. Comb. Chem.* **2,** 89–103.
4. Maniara, G., Rajamoorthi, K., Rajan, S., and Stockton, G. W. (1998) Method performance and validation for quantitative analysis by ^1H and ^{31}P NMR spectroscopy. Applications to analytical standards and agricultural chemicals. *Analyt. Chem.* **70,** 4921–4928.
5. Ellis, D. A., Martin, J. W., Muir, D. C. G, and Mabury, S. A. (2000) Development of an ^{19}F NMR method for the analysis of fluorinated acids in environmental water samples. *Analyt. Chem.* **72,** 726–731.
6. Fang, L., Wan, M., Pennacchio, M., and Pan, J. (2000) Evaluation of evaporative light-scattering detector for combinatorial library quantitation by reversed phase HPLC. *J. Comb. Chem.* **2,** 254–257.
7. Fang, L., Pan, J., and Yan, B. (2001) High-throughput determination of identity, purity and quantity of combinatorial library members using LC/MS/UV/ELSD. *Bioeng. Biotech Comb. Chem.* **71,** 162.

II

LIBRARY PURIFICATION AND SCREENING

13

Resolving Racemic Mixtures Using Parallel Combinatorial Libraries

Tingyu Li, Yan Wang, and Louis H. Bluhm

1. Introduction

As human enzymes and cell surface receptors possess handedness, the enantiomers of a racemic pair of compounds may be absorbed, activated, and degraded in different manners. In some instances, two enantiomers of a racemic drug may have different or even opposite pharmacological activities. In order to acknowledge these differing effects, the biological activity of each enantiomer often needs to be studied separately. This and other factors within the pharmaceutical industry have contributed significantly to the need for enantiomerically pure compounds *(1)*. Consequently, the need to analyze and separate racemic compounds efficiently is of significant importance in pharmaceutical research. Among the asymmetric technologies developed, chromatographic methods are widely used.

Examples of the enantioselective stationary phases that have been developed for these purposes include both monomer coated columns [Pirkle columns *(2)*, cyclodextrin columns *(3)*, macrocyclic antibiotic columns *(4)*, etc.] and polymer coated enantioselective stationary phases [polysaccharide *(5,6)*, polyacrylamide *(7)*]. Although a thorough evaluation of all enantioselective columns is impossible, the commonly used ones do seem to have their limitations. It is fair to say that despite these advances, the chiral resolution of racemic materials remains a major challenge. Resolution of a racemic sample is still a time-consuming, trial-and-error approach, often requiring experimentation with many expensive enantioselective columns. Moreover, typical separation factors achieved are less than 1.5.

To obtain more efficient enantioselective stationary phases, our laboratory has developed an efficient screening method for the development of chiral

From: *Methods in Molecular Biology, Combinatorial Library Methods and Protocols*
Edited by: L. B. English © Humana Press Inc., Totowa, NJ

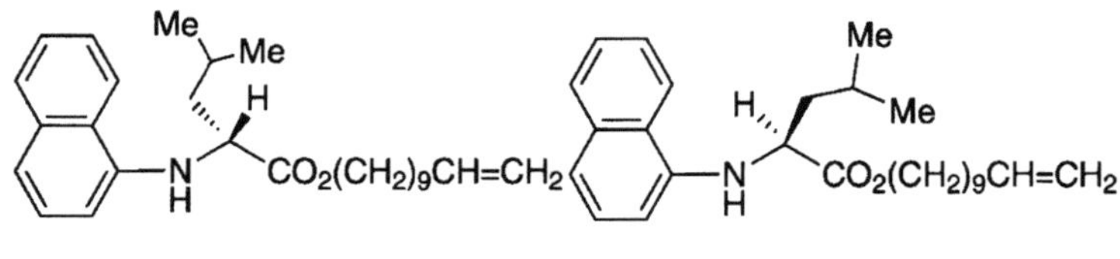

R-1 S-1

Fig. 1

selectors from parallel libraries (*see* **Note 1**) *(8,9)*. In this method, the parallel library is prepared by synthesizing all potential chiral selectors (library members) onto a solid-phase polymeric synthesis resin individually. To determine which potential chiral selector (which library member) is capable of discriminating the two enantiomers, the racemic analyte of interest in the proper solvent is allowed to equilibrate with each potential selector on the resin. The enantiomeric ratio of the analyte in the supernatant is then analyzed by circular dichroism (CD) after the equilibration period. A change in this enantiomeric ratio after equilibration implies a selective adsorption of one of the two enantiomers to the resin and thus the presence of a chiral selector. This chiral selector could then be immobilized on silica gel and the resulting stationary phase will be packed into a column and resolution of the target racemic analyte will be investigated.

The procedure is illustrated with the chiral resolution of racemic *N*-(1-naphthyl)leucine ester **1** (**Fig. 1**) with a 200-member library.

The library is a dipeptide library consisting of three modules: an end-capping module (module 1) and two amino acid modules (modules 2 and 3) (**Fig. 2**). Module 1 consists of only two components, the acetyl group (Ac) and the electron-deficient dinitrobenzoyl group (Dn). Modules 2 and 3 have identical components: both contain the same 10 amino acids. All the possible combinations of three modules yield a library containing a total of 200 members.

These modules are chosen based upon consideration of the molecular features of the target analyte. The analyte contains an electron rich, N-substituted naphthalene ring in conjunction with some hydrogen bond donor and hydrogen bond acceptor groups. According to the chiral interaction model, a minimum of three interaction sites between the chiral selector and the analyte is needed to achieve enantioselective recognition *(10,11)*. Some of these three interaction sites must be attractive, so that the analyte and the chiral selector can interact closely with each other. The electron deficient dinitrobenzoyl (Dn) group in module 1 and also in diamino propionic acid Dp of modules 2 and 3 could interact with the electron rich, N-substituted naphthalene ring through attractive pi–pi

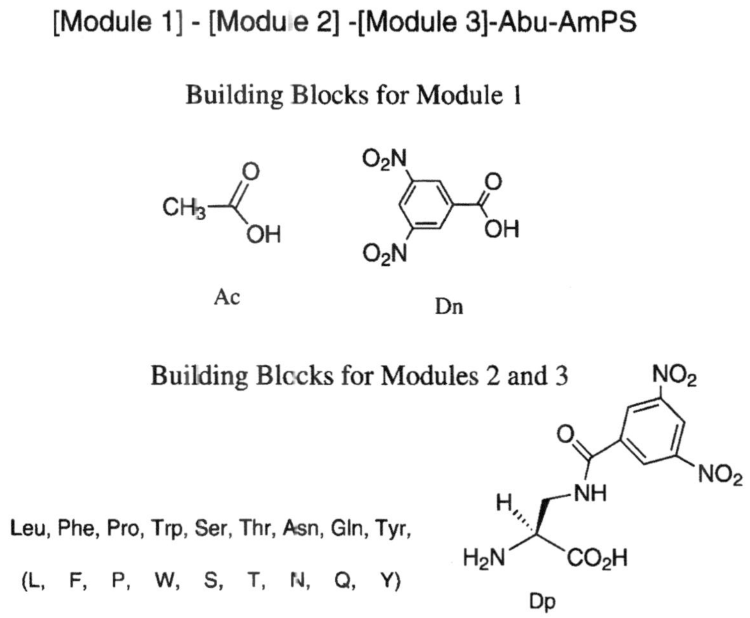

Fig. 2

interaction. Various attractive hydrogen-bonding and repulsive steric interactions are possible between the analyte and the individual library components.

All the library building blocks in this study are commercially available, with the exception of the diamino propionic acid Dp. This amino acid introduces an electron deficient aromatic ring that is absent from natural amino acids, and its Fmoc-protected form (the one needed for library synthesis) was prepared from commercially available N^2-benzyloxycarbonyl-L-asparagine (**Fig. 3**).

2. Materials

2.1. Preparation of Building Block Amino Acid Fmoc-Dp-OH (5, Fig. 3)

1. N^2-Benzyloxycarbonyl-L-asparagine (NovaBiochem, San Diego, CA).
2. [Bis(trifluoroacetoxy)iodo]benzene (Aldrich).
3. Dinitrobenzoyl chloride (Aldrich).
4. *N,N*-Diisopropylethylamine (DIPEA) (Aldrich).
5. Trimethylsilyl iodide (Aldrich).
6. 9-Fluorenylmethyloxycarbonyl-hydroxysuccinimide (Fmoc-Osu) (NovaBiochem).

2.2. Preparation of the 200-Member Library

1. 4-Aminobutyric acid (Abu) (NovaBiochem).
2. Aminomethylated polystyrene (AmPS) (NovaBiochem).

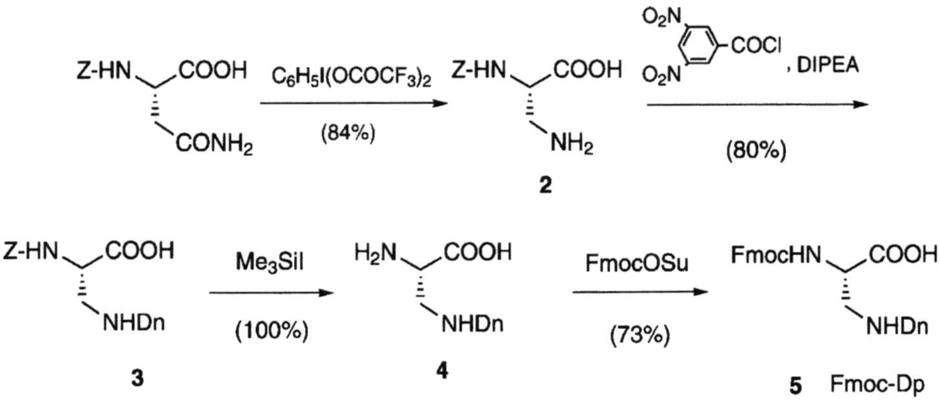

Fig. 3

3. Fmoc-Abu-OH (NovaBiochem).
4. Benzotriazolyloxy-tris[pyrrolidino]-phosphonium hexafluorophosphate (PyBop) (NovaBiochem).
5. Dichloromethane (DCM); isopropanol (IPA); piperidine (Aldrich).
6. Polyfiltronic HI-TOP manual synthesizer (Polyfiltronics, Rockland, MA).
7. Unifilter microplates (Whatman, Clifton, NJ).
8. Fmoc-T(tBu)-OH; Fmoc-L-OH (NovaBiochem).
9. Dinitrobenzoic acid (Aldrich).
10. Trifluoroacetic acid (TFA) (Aldrich).
11. Triisopropyl silane (Aldrich).

2.3. Screening of the Parallel Library with Circular Dichroism Measurement

1. Collection plates (Whatman).
2. JASCO J-720 CD spectropolarimeter (Jasco).

2.4. Preparation of DnLT (Dn-L-T-Abu-Silica Gel) Stationary Phase

1. Diisopropylcarbodiimide (DIC) (Aldrich).
2. 4-(Dimethylamino)pyridine (DMAP) (Aldrich).
3. *N*-Methylmorpholine (NMM) (Aldrich).
4. Rink acid resin (NovaBiochem).
5. 1-Hydroxybenzotriazole hydrate (HOBt) (Aldrich).
6. Silica gel (Selecto silca gel, 32–63 µm) (Fisher Scientific).
7. Acetyl chloride (Aldrich).

2.5. Packing the Column

1. HPLC column slurry packer (Alltech, Deerfield, IL).

2.6. Chromatographic Resolution of Racemic Analyte 1

1. Beckman HPLC system gold analytical gradient system (Beckman).
2. 1,3,5-Tri-t-butylbenzene (Aldrich).

3. Methods

3.1 Preparation of Building Block Amino Acid Fmoc-Dp-OH (5, Fig. 3)

3.1.1. N^2-Z-Diaminopropanoic Acid (2)

1. Add N^2-benzyloxycarbonyl-L-asparagine (2.66 g, 10 mmol) to a stirred solution of [Bis(trifluoroacetoxy)iodo]benzene (6.45 g, 15 mmol) in DMF/H_2O (80 mL, 1:1 v/v) at room temperature.
2. After 15 min, add pyridine (1.6 mL, 20 mmol), and stir for 3 h.
3. Evaporate the solvent *in vacuo* and dissolve the residue in water (100 mL).
4. Extract the solution extensively with ether and concentrate *in vacuo* to afford crude **2**, then recrystallize from ethanol/ether to give pure **2**; yield 2.01 g (84%), m.p. 228–230°C (dec.).

3.1.2. N^2-Z-N^3-Dn-Diaminopropanoic Acid (3)

1. Add dinitrobenzoyl chloride (11.5 g, 50 mmol) in THF (40 mL) slowly to a solution of **2** (3.0 g, 12.5 mmol) and DIPEA (1.61 g, 12.5 mmol) in THF (60 mL).
2. After stirring for 30 min, evaporate THF under vacuum and dissolve the residue in water (100 mL).
3. Extract the aqueous solution with EtOAc (30 mL × 3).
4. Combine the organic phase and wash it with water.
5. After drying over anhydrous Na_2SO_4, evaporate EtOAc to give pure **3** as a white solid (4.4 g, 80%), m.p. 212–213°C.

3.1.3. N^2-Fmoc-N^3-Dn-Diaminopropanoic Acid (5)

1. Add trimethylsilyl iodide (3 g, 15 mmol) to a solution of **3** (4.32 g, 10 mmol) in CH_3CN (50 mL) with stirring.
2. After 10 min, add MeOH (2 mL) to quench the reaction.
3. Evaporate the solvents in vacuum to yield the crude amino acid **4**.
4. Add the crude amino acid **4** dissolved in aqueous solution of Na_2CO_3 (9%, 25 mL) to a solution of Fmoc-OSu (5.1 g, 15 mmol) in DMF (20 mL).
5. After stirring for 20 min, pour the mixture into water (500 mL).
6. Extract the resulting aqueous solution with ether and EtOAc (100 mL × 2).
7. Adjust the pH of the aqueous solution to 1–2 with concentrated hydrochloric acid.
8. Extract the aqueous phase with EtOAc (80 mL × 3). Wash the combined EtOAc extracts with water.
9. Evaporate EtOAc under vacuum. Recrystalize the resulting crude product from CH_2Cl_2-hexane (1:1) to yield pure Fmoc-Dp-OH (**5**) as a pale yellow solid (3.80 g, 73% based on **3**). M.p. 210–211°C. ^1H NMR (DMSO-d_6) δ 3.5 (m, 2H), 4.0–4.2 (m, 4H), 7.2–7.8 (m, 9H), 8.9–9.0 (d, 3H), 9.3 (s, 1H).

3.2. Preparation of the 200-Member Library

3.2.1. Preparation of Abu-AmPS Resin

1. Add a mixture of Fmoc-Abu-OH (390 mg, 1.20 mmol), PyBop (625 mg, 1.20 mmol), and DIPEA (155 mg, 1.20 mmol) in DMF (10 mL) to 1 g (surface amino concentration, 0.40 mmol/g) of AmPS resin that was swelled first in DCM (10 min, *see* **Note 2**).
2. After agitating at room temperature for 2 h, collect the resin (Fmoc-Abu-AmPS) by filtration and wash the resin with DMF, DCM, IPA, and DCM (10 mL × 2).
3. Remove the Fmoc protecting group by treating the resin with 10 mL of 20% piperidine in DMF for 20 min.
4. Collect the deprotected resin (Abu-AmPS) by filtration and wash it with DMF, DCM, IPA, and DCM (10 mL × 2). The surface Abu concentration was determined to be 0.40 mmol/g based on the Fmoc cleavage method (*see* **Note 3**).

3.2.2. Preparation of the Parallel 200-Member Library Using the Polyfiltronic HI-TOP Manual Synthesizer

1. Add 75 mg (0.030 mmol in Abu) of the Abu-AmPS resin prepared above to each of the 200 wells out of three 96-well unifilter microplates (*see* **Note 4**).
2. Add reagents and starting materials in the proper combination to produce all 200 library members.
3. Add a mixture of Fmoc-T(tBu)-OH (16.4 mg, 0.0528 mmol), PyBOP (28.0 mg, 0.0528 mmol), and DIPEA (7.0 mg, 0.053 mmol) in 0.50 mL of DMF to one well of the 96-well unifilter microplate (*see* **Note 5**).
4. After agitating for 2 h, collect the resin by filtration and wash it with DMF, DCM, IPA, and DCM.
5. Remove the Fmoc protecting group by treatment with 0.60 mL of 20% piperidine in DMF for 20 min, followed by washing with DMF.
6. Couple Fmoc-L-OH and dinitrobenzoic acid to the resin by following **steps 3**, **4**, and **5**. In **step 3**, replace Fmoc-T(tBu)-OH with Fmoc-L-OH or dinitrobenzoic acid.
7. Remove the side chain protecting group of T by reacting with 0.6 mL of 95% TFA (2.5% water and 2.5% triisopropyl silane in TFA, *see* **Note 6**) for 1 h. Wash the resin with DMF, DCM, IPA, and DCM to yield the desired library member on the solid resin.

3.3. Screening of the Parallel Library with Circular Dichroism Measurement

1. Transfer library components on the resins from 96-well filter plates into 96-well collection plates (*see* **Note 7**).
2. Add an equal amount of the racemic *N*-(1-naphthyl)leucine ester (**1**) (1.2 mg, 0.0030 mmol) in a mixture of chloroform and heptane (2:8, 0.60 mL) to each of the 200 wells that contained 0.03 mmol selector.
3. After incubating for 3 h, transfer the supernatant in each well into a sample cell (volume 0.40 mL) of a JASCO J-720 CD spectropolarimeter.
4. Record the ellipticity (*see* **Note 8**) at 260 nm, which is the maximum CD adsorp-

tion wavelength of the enantiomerically pure *N*-(1-naphthyl)leucine ester (**1**) (*see* **Note 9**). The data obtained for all 200 wells are summarized in **Table 1**. As seen from this table, many chiral selectors were identified. From these promising selectors, the DnLT member of the library was chosen for the chromatographic purification of racemic analyte **1**.

3.4. Preparation of DnLT (Dn-L-T-Abu-Silica Gel) Stationary Phase

1. Add Fmoc-Abu-OH (3.90 g, 12 mmol), DIC (1.51 g, 12 mmol), DMAP (122 mg, 1 mmol, *see* **Note 10**), and NMM (606 mg, 6 mmol) in DCM-DMF (2:1, 30 mL) to Rink acid resin (*see* **Note 11**) (5 g, 0.6 mmol/g) preswelled with DCM (30 mL, 30 min)
2. After agitating the mixture for 5 h, collect the resin by filtration and wash the resin with DMF, IPA, and DCM (20 mL × 2). The surface concentration for Fmoc-Abu-ORink resin was determined to be 0.57 mmol/g by the Fmoc cleavage method (*see* **Note 3**).
3. Remove the Fmoc group by treatment with 20% piperidine in DMF (30 mL) for 30 min.
4. Collect the deprotected Abu-ORink resin and wash it with DMF, IPA, and DCM (20 mL × 2).
5. Add Fmoc-T(tBu)-OH (1.42 g, 3.6 mmol), PyBOP (1.25 g, 2.4 mmol), HOBt (320 mg, 2.4 mmol), and DIPEA (620 mg, 4.8 mmol) in DMF (10 mL) to the Abu-ORink resin (2 g, 1.14 mmol) prepared above.
6. After agitating for 2 h, collect the resin by filtration and wash it with DMF, IPA, and DCM (10 mL × 2). Surface concentration of Fmoc-T(tBu)-Abu-ORink was determined to be 0.55 mmol/g (*see* **Note 3**).
7. Remove the Fmoc group by the same procedures discussed in **step 3**.
8. Couple Fmoc-L-OH to the T (tBu)-Abu-ORink resin by following exactly the same procedure as described above (**steps 5–7**). The surface concentration of Fmoc-L-T(tBu)-Abu-ORink was determined to be 0.54 mmol/g (*see* **Note 3**).
9. Couple dinitrobenzoic acid to the L-T (tBu)-Abu-ORink resin by following exactly the same procedure as described above (**steps 5–6**).
10. Cleave Dn-L-T(tBu)-Abu-OH from the resin by treating the resin with 1% TFA in DCM (10 mL, 5 min, *see* **Note 12**).
11. Purify the crude product obtained by flash column chromatography on silica gel (mobile phase: 10% MeOH in DCM) to yield the desired Dn-L-T(tBu)-Abu-OH as a white solid (460 mg, 67%). ^{1}H NMR: (DMSO-d_6) δ 0.9–1.2 (m, 18H), 1.58–1.89 (m, 5H), 2.1 (t, 2H), 3.0 (m, 2H), 4.0 (m, 1H), 4.2 (m, 1H), 4.7 (m, 1H), 7.9 (br., 1H), 8.4 (br., 1H), 9.0–9.2 (m, 3H), 10.2 (br., 1H).
12. Add a mixture of Dn-L-T(tBu)-Abu-OH (330 mg, 0.58 mmol), PyBOP (302 mg, 0.58 mmol), HOBt (40 mg, 0.29 mmol), and DIPEA (115 mg, 0.87 mmol) in DMF (5 mL) to aminopropyl silica gel (0.70 g, 0.29 mmol in amino group, *see* **Note 13**).
13. After agitating for 4 h, collect the silica gel by filtration and wash it with DMF, IPA, and DCM (5 mL × 2).
14. End-cap unreacted amino groups by adding acetyl chloride (100 mg, 1.2 mmol) and DIPEA (155 mg, 1.2 mmol) in 5 mL of DMF to the silica gel prepared above.

Table 1
Ellipticities (mdeg) Measured at 260 nm for Each Member of the Library[a]

DnLL	DnLF	DnLP	DnLW	DnLS	DnLT	DnLN	DnLQ	DnLY	DnLDp
+4.0	+3.2	+10.2	+11	+6.5	+16.8	+6.8	+2.4	+11.2	+4.0
DnFL	DnFF	DnFP	DnFW	DnFS	DnFT	DnFN	DnFQ	DnFY	DnFDp
+5.8	+5.5	+9.1	+4.8	+6.0	+6.0	+2.7	+6.7	+11	+5.1
DnPL	DnPF	DnPP	DnPW	DnPS	DnPT	DnPN	DnPQ	DnPY	DnPDp
+1.5	+1.2	–0.25	+1.5	0	+1.3	+0.1	–0.4	0	+1.3
DnWL	DNWF	DnWP	DnWW	DnWS	DnWT	DnWN	DnWQ	DnWY	DnWDp
+10.0	+4.8	+5.4	+11	+12.4	+24.0	+10.3	+9.0	+14	+5.6
DnSL	DnSF	DnSP	DnSW	DnSS	DnST	DnSN	DnSQ	DnSY	DnSDp
+5.7	+1.9	+3.1	+4.7	+3.7	+7.5	+3.5	+3.0	+5.5	+2.4
DnTL	DnTF	DnTP	DnTW	DnTS	DnTT	DnTN	DnTQ	DnTY	DnTDp
+3.9	+3.9	+2.5	+2.0	+2.4	+6.1	+3.3	+3.1	+8.5	+3.0
DnNL	DnNF	DnNP	DnNW	DnNS	DnNT	DnNN	DnNQ	DnNY	DnNDp
+22.6	+14.1	+1.0	+10.0	+15.0	+19.0	+11.0	+5.2	+8.5	+6.3
DnQL	DnQF	DnQP	DnQW	DnQS	DnQT	DnQN	DnQQ	DnQY	DnQDp
+4.1	+2.4	+0.7	+2.7	+4.0	+6.7	+2.9	+1.9	+5.4	+2.9
DnYL	DnYF	DnYP	DnYW	DnYS	DnYT	DnYN	DnYQ	DnYY	DnYDp
+8.6	+5.2	+2.8	+14.4	+13.8	+23	+11.9	+9.1	+18.6	+9.6
DnDpL	DnDpF	DnDpP	DnDpW	DnDpS	DnDpT	DnDpN	DnDpQ	DnDpY	DnDpDp,
+3.5	+0.75	+3.0	+3.7	+3.2	+5.0	+1.6	+2.0	+3.0	+2.3
AcLL	AcLF	AcLP	AcLW	AcLS	AcLT	AcLN	AcLQ	AcLY	AcLDp
+2.4	–0.5	+0.6	+0.2	+0	–0.4	–0.1	–0.5	+0.6	0
AcFL	AcFF	AcFP	AcFW	AcFS	AcFT	AcFN	AcFQ	AcFY	AcFDp
0	0	0	–0.4	+0.4	–0.4	+0.25	–0.4	–0.3	–0.6
AcPL	AcPF	AcPP	AcPW	AcPS	AcPT	AcPN	AcPQ	AcPY	AcPDp
+0.3	–0.36	+0.6	–0.1	+0.8	–0.2	–0.15	–0.4	+0.15	+1.5
AcWL	ACWF	AcWP	AcWW	AcWS	AcWT	AcWN	AcWQ	AcWY	AcWDp
+0.2	–0.3	+0.6	–0.1	+0.3	+0.4	+0.5	+0.6	+0.3	+0.4
AcSL	AcSF	AcSP	AcSW	AcSS	AcST	AcSN	AcSQ	AcSY	AcSDp
+1.0	0	+0.3	–0.9	+0.4	0	–0.1	+0.5	–0.8	–0.14
AcTL	AcTF	AcTP	AcTW	AcTS	AcTT	AcTN	AcTQ	AcTY	AcTDp
–0.11	+1.1	+0.4	+0.7	+1.4	+1.2	+0.3	0	0	+0.7
AcNL	AcNF	AcNP	AcNW	AcNS	AcNT	AcNN	AcNQ	AcNY	AcNDp
+1.5	+0.7	–0.25	–0.23	–0.3	+0.2	0	+0.4	+0.6	0
AcQL	AcQF	AcQP	AcQW	AcQS	AcQT	AcQN	AcQQ	AcQY	AcQDp
+0.5	+1.6	+0.4	–0.4	–0.4	+0.6	+2.4	0	0	+0.5
AcYL	AcYF	AcYP	AcYW	AcYS	AcYT	AcYN	AcYQ	AcYY	AcYDp
–0.1	+0.5	+1.2	+0.3	+0.5	+1.0	–0.2	+0.3	+0.7	–0.25
AcDpL	AcDpF	AcDpP	AcDpW	AcDpS	AcDpT	AcDpN	AcDpQ	AcDpY	AcDpDp
+0.1	+0.4	+0.6	+1.1	–0.8	+1.5	+0.1	+0.7	+1.0	–0.3

[a] Equilibration solvents: $CHCl_3$/heptane (2/8).

15. After 20 min, collect the silica gel by filtration.
16. Remove the t-Bu protecting group of Thr(T) by reacting with 5 mL of 95% TFA with 2.5% water and 2.5% triisopropyl silane for 1 h.
17. Collect the desired chiral stationary phase and wash it with DCM, DMF, IPA, methanol.
18. Dry the desired chiral stationary phase at 80°C for 5 h.

3.5. Packing the Column

1. Slurry pack the stationary phase prepared above into a 50 × 4.6 mm stainless steel column using an Alltech HPLC column slurry packer *(12)*. Slurry solvent: 1/1 (v/v) mixture of toluene and cyclohexanol. Pressurizing solvent: heptane. Packing pressure: 6000 psi.

3.6. Chromatographic Resolution of Racemic Analyte 1

1. Inject racemic *N*-(1-naphthyl)leucine ester **1** onto this column connected to a Beckman HPLC analytical gradient system. Chromatographic conditions: mobile phase, 20% chloroform in heptane; flow rate, 1.2 mL/min; UV detection at 254 nm.
2. Measure the dead time by injecting 1,3,5-tri-*t*-butylbenzene, a void volume marker *(13)*, onto the column. t_0 equals 0.43 min.
3. Calculate the separation factor. The retention factor (k) equals $(t_r-t_0)/t_0$, in which t_r is the retention time and t_0 is the dead time. For the lesser retained *R* enantiomer, its retention factor (k_r) is 1.7. For the more strongly retained *S* enantiomer, its retention factor (k_s) is 29. Thus, the separation factor for the resolution of racemic naphthyl leucine ester **1**, which equals k_s/k_r, is 17. Excellent separation was therefore achieved (*see* **Note 14**).

4. Notes

1. For other related work in this area, see the references cited in refs. *8* and *9*.
2. Thorough swelling of the resin is crucial for the success of the synthesis. With AmPS resin, both DCM and $CHCl_3$ are excellent swelling solvents.
3. To about 20 mg of resin was added 3 mL of 20% piperidine in DMF in a quartz UV cuvet. After the mixture was gently agitated for 3–5 min, the resin was allowed to settle to the bottom of the cuvet. The cuvet was then placed into the UV spectrophotometer, and the absorbance of the sample at 290 nm was recorded with a solution of 20% piperidine in DMF as the reference cell. The amount of the Fmoc group on the resin was then determined by comparing the UV absorbance with a calibration curve generated by cleaving known amounts of Fmoc-Gly-OH following similar procedures.
4. 800 µL Whatman uni-filter PKP multi-chem plates were used. The PKP filter, a glass microfiber media treated to be oliophobic, enables the retention of most organic solutions in the well, thus ensuring they will not drip prematurely.
5. The specific experimental procedure for the synthesis of the DnLT member of the library is shown here. Other library members were prepared by following this identical sequence.
6. Triisopropyl silane serves as a cationic quencher. Without it, side chains of amino

acids may undergo side reactions under the acidic reaction conditions.

7. Resin transfer can be effected by placing a collection plate face to face over the plate that contains thoroughly dried resins and then flipping the plates. If desired, the residual amount of the resin left in the original plate could be transferred by adding 0.5 mL of DCM to the original plate and then pipetting the slurry to the collection plate. DCM will evaporate upon standing in a fume hood overnight.
8. Ellipticity is the intensity of the circular dichroism spectra. The higher the ellipticity, generally the higher the enantiomeric excess.
9. For CD measurement, a syringe with a fine needle or a plastic disposable micropipet is used to transfer the supernatant from the incubation plate to the sample cell of a JASCO J-720 CD spectropolarimeter.
10. DMAP is crucial for the successful coupling of the amino acid to the resin, although a small degree of reaction will occur in its absence.
11. Rink acid resin is highly sensitive to acid. Great care must be exercised not to introduce acid during the synthesis to prevent premature cleavage.
12. Careful control of the reaction time is very important. Prolonged reaction time will lead to the premature cleavage of the side chain protecting groups.
13. Aminopropyl silica gel was prepared by the following procedure: 8.44 g of HPLC grade Allsphere silica gel (surface area 220 m^2/g, pore size 80 Å, particle size 5 μm) was placed in a round-bottom flask equipped with a Dean-Stark trap and reflux condenser. After adding 40 mL toluene, the mixture was refluxed for 1 h to remove surface-adsorbed water. The reaction was stopped temporarily to introduce 6.8 mL 3-aminopropyltriethoxysilane and allowed to continue overnight. The silica was collected by filtration, rinsed with toluene, EtOH, and DCM (2 × 30 mL each), and placed in the oven at 60°C overnight. Elemental analysis indicated C = 2.51%, H = 0.65%, N = 0.78%; the surface amino coverage based on %N was determined to be 0.55 mmol/g.
14. In chiral separations, a separation factor over 1.5 is generally considered an excellent separation.

References

1. Stinson, S. C. (1995) Chiral Drugs. *Chem. Eng. News* **73(41),** Oct. 9, 44–74.
2. Pirkle, W. H. and Welch, C. J. (1994) Chromatographic and 1H NMR support for a proposed chiral recognition model. *J. Chromatogr. A* **683,** 347–353.
3. Armstrong, D. W., Ward, T. J., Armstrong, R. D., and Beesley, T. E. (1986) Separation of drug stereoisomers by the formation of cyclodextrin inclusion complexes. *Science* **232,** 1132–1135.
4. Berthod, A., Liu, Y., Bagwill, C., and Armstrong, D. W. (1996) Facile LC enantioresolution of native amino acids and peptides using a teicoplanin chiral stationary phase *J. Chromatogr. A* **731,** 123–137.
5. Yashima, E., Yamamoto, C., and Okamoto, Y. (1996) NMR studies of chiral discrimination relevant to the liquid chromatographic enantioseparation by a cellulose phenylcarbamate derivative. *J. Am. Chem. Soc.* **118,** 4036–4048.

6. Oguni, K., Oda, H., and Ichida, A. (1995) Development of chiral stationary phases consisting of polysaccharide derivatives. *J. Chromatogr. A*, **694,** 91–100.
7. Blaschke, G. (1988) Substituted polyacrylamides as chiral phases for the resolution of drugs, in *Chromatographic Chiral Separations* (Zief, M. and Crane, L. J., eds.), Marcel Dekker, Inc., New York, pp. 179–198.
8. Wang, Y. and Li, T. (1999) Screening of a parallel combinatorial library for selectors for chiral chromatography. *Anal. Chem.* **71,** 4178–4182.
9. Wang, Y., Bluhm, L., and Li, T. (2000) Identification of chiral selectors from a 200-member parallel combinatorial library. *Anal. Chem.* **72,** 5459–5465.
10. Pirkle, W. H. (1997) On the minimum requirements for chiral recognition. *Chirality* **9,** 103–104.
11. Wainer, I. W. and Caldwell, J. (1997) The 1996 Chirality debate. *Chirality* **9,** 95–96.
12. Poole, C. F. and Poole, S. K. (1991) *Chromatography today*. Elsevier, New York, pp. 350–353.
13. Pirkle, W. H. and Welch, C. J. (1991) A convenient void volume marker for several chiral HPLC columns. *J. Liq. Chromatogr.* **14,** 1–8.

14

Ligand Libraries for the Extraction of Metal Ions

Dynamic Combinatorial and High-Throughput Screening Methods

Seema Choudhary and Janet R. Morrow

1. Introduction

Combinatorial methods have been applied with a measure of success in the field of metal ion coordination chemistry *(1–5)*. High-throughput screening methods have been used to facilitate the discovery of new catalysts and to design ligands for binding metal ions *(6–9)*. Metal ion coordination complexes have been central to the development of dynamic combinatorial methods as well. Dynamic combinatorial methods utilize a library whose composition is controlled through reversible interactions with a target molecule *(10,11)*. Metal–ligand bonds are often labile enough to undergo reversible bond formation under mild conditions, making metal ions ideal components for dynamic libraries. Such metal ion complexes are used in dynamic combinatorial libraries for the purpose of designing compounds with unusual shapes or recognition properties *(12,13)*.

An important area of research in metal ion coordination chemistry that has rarely been studied by combinatorial methods is the liquid–liquid extraction of metal ions in the presence of ligands. Such processes are extremely important in metal ion separations, remediation, and purification *(14,15)*. Metal ion extraction from aqueous to organic solution is one of the most commonly studied types of separations *(16,17)*. In this process, a ligand binds to a water-soluble metal ion and the lipophilic metal ion complex is transferred into the organic layer. Successful extraction is observed for ligands that effectively bind the metal ion and that promote partitioning of the complex into the organic solvent layer. However, many important variables in extraction are not readily predicted, including metal ion complex solubility and stability. It is frequently

From: *Methods in Molecular Biology, Combinatorial Library Methods and Protocols*
Edited by. L. B. English © Humana Press Inc., Totowa, NJ

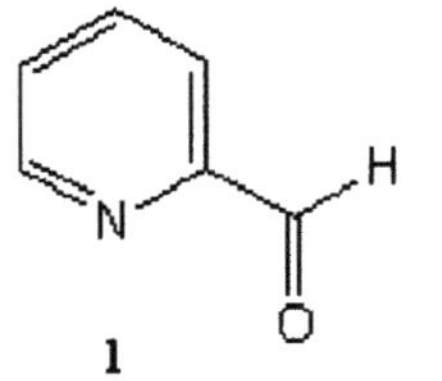

1

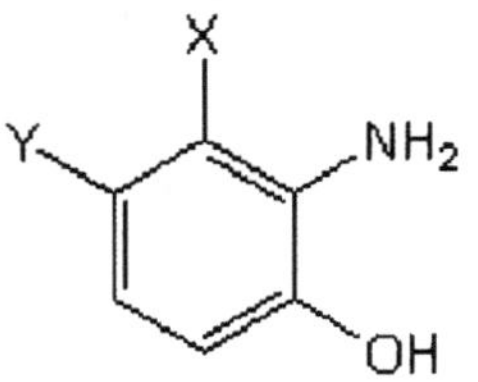

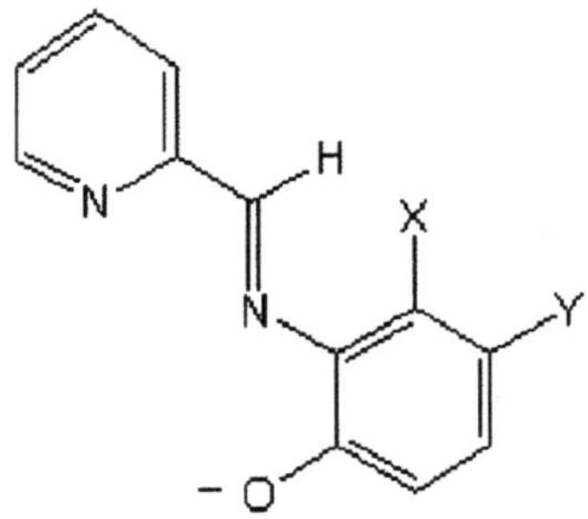

2 X = Y = H
3 X = CH_3, Y = H
4 X = H, Y = CH_3
5 X = H, Y = $C(CH_3)_3$
6 X = H, Y = Cl
7 X = H, Y = C_6H_5

SB12 X = Y = H
SB13 X = CH_3, Y = H
SB14 X = H, Y = CH_3
SB15 X = H, Y = $C(CH_3)_3$
SB16 X = H, Y = Cl
SB17 X = H, Y = C_6H_5

Fig. 1. Aldehyde, aminophenols, and Schiff-bases.

observed that two different ligands combine to promote extraction more effectively than would either alone. Such synergistic effects are not well understood. In order to optimize extraction, it is a common approach to synthesize several derivatives of the type of ligand under study. Thus, combinatorial methods that facilitate the study of large numbers of ligands and ligand combinations are likely to accelerate the discovery of new ligand systems for extraction. In addition, the selection of the most stable and soluble metal ion complexes from ligand libraries may lead to the discovery of new principles of coordination chemistry that facilitate other areas of research including the arena of metallodrug design.

We recently reported on the application of combinatorial methods to selection of stable and chloroform-soluble Schiff-base metal complexes from a dynamic library in order to improve the extraction of metal ions *(18)*. In this study, we built a dynamic library of Schiff-base complexes to capitalize on the reversible nature of metal-catalyzed imine bond formation. Library components were chosen to optimize stability and chloroform solubility of metal complexes. By utilizing the template effect, metal ions promote the assembly of a group of ligands, which then undergo a condensation reaction to form a metal complex. Therefore, aldehydes containing a metal ion-binding site were used to facilitate metal ion templated formation of the Schiff-base ligands (**Fig. 1**).

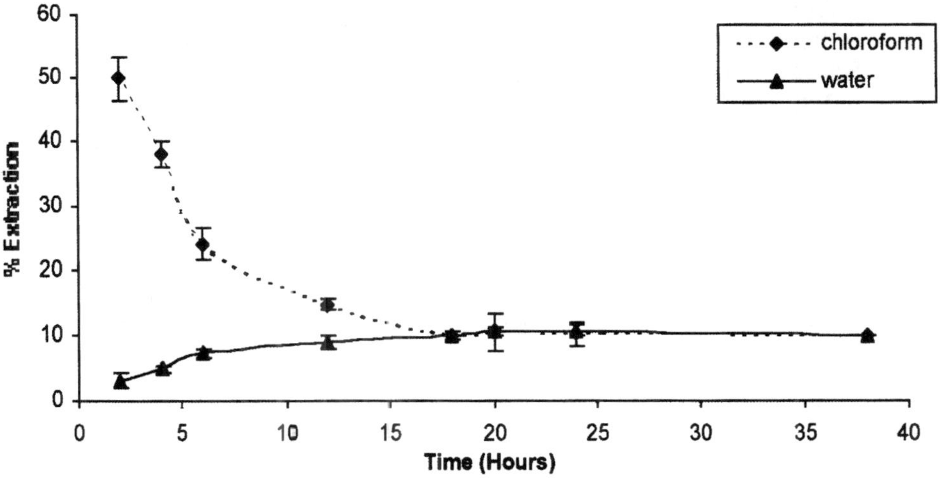

Fig. 2. Timed extraction study for Cd(II) Schiff-base. Extraction started with either 0.0500 m*M* Cd(PCA-**2**)$_2$ in chloroform and 5.00 m*M* CHES (pH = 8.5) and 0.100 *M* NaCl in water (chloroform) or 0.100 *M* NaCl, 5.00 m*M* buffer (pH = 8.5), 0.0500 m*M* Cd^{2+} and 0.100 m*M* PCA and 2AP in water followed by chloroform addition (water).

Deprotonation of the phenol group of the Schiff-base gives an anionic ligand that facilitates the extraction of cationic metal ions. Schiff-bases are good ligands for metal ions, especially for Zn(II) and Cd(II) ***(19)***. As these metal ions are d^{10} and have no crystal field stabilization energy, their flexible coordination geometries may be influenced by subtle interligand interactions. Also, ligand exchange kinetics are typically quite rapid, facilitating the rapid equilibration of metal complex libraries. Given that both metal–ligand and imine bond formation are reversible, these extraction systems are thus a rare example of double-level orthogonal dynamic libraries.

All extraction studies were conducted under conditions where the compounds formed by these interacting library members are at equilibrium with each other. In order to confirm that our extraction systems functioned as a dynamic library, it was necessary to ensure that the two-phase system was at equilibrium. To demonstrate this, we compared the extent of metal ion extraction with reagents initially dissolved in either the chloroform or the water layer. A dithizone assay as described herein was used to measure extracted metal ion concentration. As shown in **Fig. 2**, Cd(II) concentrations distributed into chloroform after 24 or 40 h were identical within experimental error for both sets of conditions, suggesting that the equilibrium was established between the two layers within 24 h. Thus the percentage of metal ion in the organic phase was the same whether the complex extracts from aqueous phase or organic phase,

Table 1
Extraction of Zn(II) or Cd(II) (0.0500 m*M*) from Aqueous Solutions Containing 1 (0.100 m*M*), 0.100 m*M* aminophenol, 5.0 m*M* buffer, 0.10 *M* NaCl, into 1 mL of $CHCl_3$[c]

Aminophenol	% Zn[a]	%Zn[b]	%Cd[a]
2	29±2	15±0.7	9.3±2
3	46±0.4	0	1.4±0.7
4	47±0.9	7.6±0.3	23±3
5	39±3	16±1	37±2

[a] pH = 8.5, CHES buffer.
[b] pH = 6.5, cacodylate buffer.
[c]Reprinted with permission from **ref. *18***.

confirming that our system was at equilibrium. All further experiments were stirred for a 24-h equilibration time.

Different conditions were examined in order to characterize the extraction system. Effects of pH, counterions, the concentration of aldehyde and aminophenol components, and ligand synergy in multicomponent systems are all potentially important. Extraction is dependent on pH but not on counterion, suggesting that the counterion does not participate in extraction. Less Zn(II) and no Cd(II) extracts at pH 6.5 compared to pH 8.5, consistent with the deprotonation of the phenol group of the Schiff-base to give an anionic ligand that facilitates the extraction (**Table 1**) ***(18)***. Aminophenol and aldehyde are both important for extraction as no extraction was observed in absence of either component. An exception was found for Zn(II) at pH 8.5 where 19% extraction was observed with **5** even in the absence of aldehyde. Note that even in mixtures containing a single aminophenol and aldehyde there are several possible neutral metal complexes that may extract including the bis-Schiff–base complex, a complex containing one aminophenol and one Schiff-base ligand, or a Schiff-base-hydroxide complex. Extraction solutions analyzed by ^1H NMR and by mass spectroscopy support the extraction of the bis-Schiff-base complex $[Zn(SB)_2]$ as the predominant extracted species. The crystal structures of such bis-Schiff-base complexes of Zn(II) and Cd(II) demonstrate that small changes in ligand substituents have a dramatic effect on the geometry and structure of the complex ***(18)***.

Mixtures containing two different types of aminophenols were prepared to study the synergistic effect of ligands in a dynamic library. To calculate the expected amount of $CHCl_3$ extraction of metal ions by mixtures of aminophenols, it was assumed that the bis-Schiff-base complex was the predominant species. The calculated and experimental values for Cd(II), differ little and no ligand synergy was observed. For Zn(II), several combinations were promis-

Table 2
Extraction of Zn(II) (0.0500 m*M*) from Aqueous Solutions Containing 1 (0.100 m*M*) and Two Different Aminophenols (0.100 m*M* each), pH = 8.5, 5.0 m*M* CHES Buffer, 0.10 *M* NaCl into 1 mL of $CHCl_3$[a]

Aminophenol	% Zn Extracted	% Zn Calculated
2+3	39 ± 0.8	30
2+4	63 ± 10	53
2+5	67 ± 9	47
3+4	48 ± 1	47
3+5	60 ± 2	39
4+5	75 ± 6	57

[a]Reprinted with permission from **ref.** ***18***.

ing (**Table 2**). A mixture containing **3**, **5**, and PCA showed a strong synergistic effect for Zn(II) extraction. An additional 19% extraction above the calculated percent extraction was observed with these two aminophenols together.

Traditional combinatorial chemistry methods utilize high-throughput screening methods to facilitate the study of large numbers of compounds. Such an approach can be used to facilitate metal ion extraction studies as well. Our screening assays use a reactor that combines the advantages of the 48-well format with large reaction volumes. The Miniblocks synthesizer used here consists of two complementary 48-position blocks, which allow reaction products to be collected into standard microplate formats *(20)*. The valve mechanism in this synthesizer allows all the reaction tubes to be closed and opened at the same time. In these assays, dithizone dye was used as an indicator of metal ion extraction.

Dithizone, a highly sensitive extraction photometric reagent *(21)* for metal ions such as Zn^{2+}, Cd^{2+}, Cu^{2+}, and Hg^{2+}, was used to screen the extraction of metal ion in the extracted complex. Dithizone was titrated into an excess of the metal complex extracted in chloroform. The chloroform solution color appeared as green, blue, orange, or pink, depending on the concentration of metal ion. **Fig. 3** shows the reactor and screening assay used. In these studies, we have shown that high-throughput screening can be combined with dynamic combinatorial methods for the optimization of metal ion extraction.

2. Materials

2.1. Chemicals and Instrumentation

1. 2-Pyridinecarboxaldehyde, (PCA), aminophenols, (AP), including 2-aminophenol, 2-amino-m-cresol, 2-amino-p-cresol, 2-amino-4-tert-butylphenol, 2-amino-chlorophenol, 2-aminophenylphenol, were purchased from Aldrich (Milwaukee, WI) and used without further purification.

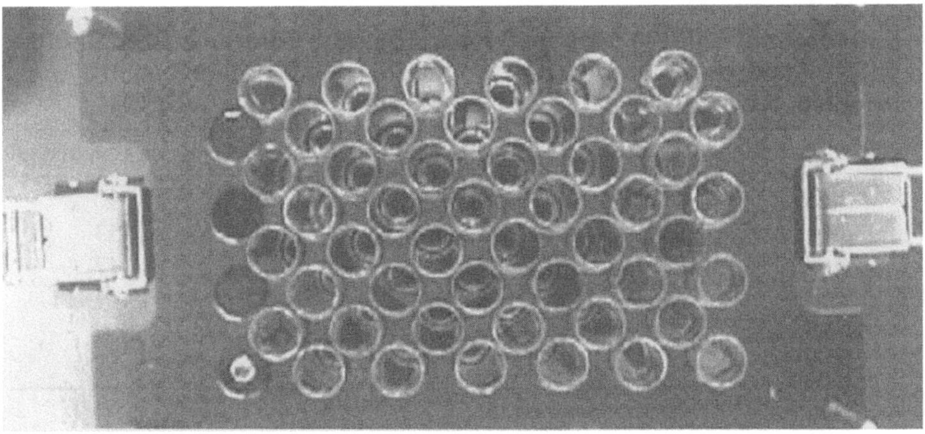

Fig. 3. High-throughput screening using the Bohdan Miniblock.

2. 2-(*N*-Cyclohexylamino)ethanesulfonic acid (CHES) and 2-(*N*-morpholino)ethanesulfonic acid (MES) buffers, chloroform ($CHCl_3$), sodium chloride (NaCl) , zinc nitrate, cadmium nitrate, and dithizone
3. Millipore MILLI-Q purified water was used for all the experiments.
4. An Orion Research ion analyzer/501 with an Orion Ross combination 81115BN pH electrode was used to adjust the pH of all solutions.
5. A Hewlett Packard diode array 8452A spectrophotometer was used to carry out all UV-Vis spectra and kinetic measurements.
6. A Bohdan Miniblock reactor with compact washing and shaking station was used for high-throughput screening with 5 mL polypropylene tubes. Vacuum collection base with microplate format rack was used to collect the samples.

2.2. Stock Solutions

1. 1.00 *M* NaCl to maintain the ionic strength of solutions.
2. 0.100 *M* Buffer solutions [MES (pH 6.5), CHES (pH 8.5)].
3. 10.0 m*M* 2-pyridinecarboxaldehyde aqueous solution.
4. Solutions of 0.500 *M* and 30.0 m*M* aminophenol in DMSO; solutions must be freshly prepared.
5. A solution containing 3.90 mg of dithizone in 15–20 mL of chloroform; freshly prepared.

3. Methods

3.1. Extraction Study

1. Add to a 20 mL vial with aluminum-lined Teflon crimp-cap (*see* **Note 1**), 10 mL of a solution containing 0.100 *M* NaCl, 0.0500 m*M* metal, 5 m*M* buffer, and 0.1 m*M* PCA in water.

2. In order to prevent the oxidation of aminophenols (AP), degas the solution with nitrogen gas before adding 2 µL of 0.500 *M* AP solution to give a final concentration of 0.100 m*M* AP.
3. Equip the vials with magnetic stir bars, seal immediately, and degas the solutions again for a few minutes.
4. Add 1 mL of degassed $CHCl_3$ by syringe to each vial.
5. Stir the solutions in the vials vigorously for 24 h on a magnetic stir plate.
6. Watch for a change in color of $CHCl_3$ over a period of time from transparent to hues of red.
7. Take aliquots from the organic phase and analyze the samples for metal ion concentration by the dithizone assay.

3.2. Equilibrium Study

1. Prepare the solutions as described above in **Subheading 3.1.** (*see* **Note 2**) for water–chloroform extraction for different periods of time.
2. Prepare a $CHCl_3$ solution containing 0.500 mmol of synthesized metal complex [i.e., Cd(PCA-**2**)$_2$] *(18)* and degas the solution.
3. Add 1 mL of this solution containing 0.0500 m*M* metal complex to 10 mL of degassed solution containing 5.00 m*M* CHES and 0.100 *M* NaCl in a 20 mL glass vial equipped with a magnetic stir bar.
4. Allow this solution to equilibrate for different periods of time and analyze the amount of metal ion by dithizone assay.
5. Compare the results for both extractions (**Fig. 2**).

3.3. High-Throughput Screening

1. Load the Miniblock reactor with 5 mL polypropylene pinch tubes by placing it into tube inserts (*see* **Note 3**). Close the valve mechanism before pouring any reagents in the tube.
2. Use different aminophenols (we used six) with PCA in the presence of either Zn(II) or Cd(II) at different pH values.
3. Add 2 mL of solutions containing 0.100 *M* NaCl, 5.00 m*M* buffer, 0.0500 m*M* metal ion, and 0.100 m*M* PCA in water to these tubes.
4. To this solution, add 60 µL of 30.0 m*M* AP solution to give a final concentration of 0.100 m*M* AP.
5. Add 2 mL of $CHCl_3$ to this solution.
6. Agitate the Miniblock reactor at shaking speed of 600–700 rpm on a shaker for 24 h.
7. After shaking, place the Miniblock on vacuum collection base. Open the valve of the Miniblock reactor and collect the chloroform from polypropylene tubes to glass test tubes placed on a microplate format rack.
8. To 500 µL of collected chloroform, add 3 mL total of dithizone solution in three aliquots.
9. Observe the different solution colors for extracted metal Schiff-base complexes ranging from blue to green to orange and pink according to the concentration of metal ion extracted.

3.4. Dithizone Assay

1. To a standard 1.0 cm quartz cuvet, add 3 mL of $CHCl_3$ and measure a blank on the UV-Vis spectrophotometer. Add the dithizone solution in $CHCl_3$ and adjust the concentration such that the absorbance of dithizone ranges between 0.7 and 0.9 absorbance units. Always prepare the dithizone solution fresh (*see* **Note 4**).
2. To monitor metal ion extractions (*see* **Note 5**), add a known volume from the extracted $CHCl_3$ phase of the sample complex to the dithizone solution and record the change in absorbance of free dithizone at 606 nm (*see* **Note 6**).
3. Make a minimum of three subsequent additions to determine an average change in absorbance upon addition of the chloroform extract.
4. Calculate the number of moles of free dithizone from Beer's law using Eq. (1) for a 1 cm path length cell.
5. By using the difference in concentration of free dithizone due to added extractant, calculate the number of moles of metal ion by using Eq. (2).

$$n_{\mathrm{Dz}} = A_{606} \times V_{\mathrm{total}}/\varepsilon_{606(\mathrm{Dz})} \quad (1)$$

$$n_{\mathrm{metal}} = (\Delta n_{\mathrm{Dz}} \times C/2 \quad (2)$$

where n_{Dz} is number of moles of free Dz in solution, C is the dilution factor, V_{total} is the total volume of the solution in L, A_{606} is the absorbance at 606 nm in absorbance units, ε_{606} is the molar extinction coefficient at 606 nm, which is $4.15 \times 10^4 M^{-1}\,\mathrm{cm}^{-1}$ for dithizone, n_{metal} is the number of moles of metal ion, and Δn_{Dz} is the difference in the moles of free dithizone due to added extractant (*see* **Note 7**).

3.5. Calculation of Extraction Constants and Percentage Extraction

1. Calculate the extraction constant (K) for every aminophenol by using Eqs. (3) and (4). We assume that the bis-Schiff-base complex is the predominant extracted species.

$$2\mathrm{PCA}+2\mathrm{AP}+\mathrm{M}^{2+} \rightleftharpoons \mathrm{M(PCA-2AP)_2} \quad (3)$$

$$K = x/[(a-2x)^2(b-2x)^2(c-x)] \quad (4)$$

2. The variable x is the concentration of metal ion extracted, which is equal to the concentration of complex extracted. This is determined experimentally by the dithizone assay. Here a is the total concentration of PCA (0.0001 M), b is the total concentration of aminophenol, AP (0.0001 M) and c is the total concentration of metal ion, M^{2+} (0.00005 M).
3. Using these values of K, calculate the expected amount of metal ion extracted for two different types of aminophenols in a mixture containing PCA.

$$2\mathrm{PCA} + 2\mathrm{AP1} + 2\mathrm{AP2} + \mathrm{M}^{2+} \rightleftharpoons \mathrm{M(PCA-AP1)_2} + \mathrm{M(PCA-AP2)_2} \quad (5)$$

4. Defined d as the total concentration of aldehyde, PCA (0.0001 M), e as total concentration of aminophenol 1, AP1 (0.0001 M), f as total concentration of aminophenol 2, AP2 (0.0001 M), and g as total concentration of metal ion, M (0.00005 M).

5. Assuming that the two aminophenols behave independently and using the value of the extraction constant determined for the two different aminophenol complexes separately ($K1$ and $K2$), calculate y and z, which are the concentrations of the two different bis-Schiff-base complexes. Each of these complexes contains two of the same Schiff-base ligands. Any increased amount of extraction above that calculated by this method is assumed to arise from the formation of mixed Schiff-base complexes.

$$[(d - 2y - 2z)^2(e - 2y)^2(g - y - z)] = y/K1 \quad (6)$$

$$[(d - 2y - 2z)^2(f - 2z)^2(g - y - z)] = z/K2 \quad (7)$$

6. Solve Eqs. (6) and (7) iteratively by using an initial guess of y and z of 0.00001. Calculate the amount of metal by taking the sum of y and z. Compare this value with the experimental value of metal ion extracted with a mixture containing two aminophenols and PCA (**Table 2**).

4. Notes

1. The use of vials with aluminum-lined Teflon crimp-caps minimizes loss of chloroform through evaporation and facilitated the use of an inert atmosphere to minimize oxidation of the aminophenols. Although best results were obtained under nitrogen, most aminophenols had similar extraction properties without an inert atmosphere. The susceptibility toward oxidation varies with aminophenol substituent *(22)*.
2. In order for this system to function as a dynamic library, it was necessary to ensure that the two-phase system was at equilibrium. Studies were done to determine the time required for the two-phase system to reach equilibrium. For comparison, extraction was performed using reagents that were initially in either the organic or aqueous phase to determine the time needed for the system to equilibrate between the two phases.
3. Make sure that the valve is open before inserting the reaction tube. Failing to do this may result in an inoperable or damaged pinch tube.
4. Dithizone has 16.8 g/L solubility at 20°C in chloroform with two absorbance bands at 440 nm ($\varepsilon = 1.59 \times 10^4\ M^{-1}\ cm^{-1}$) and 606 nm ($\varepsilon = 4.14 \times 10^4\ M^{-1}\ cm^{-1}$).
5. There are two overlapping absorbance bands for dithizone (Dz) metal complexes and free dithizone. The $Cd(Dz)_2$ complex absorbs strongly at 520 nm but very weakly at 606 nm (ε ~100 $M^{-1}\ cm^{-1}$ at 606 nm). Thus, the 606 nm absorbance was used to monitor free dithizone.
6. Solutions containing known amounts of metal bis-Schiff-base complex were analyzed in the dithizone assay. The expected decrease in free dithizone absorbance was observed, confirming that there was no interference from the metal Schiff-base complexes for this assay.
7. There are two dithizone molecules per metal ion because dithizone forms a bis-complex with Cd(II) and Zn(II) ions.

References

1. Still, W. C., Hauck, P., and Kempf, D. (1987) Stereochemical studies of lasalocid epimers. Ion-driven epimerizations. *Tetrahedron Lett.* **28,** 2817–2820.
2. Hasenknopf, B., Lehn, J.-M., Kneisel, B. O., Baum, G., and Fenske, D. (1996) Self-assembly of a circular double helicate. *Angew. Chem. Int. Ed. Engl.* **35,** 1838–1840.
3. Hill, C. L. and Zhang, X. (1995) A smart catalyst that self-assembles under turnover conditions. *Nature* **373,** 324–326.
4. Huc, I., Krische, M. J., Funeriu, D. P., and Lehn, J.-M. (1999) Dynamic combinatorial chemistry: substrate H-bonding directed assembly of receptors based on bipyridine-metal complexes. *Eur. J. Inorg. Chem.* 1415–1420.
5. Albrecht, M., Blau, O., and Frohlich, R. (1999) An expansible metalla-cryptand as a component of a supramolecular combinatorial library formed from Di(8-hydroxyquinoline) ligands and gallium (III) or zinc (II) ions. *Chem. Eur. J.* **5,** 48–56.
6. Burger, M. T. and Still, W. C. (1995) Synthetic ionophores. Encoded combinatorial libraries of cyclen-based receptors for Cu^{2+} and Co^{2+}. *J. Org. Chem.* **60,** 7382–7383.
7. Francis, M. B., Finney, N. S., and Jacobsen, E.N. (1996) Combinatorial approach to the discovery of novel coordination complexes. *J. Am. Chem. Soc.* **118,** 8983–8984.
8. Hoveyda, A. H. (1998) Catalyst discovery through combinatorial chemistry. *Chem. Bio.* **5,** R187–R191.
9. Francis, M. B., Jamison, T. F., and Jacobsen, E,N. (1998) Combinatorial libraries of transition-metal complexes, catalysts and materials. *Curr. Opion. Chem. Bio.* **2,** 422–428.
10. Lehn, J. M. (1999) Dynamic combinatorial chemistry and virtual combinatorial libraries. *Chem. Eur. J.* **5,** 2455–2463.
11. Goral, V., Nelen, M., Eliseev, A. V., and Lehn, J.-M. (2001) Double-level "orthogonal" dynamic combinatorial libraries on transition metal template. *Proc. Natl. Acad. Sci.* **98,** 1347–1352.
12. Klekota, B., Hammond, M. H., and Miller, B. L. (1997) Generation of novel DNA-binding compounds by selection and amplification from self-assembled combinatorial libraries. *Tetrahedron Lett.* **38,** 8639–8642.
13. Goodman, M. S., Jubian, V., Linton, B., and Hamilton, A. D. (1995) A combinatorial library approach to artificial receptor design. *J. Am. Chem. Soc.* **117,** 11610–11611.
14. Yordanov, A. and Roundhill, M. (1998) Solution extraction of transition and post-transition heavy and precious metals by chelate and macrocyclic ligands. *Coord. Chem. Rev.* **170,** 93–124.
15. Nash, K. L. and Choppin, G. R. (1997) Separations chemistry for actinide elements: recent developments and historical perspective. *Separation Science and Technology* **32,** 255–274.
16. Rydberg, J., Musikas, C., and Choppin, G. R. (eds.) (1992) *Principles and Practices of Solvent Extraction*, Marcel Dekker, New York.
17. Cox, B. G. and Schneider, H. (1992) *Coordination and Transport Properties of Macrocyclic Compounds in Solution,* Elsevier, New York.

18. Epstein, D. M., Choudhary, S., Churchill, M. R., Keil, K. M., Eliseev, A. V., and Morrow, J. R. (2001) Chloroform-soluble Schiff-base Zn(II) or Cd(II) complexes from a dynamic combinatorial library *Inorg. Chem.* **40,** 1591–1596.
19. Gillard, R. D. and McCleverty, J. A. (eds.) (1987) Zinc and cadmium, In *Comprehensive Coordination Chemistry*, 1st ed, Pergamon Press, Oxford, England, vol. 5, p. 940.
20. http://www.bohdan.com.
21. Cheng, K. L., Veno, K., and Imammura, T. *Handbook of Organic and Analytical Reagents*, CRC Press, Inc., Boca Raton, FL, 1982.
22. Epstein, D. M. (2000) Cleavage of phosphodiesters and 5'-CAP of m-RNA by lanthanide (II) macrocyclic complexes and the selection of Schiff-base transition metal complexes from a dynamically generated combinatorial library *Ph.D. Thesis*, State University of New York at Buffalo.

15

Automated Liquid–Liquid Extraction and Ion-Exchange Solid-Phase Extraction for Initial Purification

Sean X. Peng and Charles Henson

1. Introduction

The development of combinatorial chemistry has made a profound impact on the pharmaceutical industry by producing a large number of structurally diverse compounds in a very short amount of time. Combined with high-throughput screening, bioinformatics, and laboratory automation, the combinatorial chemistry approach has led to a significantly accelerated drug discovery process compared to a traditional one-compound-at-a-time approach *(1)*. With a potential explosion of the number of biological targets available for various diseases from genomics and proteomics, combinatorial chemistry and parallel synthesis will be further embraced by the pharmaceutical industry, leading the way to rapidly generate a large number of chemical libraries to screen for different disease targets. Currently, thousands of potentially bioactive compounds are made every week by combinatorial chemistry synthesis. Subsequently, these compounds go through various high-throughput biological screens in different therapeutic areas to find biologically active compounds or hits. However, in order to ensure true hits from these various screening assays, initial sample purification to remove assay-interfering components is required to prevent false positive or negative results. This initial sample cleanup step poses a great challenge to synthetic and analytical chemists, because purification in a high-throughput and fully automated fashion is needed to keep up with the fast pace of combinatorial synthesis and high-throughput biological screening.

In general, combinatorial chemistry employs either solid-phase or solution-phase synthesis *(2)*. The solid-phase synthesis has the advantage of generating cleaner samples, because the solid support material, i.e., resin beads, can be

From: *Methods in Molecular Biology, Combinatorial Library Methods and Protocols*
Edited by: L. B. English © Humana Press Inc., Totowa, NJ

filtered and washed after the reaction is complete and purer reaction product is obtained upon cleavage of the linker on the resin. However, excess cleaving reagents are often needed to achieve complete cleavage in a short amount of time because many cleavage reactions are slow with stoichiometrical amounts. For solution-phase synthesis, excess starting reagents and by-products often remain in the intermediate and final product and are difficult to remove. When these excess reagents and by-products interfere with biological screening assays of interest, the screening results become unusable. To address these problems, tremendous ongoing efforts have been made in the purification of combinatorial library products. Several different approaches have been developed and employed for initial purification of crude library samples. These include liquid–liquid extraction (LLE) ***(3–5)***, various modes of solid-phase extraction (SPE) ***(6–8)***, solid-phase scavenging (SPS) ***(9–12)***, and parallel-column preparative high-performance liquid chromatography (HPLC) ***(13)***. LLE, SPE, and SPS are generally employed for initial or primary purification to remove unspent starting reagents, by-products, and impurities from reaction mixtures. Because these methods can be fully automated in a high-throughput fashion such as in a 96-well format, they are generally employed as the first purification step immediately after crude library samples are made. After this initial purification, the purified samples pass through high-throughput biological screens. When a hit sample is found, a parallel-column preparative HPLC method is typically employed to isolate the hit compound for an additional confirmatory biological assay. Preparative HPLC can also be used for initial purification to provide high purity samples for biological screening. Because of its limited throughput, this HPLC method may be best suited for purification of hit samples for hit identification and confirmation as the number of samples to be purified is significantly reduced after initial biological screening. For the initial biological screening, however, the assay-interfering components are typically removed using LLE, SPE, and SPS as these approaches offer higher sample throughput, faster speed, lower cost, and easier adaptation to automation. While the LLE method exploits the differences in partitioning between the desired product and unwanted components in two immiscible organic and aqueous phases, the SPE and SPS methods take advantage of the differences in interactions (reversible and non-covalent interaction in SPE versus irreversible and covalent bonding in SPS) between the solid-phase resins or scavengers and the desired products or unwanted components. Currently, automated LLE and ion-exchange SPE are the two commonly used methodologies for initial purification of crude combinatorial library samples.

LLE is a well established and widely utilized extraction technique in organic synthesis. It is simple, rapid, and convenient, producing extremely clean extracts with high product recovery. Recent development of a fully automated

96-well LLE methodology has made LLE a very attractive and practical method for post-reaction purification of combinatorial libraries for high-throughput screening *(3)*. The automated LLE method selectively removes unreacted starting materials, by-products, and impurities from reaction mixtures. Automated ion-exchange SPE has also been commonly employed to separate ionizable by-products and impurities from neutral target compounds, or to remove unwanted neutral components from ionizable target compounds, for purification of small molecule libraries *(7,8)*. In ion-exchange SPE, two strategies are commonly employed: either the desired final products or the unwanted by-products and excess reagents are extracted onto the ion-exchange resins. Typically, cationic ion-exchange SPE is employed to remove cationic by-products and excess reagents from neutral target molecules by extracting them onto the cationic resins or to selectively extract cationic target compounds and then elute them from the resins. Conversely, anionic ion-exchange SPE selectively removes unwanted anionic components in the reaction mixture by extracting them onto the anionic resins or selectively attaches anionic target compounds onto the resins and then washes them off the resins. Because most of the starting reagents for combinatorial library synthesis are acids such as acid chlorides or bases such as amines, the resulting unwanted by-products or unreacted excess reagents are typically ionizable components, such as carboxylic acids and primary and secondary amines, which can be readily removed by either cationic (e.g., for amine removal) or anionic (e.g., for carboxylic acid removal) ion-exchange resins. Here, we describe the protocols for post-reaction purification by automated 96-well LLE and ion-exchange SPE, the two most commonly used techniques for removal of unwanted or assay-interfering components from crude reaction mixtures in combinatorial library synthesis. In these protocols, a robotic liquid handler and various types of 96-well plates are used to facilitate automation, increase sample throughput, and reduce solvent consumption. A schematic representation of each purification technique is shown in **Figs. 1** and **2**.

2. Materials

2.1. Chemicals for LLE

1. Butylacetate (Aldrich, Milwaukee, WI).
2. Hydrochloric acid (Aldrich).
3. Sodium hydroxide (Aldrich).
4. Dimethylsulfoxide (Aldrich).
5. Hydromatrix diatomaceous earth material (Varian, Harbor City, CA, USA).

2.2. Apparatus for LLE

1. 96-well (2-mL capacity) hydrophobic GF/C glass fiber filter plate (Whatman, Clifton, NJ).

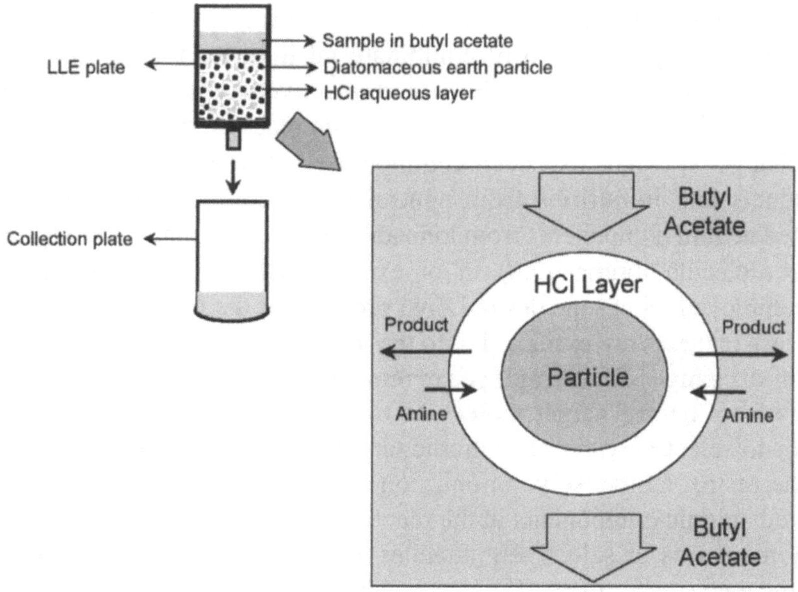

Fig. 1. Schematic representation of automated 96-well liquid–liquid extraction for the removal of basic and water soluble components such as amines from crude library samples.

2. 96-well (1- and 2-mL capacity) collection plate (Varian).
3. Polypropylene reagent reservoir with 2 baffles (Tomtec, Hamden, CT).
4. Manifold disposable waste tray (Waters, Milford, MA).
5. Quadra 96 model 320 96-channel liquid handler (Tomtec).
6. SPE Dry-96 96-well plate sample concentrator (Jones Chromatography, Lakewood, CO).

2.3. Chemicals for Ion-Exchange SPE

1. Methanol (J. T. Baker, Phillipsburg, NJ).
2. Acetonitrile (J. T. Baker).
3. Ammonia, 2 *M* in methanol (Aldrich).
4. Sodium hydroxide (Aldrich).
5. Trifluoroacetic acid (J. T. Baker).
6. Dimethylsulfoxide (Aldrich).
7. Dowex strong cation exchanger 50WX8-100 (*see* **Note 1**), with sulfonic acid as the functional group on polystyrene-based resins (Supelco, Bellefonte, PA).
8. Dowex strong anion exchanger 1X8-100 (*see* **Note 1**), with trimethylbenzyl ammonium as the functional group on polystyrene-based resins (Supelco).

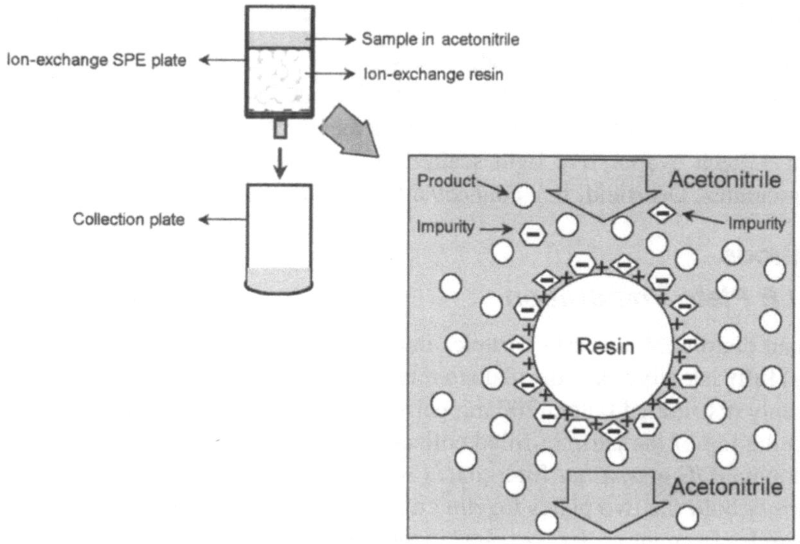

Fig. 2. Schematic representation of automated 96-well ion-exchange solid-phase extraction for the removal of anionic components such as carboxylic acids from crude library samples.

2.4. Apparatus for Ion-Exchange SPE

1. Empty 96-well plate (2-mL capacity) with bottom frits (Varian).
2. 96-well (2-mL capacity) collection plate (Varian).
3. Polypropylene reagent reservoir with two baffles (Tomtec).
4. Manifold disposable waste tray (Waters).
5. Quadra 96 model 320 96-channel liquid handler (Tomtec).
6. SPE Dry-96 96-well plate sample concentrator (Jones Chromatography, Lakewood, CO).

2.5. Chemicals for Purity Determination

1. HPLC-grade methanol (J. T. Baker).
2. HPLC-grade acetonitrile (J. T. Baker).
3. Formic acid (J. T. Baker).
4. Triethylamine (Aldrich).
5. Heptafluorobutyric acid (Aldrich).
6. Tetrabutylammonium dihydrogenphosphate (Aldrich).
7. Monobasic sodium phosphate (Aldrich).
8. Dibasic sodium phosphate (Aldrich).
9. Phenyl isothiocyanate (Aldrich) (*see* **Note 2**).
10. Phenyl isocyanate (Aldrich) (*see* **Note 2**).
11. p-Bromophenacyl bromide (Aldrich) (*see* **Note 3**).

2.6. HPLC Instrumentation for Purity Determination (see Note 4).

1. A Waters 2690 separation module (Waters).
2. A Waters Symmetry C18 column (150 × 3.9 mm, 5 mm particle size) (Waters).
3. A Waters 996 photodiode array (PDA) detector (Waters).
4. An Alltech evaporative light scattering detector (ELSD) ELSD2000 (Alltech Associates, Deerfield, IL), connected to the outlet of a PDA detector.

3. Methods

3.1. LLE Plate Preparation

1. Load diatomaceous earth particles into a 96-well collection plate (1-mL capacity) fully; slowly shake the plate to remove extra particles so that the particles are evenly distributed (about 200 mg per well) in each well (*see* **Note 5**).
2. On the top of the particle-filled collection plate, place an empty 96-well hydrophobic GF/C glass fiber filter plate (2-mL capacity) upside down.
3. Firmly hold the two plates together and invert both plates quickly to unload the particles from the collection plate to the filter plate; tap the bottom of the collection plate lightly to ensure that all particles in the collection plate are transferred.
4. Remove the collection plate and cover the LLE plate (i.e., the particle-filled filter plate) with a sheet of aluminum foil.
5. Repeat steps 1–3 to prepare additional LLE plates as necessary.

3.2. LLE Extraction

1. Place an LLE plate in the vacuum manifold and a collection plate inside of the manifold.
2. Program the Quadra 96 liquid handler to execute the following sequentially (*see* **Note 6**):
   a. Aspirate 800 µL of butylacetate and dispense it into the sample plate containing combinatorial samples.
   b. Perform six aspirate–dispense mixing cycles to dissolve the combinatorial samples in butylacetate.
   c. Aspirate 800 µL of 2 *N* hydrochloric acid (for removal of basic components) or 2 *N* sodium hydroxide (for removal of acidic components) and dispense it into the LLE plate.
   d. Pull 800 µL of combinatorial samples from the sample plate and place it into the LLE plate.
   e. Wait for 3–5 min and then apply gentle vacuum (< 3 in. Hg) to the vacuum manifold.
   f. Load 800 µL of butylacetate into the LLE plate to elute the sample from the LLE plate to the collection plate.
3. Take the collection plate and place it in an SPE Dry-96 96-well plate sample concentrator; evaporate the solvent completely under a stream of heated nitrogen (45°C).

4. Store the collection plate containing the dry sample or redissolve the sample in dimethylsulfoxide (DMSO) using the Quadra 96 for storage or for purity determination by HPLC/UV/ELSD.

3.3. Ion-Exchange SPE Plate Preparation

1. Load ion-exchange resins, either cationic or anionic ion exchangers, into an empty 96-well plate with frits (about 600 mg per well), shake the plate until the particles are evenly distributed in each well (*see* **Note 7**).
2. Cover the ion-exchange SPE plate with a sheet of aluminum foil.
3. Repeat steps 1–3 to prepare additional ion-exchange SPE plates as necessary.

3.4. Cationic Ion-Exchange SPE Extraction

1. Place an ion-exchange SPE plate filled with cationic exchange resins in the vacuum manifold and a waste tray inside of the manifold.
2. Program the Quadra 96 liquid handler to execute the following sequentially (*see* **Note 8**):
   a. Aspirate 800 µL of acetonitrile and dispense it into the sample plate containing combinatorial samples.
   b. Perform six aspirate–dispense mixing cycles to dissolve the combinatorial samples in acetonitrile.
   c. Aspirate 600 µL of water and dispense it into the ion-exchange SPE plate.
   d. Aspirate 600 µL of acetonitrile and dispense it into the ion-exchange SPE plate.
   e. Remove the waste tray from the vacuum manifold, discard the collected eluent, and place a 96-well collection plate inside the vacuum manifold.
   f. Pull 800 µL of combinatorial samples from the sample plate and place it into the ion-exchange SPE plate.
   g. Wait for 1–2 min and then apply gentle vacuum (< 3 in. Hg) to the vacuum manifold to elute the sample from the resins (*see* **Note 9**).
3. Take the collection plate and place it in an SPE Dry-96 96-well plate sample concentrator; evaporate the solvent completely under a stream of heated nitrogen (45°C).
4. Store the collection plate containing the dry sample or redissolve the sample in DMSO using the Quadra 96 for storage or for purity determination by HPLC/UV/ELSD.

3.5. Anionic Ion-Exchange SPE Extraction

1. Place an ion-exchange SPE plate filled with anionic exchange resins in the vacuum manifold.
2. Program the Quadra 96 liquid handler to execute the following sequentially (*see* **Note 10**):
   a. Aspirate 800 µL of acetonitrile and dispense it into the sample plate containing combinatorial samples.
   b. Perform six aspirate–dispense mixing cycles to dissolve the combinatorial samples in acetonitrile.

c. Aspirate 800 μL of water and dispense it into the ion-exchange SPE plate, drain it with low vacuum applied.
d. Add 5 mL of 4 *N* sodium hydroxide to the ion-exchange plate and drain it with low vacuum applied (*see* **Note 11**).
e. Add 4 mL of water to the ion-exchange plate and drain it with low vacuum applied.
f. Aspirate 800 μL of acetonitrile and dispense it into the ion-exchange SPE plate.
g. Remove the waste tray from the vacuum manifold, discard the collected eluent, and place a 96-well collection plate inside the vacuum manifold.
h. Pull 800 μL of combinatorial sample from the sample plate and load it into the ion-exchange SPE plate.
i. Wait for 1–2 min and then apply gentle vacuum (< 3 in. Hg) to the vacuum manifold to elute the sample from the resins (*see* **Note 9**).

3. Take the collection plate and place it in an SPE Dry-96 96-well plate sample concentrator; evaporate the solvent completely under a stream of heated nitrogen (45°C).
4. Store the collection plate containing the dry sample or redissolve the sample in DMSO using the Quadra 96 for storage or for purity determination by HPLC/UV/ELSD.

3.6. Purity Determination

Purity of the samples extracted by the automated LLE or ion-exchange SPE can be determined by HPLC/UV/ELSD. The UV wavelength in a UV or PDA detector is typically set at 210 nm to observe both strong and weak UV-absorbing components. The ELSD detector is used to monitor non-volatile components that do not contain strong UV-chromophores.

1. Sample preparation: An appropriate amount of the dry sample is dissolved in DMSO to give a concentration of about 100 mg/mL. The sample can then be directly injected into the HPLC system for the determination of product purity and recovery. However, when extraction efficiency (e.g., removal of certain components such as amines and carboxylic acids) needs to be evaluated and higher detection sensitivity is required for small amounts of impurities that lack strong UV-chromophores, chemical derivatization can be employed to improve UV detection sensitivity and retention. Amines can be derivatized using phenyl isothiocyanate (PITC) or phenyl isocyanate (PIC) as follows (*see* **Note 2**): 5 μL of PITC or PIC reagent is added to 100 μL of the extracted sample dissolved in 1 mL acetonitrile to form UV-absorbing phenylthiocarbamoyl or phenylcarbamoyl derivatives (reaction at room temperature for 10 min). Subsequently, 10-μL aliquots of the resulting solutions are injected onto the HPLC system. For carboxylic acids, they can be derivatized using *p*-bromophenacyl bromide as follows (*see* **Note 3**): 10 μL of *p*-bromophenacyl bromide reagent is added to 100 μL of the extracted sample dissolved in 1 mL acetonitrile, followed by addition of 10-μL aliquots of triethylamine (as a catalyst), to form UV-absorbing ester deriv-

atives (reaction at 50°C for 20 min). Subsequently, 10-μL aliquots of the resulting solutions are injected onto the HPLC system.

2. HPLC conditions: A generic 15-min linear gradient method is used where 20 μL of sample is injected into a Waters Symmetry C18 column (150 × 3.9 mm ID, 5 mm) and eluted from mobile phase A (acetonitrile:water:formic acid; 5:95:0.1; v/v/v) to mobile phase B (acetonitrile:water:formic acid; 95:5:0.1; v/v/v) with a flow rate of 1.0 mL/min (*see* **Note 12**). A PDA detector is connected with an ELSD detector in series to monitor both UV and light scattering signals.
3. PDA UV detector settings: wavelength 200–400 nm with the monitoring wavelength at 210 nm and resolution at 1.2 nm.
4. ELSD detector settings: in the "Impactor On" mode—drift tube temperature at 40°C and a nitrogen flow rate at 1.5 L/min; in the "Impactor Off" mode—drift tube temperature at 95°C and a nitrogen flow rate at 2.2 L/min.
5. HPLC assay: The purity of the final combinatorial products is determined by the aforementioned gradient HPLC method. The library samples before and after extraction are injected onto the HPLC system. The product and impurity peak areas are then compared to assess the product purity and recovery. The product purity is determined by the percent peak area of the product of interest in the post-extraction sample solution. The product recovery and the removal of the excess reagents, by-products, or impurities are evaluated by the ratio of the peak areas of the respective component before and after extraction.

4. Notes

1. Any other equivalent (cation or anion) ion-exchange resins from different manufacturers may be used. Other types of products may be better in some situations.
2. Both phenyl isocyanate (PIC) and phenyl isothiocyanate (PITC) are used as derivatization reagents for aliphatic amines that lack strong UV-chromaphores to improve detectability and retention on a reversed-phase HPLC column. PIC is for both primary and secondary amines, while PITC is mainly for primary amines ***(14)***. They are used to determine extraction efficiencies of individual amines.
3. It is used as a derivatization reagent for aliphatic carboxylic acids that lack strong UV-chromaphores to improve detectability and retention on a reversed-phase HPLC column ***(14)***. It is used to determine extraction efficiencies of individual carboxylic acids.
4. Any equivalent analytical HPLC system can be used. If no ELSD detector is available, a UV detector may be sufficient enough for purity determination. An ELSD detector is preferred over a UV detector because it provides more uniform responses independent of the optical properties of compounds.
5. A 1-mL 96-well collection plate is used only as a template to facilitate the packing of an LLE plate so that an equal amount of particles can be easily transferred to each well of a 96-well filter plate.
6. The Tomtec Quadra 96 liquid handler can be replaced with another robotic liquid handler with 96-, 12-, or 8-channels. Each aspiration step by the Quadra 96 is generally proceeded by 50 μL of air gap to aid in complete and accurate dispensing.

7. When packing dry ion-exchange resins, use a 1-mL 96-well collection plate as a template and load the resins into the collection plate; then place the filter plate on the top of the collection plate upside down and invert both plates together to transfer the resins from the collection plate to the filter plate.
8. Each aspiration step by the Quadra 96 is generally proceeded by 50 μL of air gap to aid in complete and accurate dispensing. This extraction procedure is used to remove excess amines or other basic by-products and impurities in the final combinatorial products. It takes advantage of the difference in basicity between the impurities such as amines (pK_a 9.5–10.5 for most of their conjugate acids or amonium ions) and the desired product ($pK_a < 0$ for most of their conjugate acids). Therefore, amines are protonated cations at a neutral pH and retained by the cation exchange SPE resins, while the desired product is not. If basic ionizable final products (e.g., amines) are to be purified and nonionizable impurities are to be removed, then an additional final elution step with a basic solvent (e.g., 1 mL of 2 *M* ammonia in methanol) is needed to obtain the desired final product that is retained on the resins after step g.
9. To further improve product recovery, an additional 200 μL of acetonitrile may be added into the ion-exchange SPE plate to elute the residual product remained in the SPE plate.
10. Each aspiration step by the Quadra 96 is generally proceeded by 50 μL of air gap to aid in complete and accurate dispensing. This extraction procedure is used to remove carboxylic acids or other acidic by-products and impurities in the final combinatorial products. It takes advantage of the difference in acidity between the impurities, such as acids ($pK_a < 5$, deprotonated anions at neutral pH) and the desired product (unionized form at neutral pH). Therefore, acids are retained by the anion exchange SPE resins, while the desired product is not. If acidic ionizable final products (e.g., acids) are to be purified and nonionizable impurities are to be removed, then an additional final elution step with an elution solvent (e.g., 1 mL of 1–2 *M* trifluoroacetic acid in methanol) is needed to obtain the desired final product that is retained on the resins after step g.
11. A 4 *N* sodium hydroxide solution is used to change the counter-ion on the resin from chloride (higher affinity anion) to hydroxide (lower affinity anion) for better retention of carboxylic acids.
12. For the library samples that contain very polar by-products, excess reagents, or impurities, an ion-pair reagent may be added to the mobile phase to improve retention of the polar components on the reversed-phase HPLC column. For small polar basic components such as amines, 0.1–1% of heptafluorobutyric acid can be added into the mobile phase containing acetonitrile or methanol in water. For small polar acidic components such as carboxylic acids, 5–20 m*M* of tetrabutylammonium dihydrogenphosphate can be added into the mobile phase containing acetonitrile or methanol in 50 m*M* phosphate buffer (pH 7.4).

References

1. Dolle, R. E. (2000) Comprehensive survey of combinatorial library synthesis: 1999. *J. Comb. Chem.* **2,** 383–433.

2. Czarnik, A. W. (1998) Combinatorial chemistry. *Anal. Chem.* **70,** 378A-386A.
3. Peng, S. X., Henson, C., Strojnowski, M. J., Golebiowski, A., and Klopfenstein, S. R. (2000) Automated high-throughput liquid-liquid extraction for initial purification of combinatorial libraries. *Anal. Chem.* **72,** 261–266.
4. Johnson, C. R., Zhang, B., Fantauzzi, P., Hocker, M., and Yager, K. M. (1998) Libraries of N-alkylaminoheterocycles from nucleophilic aromatic substitution with purification by solid supported liquid extraction. *Tetrahedron* **54,** 4097–4106.
5. Cheng, S., Comer, D. D., Williams, J. P., Myers, P. L., and Boger, D. L. (1996) Novel solution phase strategy for the synthesis of chemical libraries containing small organic molecules. *J. Amer. Chem. Soc.* **118,** 2567–2573.
6. Nilsson, U. J. (2000) Solid-phase extraction for combinatorial libraries. *J. Chromatogr. A* **885,** 305–319.
7. Gayo, L. M. and Suto, M. J. (1997) Ion-exchange resins for solution phase parallel synthesis of chemical libraries. *Tetrahedron Lett.* **38,** 513–516.
8. Siegel, M. G., Hahn, P. J., Dressman, B. A., Fritz, J. E., Grunwell, J. R., and Kaldor, S. W. (1997) Rapid purification of small molecule libraries by ion exchange chromatography. *Tetrahedron Lett.* **38,** 3357–3360.
9. Coates, S. W., Kirkland, J. J., Langlois, T., Majors, R. E., Szafranski, C. A., Thompson, L. A., and Wang, Q. (2000) New solid-phase scavengers improve recovery and speed throughput in parallel and related synthesis. *LC-GC* **18,** S30–S34.
10. Booth, R. J. and Hodges, J. C. (1997) Polymer-supported quenching reagents for parallel purification. *J Am. Chem. Soc.* **119,** 4882–4886.
11. Kaldor, S. W., Siegel, M. G., Fritz, J. E., Dressman, B. A., and Hahn, P. J. (1996) Use of solid supported nucleophiles and electrophiles for the purification of non-peptide small molecule libraries. *Tetrahedron Lett.* **37,** 7193–7196.
12. Parlow, J. J., Devraj, R. V., and South, M. S. (1999) Solution-phase chemical library synthesis using polymer-assisted purification techniques. *Curr. Opin. Chem. Biol.* **3,** 320–336.
13. Zeng, L. and Kassel, D. B. (1998) Development of a fully automated parallel HPLC/mass spectrometry system for the analytical characterization and preparative purification of combinatorial libraries. *Anal. Chem.* **70,** 4380–4388.
14. Li, F. and Lim, C. K. (1993) Colored and UV-absorbing derivatives, in Handbook of Derivatives for Chromatography, second edition (Blau, K. and Halket, J., eds.), John Wiley & Sons, pp. 157–174.

16

On-Bead and Solution Screening Approaches for Genomically Derived Targets

Discovery of Surrogate Ligands and Substrates Using Combinatorial Chemistry Libraries

David S. Thorpe, Gérard Rossé, Helen Yeoman, Sydney Wilson, Patti Willson, Greg Harlow, Anna Robinson, and Kenneth F. Wertman

1. Introduction

Sequencing of the human and other genomes has revolutionized drug discovery. Modern functional genomics methods have brought a wealth of drug discovery targets. Bioinformatic tools are classifying candidate drug targets into different mechanistic classes. Many classes are apparent, but pharmaceutical companies are clearly focusing on proteases, protein kinases, and G-protein coupled receptors (GPCRs). This is perhaps because of the relatively direct nature of assigning these sequences to gene families, especially when the family features are so distinctive, e.g., seven transmembrane segments as a signature for GPCRs. Moreover, two of these are proven pharmaceutical targets with large drug markets (proteases and GPCR). Even though most target genes will fit into mechanistic families as determined by bioinformatic tools, others will fail to fit. Also, some fits will be misleading—for example, the sequences of membrane forms of guanylyl cyclases are homologous to protein kinases *(1)*. Thus, there is a need for tools to find surrogate ligands and substrates to serve as reagents to measure inferred bioactivities and discover leads to both pharmacologically validate novel drug targets and provide entry for lead optimization to obtain new drugs for unmet medical needs.

The focus of this chapter is on the methods of screening combinatorial libraries for genomically derived targets. Most methods are based on a presumption

From: *Methods in Molecular Biology, Combinatorial Library Methods and Protocols*
Edited by: L. B. English © Humana Press Inc., Totowa, NJ

of the role played by the genomically derived protein target. However, one method, on-bead protein staining, does not assume the function of a given protein *a priori* in order to find surrogate ligands. It is limited, however, by the requirement of a soluble protein target, the synthetic chemistry employed in the production of the library, and the methods to identify the ligands on the beads that bind the target. We will not review the synthesis of split mix peptide libraries, or other combinatorial chemical libraries, because these have been described extensively in this book and elsewhere ***(2–4)***.

The balance of the chapter is divided into two broad categories: assays where the combinatorial compounds are displayed on the surface of solid supports (on-bead methods) and an assay where the combinatorial compounds are free of supports and soluble in aqueous media to test against GPCRs. The on-bead section is in turn comprised of three parts. The first describes screens to detect the binding of protein targets to beads with ligands. The second on-bead part discusses protease screens for the discovery of surrogate substrates for enzymes that cleave substrates. The third aspect of on-bead techniques is on protein kinase screens to discover surrogate substrates for enzymes that add or transfer traceable moieties to a substrate. Finally, only one aspect of screening soluble combinatorial libraries will be described. A brief discussion of exotic Fluorescent Imaging Plate Reader (FLIPR) technology on the pharmaceutically critical class of GPCRs will finish out the chapter. As a result, this chapter will discuss a total of six separate protocols, which are now briefly introduced.

1.1. On-Bead Noncovalent Binding of Protein Targets to Beads Displaying Combinatorial Molecules

These methods are refined techniques based on the methods first explored by Lam and coworkers ***(3)***. Two protocols will be provided to describe the histochemical identification of beads that bind protein targets. The first protocol is the binding of a biotinylated target protein, and the second protocol is the binding of a protein target that is detected using a monoclonal antibody. The library beads are either in a tube (e.g., 50 mL) and spun in a centrifuge for washing steps (e.g., Beckman TJ6 at approx 1,800*g* for 3 min followed by aspirating the supernatant fluid), or the resin is inside a polyethylene fritted polypropylene syringe, which facilitates rapid washing steps. Clearly, the buffers and cofactors used for a given genomically derived target may need to be different from those used in the illustrative protocols given in this chapter. Basic principles of biochemistry and enzymology or pharmacology should guide the investigation. Although many histochemical methods are available—e.g., fluorescence activated cell (or bead) sorting using Bodipy- or Cy5-labeled protein targets—the simplest and least expensive methods have relied on enzymatic staining. We have used both biotin and antibodies to provide convenient molecular tags for our protein targets to identify the beads that have bound them. Both of these will be reviewed in detail here using illustrative test cases. These cases are the catalytic subunit of protein kinase A (PKA-C) and the protease called clotting factor Xa. For most screening projects, the concentrations of protein targets, antibodies, and conjugates with alkaline phosphatase are generally within about fivefold of the concentrations given in these protocols and are fairly robust across several different target types *(5)*. All steps are performed at room temperature (RT, generally 20–25°C) unless specified.

1.2. Screening for Ligands Using a Monoclonal Antibody to the Protein Target: Xa

Polyclonal antibodies have been used successfully; however, we have only a limited amount of experience with these. The specificity of the antibody handle becomes more critical as less pure preparations of target protein are used. For example, lysates of membrane preparations have been used with success as a source of the target protein provided a sufficiently selective antibody is available. Clotting factor Xa is the example here because it is the best understood system at Selectide. Screening using biotinylated Xa has been reported already *(6)*.

1.3. On-Bead Libraries for the Discovery of Substrates for Proteases Using Pure Enzyme: Escherichia coli Leader Peptidase

Proteases have proven to be excellent drug targets. Based on the experiences from angiotensin converting enzyme to HIV protease inhibitors, many therapeutics will be targeted to proteases found from analysis of microbial and human genomes. Although BLAST analyses will often place a candidate gene within certain families (e.g., aspartyl proteases), it is still sometimes difficult to find satisfactory surrogate substrates *a priori*. A powerful approach to rapidly discover substrates is described here *(7)*. It is based on the principle of Fluorescent Resonance Energy Transfer, FRET (**Fig. 1**). In this case, Lucifer yellow is paired with Dabsyl in a FRET quenching strategy to discover peptide sequences that can be cleaved to disrupt the FRET quenching phenomenon, and the beads with such substrates become fluorescent. We expect that the scope of this approach should include other hydrolases and substrate classes amenable to combinatorial methods and structural determination. The method described here uses a purified protease for screening.

1.4. On-Bead Libraries for the Discovery of Substrates for Proteases from Unpurified Cellular Lysates: Napsin A

There are times and circumstances where purified proteases are not available or desirable for the research strategy. An important adaptation of the

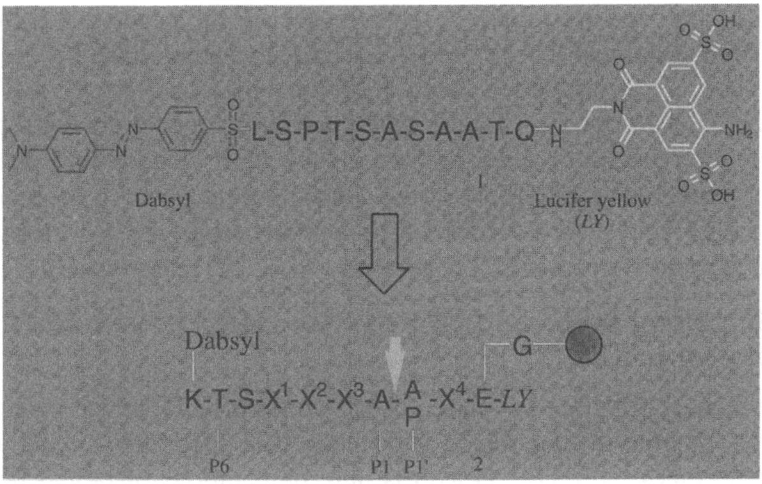

Fig. 1. A substrate for *E. coli* leader peptidase is shown, **1**, with a quenching Dabsyl group and a fluorescent lucifer yellow attached suitable for FRET-based assay. This substrate was a template for design of a peptide library as shown in structure **2**. The arrow indicates the position of cleavage for the peptidase between Ala-Ala/Pro. X represents randomized positions in which any of 19 different natural amino acids are coupled (all except Cys). Note that Dabsyl is on the epsilon amino group of the N-terminal lysyl residue, K.

Meldal *(7)* proteolytic screening approach is the use of crude lysates from genetically engineered HEK293 cells producing a protease encoded by an expression plasmid *(8,9)*. An example is provided for a genomically derived protease—napsin A.

1.5. Discovery of Surrogate Substrates for Protein Kinases Using OBOC Peptide Libraries

The human genome is projected to encode a large number of protein kinases, which serve important roles in signal transduction required for a variety of cellular processes such as differentiation, growth, division, and reaction and adaptation to external stimuli. A growing number of human diseases are known to involve mutations, overproduction, or inappropriate expression of protein kinases or their associated regulators and effectors. There are many protein kinases being pursued as therapeutic targets by the pharmaceutical industry. Presumably, many more will emerge in the near future with the completion of sequencing of the human genome. In addition, protein kinases from many pathogenic organisms are pharmaceutical targets. The hope is that combinatorial chemistry will be able to search pharmacophoric space with sufficient resolution as to yield reasonably selective inhibitors in a target family famous for

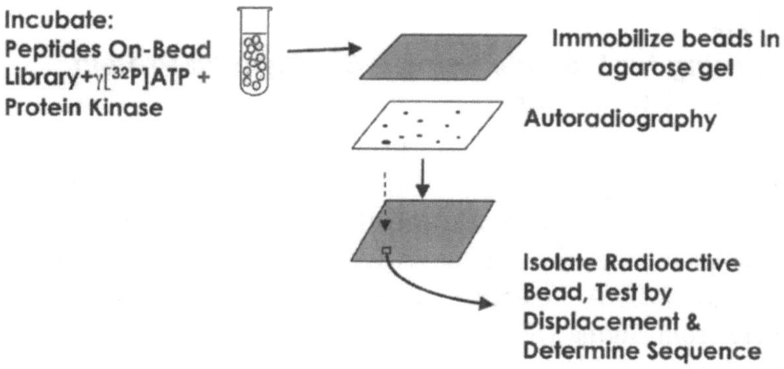

Fig. 2. Schematic depicting the method to discover substrates for protein kinases. Beads with unique peptide sequences are incubated with a protein kinase and $\gamma[^{32}P]$ATP. After washing, the beads are spread out in agarose and autoradiographed. Radioactive beads are isolated, tested by quantitative methods, and the presumptive phosphopeptides are sequenced.

its extreme degree of conservation *(10)*. Interestingly, in spite of extensive sequence homology within the protein kinase family, they retain remarkable specificity for their protein substrates. This may prove to have important therapeutic consequences and suggests alternate strategies for making inhibitors. A major challenge in the study of protein kinases is how to identify their specific substrates, both for the purpose of assay development and multiplexing, and for fragment analysis to develop specific kinase inhibitors by medicinal and combinatorial chemistry approaches. A number of combinatorial methods have been developed to identify protein kinase substrates *(11)*. Some methods rapidly find motifs for protein kinases, however, we prefer to rapidly obtain distinct sequences. The technology we employ is based on the method first described by Lam and coworkers *(12)*, and it is schematized in **Fig. 2**. The protocol described in this chapter was validated for cAMP dependent protein kinase (also known as protein kinase A, or PKA) and a proprietary protein kinase. In the case of PKA, the expected sequence motif was found: RRXS *(13)*.

1.6. Cellular Bioassay of Soluble Combinatorial Libraries for Putative GPCRs Using FLIPR Technology

Most drugs act through G-protein coupled receptors (GPCRs). Estimates vary wildly; however, the human genome is projected to encode about 1000 different GPCRs. Many methods are available for finding surrogate agonists for putative GPCRs discovered through genomics *(11)*. Combinatorial libraries are being screened to both find surrogate ligands for these genomically derived targets and discover leads for drug development.

2. Materials

2.1. Screening for Ligands Using a Biotinylated Protein as Probe: PKA-C Subunit

Conjugated Streptavidin-Alkaline Phosphatase (SA-AP; cat. no. 21323), BCIP (cat. no. 34040), and EZ link NHS-LC-biotin (cat. no. 21335) were from Pierce Chemical Company (Rockford, IL). Sometimes, conjugated avidin-alkaline phosphatase (Sigma, cat. no. A-2527; 1:1 ratio of avidin and enzyme) is used instead of SA-AP. General reagents such as glycine and dithiothreitol (DTT) were from Sigma (St. Louis, MO) including the catalytic subunit of bovine cardiac protein kinase A (cat. no. P-2645). TBS is 150 m*M* NaCl and 50 m*M* Tris, pH 7.5. HBS is 150 m*M* NaCl and 50 m*M* Hepes, pH 9. A convenient biotinylation kit is available from Boehringer-Mannheim (cat. no. 1418-165), and it includes desalting columns. Alternatively, NAP 5 desalting columns (cat. no. 17-0853-02) are available from Pharmacia Biotech.

2.2. Screening for Ligands Using a Monoclonal Antibody to the Protein Target: Xa

Conjugated goat anti-mouse alkaline phosphatase, GAM-AP, was from American Qualex (San Clemente, CA, cat. no. A106AN). BCIP was from Pierce (Rockford, IL). Human clotting factor Xa was from Calbiochem (San Diego, CA, cat. no. 233526). The anti-Xa mAb cat. no. 5295 was from American Diagnostica (Greenwich, CT). All steps are performed at room temperature (RT) (20–25°C) unless stated otherwise.

2.3. Substrate Library Screen Using E. coli Leader Peptidase

The resin used to make the library was $PEGA_{1900}$ from Polymer Laboratories, Shropshire, UK (*see* **Note 1**) with a loading of 0.2 mmol/g, batch cat. no. PEGA34. *E. coli* leader peptidase was cloned and purified using modifications of published procedures *(8)*. The specific activity of the leader peptidase was not known. A stock solution of *E. coli* leader peptidase in 20 m*M* Tris-HCl buffer, pH 7.4, 5 m*M* MgC12, 0.5% Triton X-100, and 50% glycerol was used. General reagents such as morpholino ethane sulfonic acid (MES) were from Sigma (St. Louis).

2.4. Substrate Library Screen for Napsin A Using an Unpurified Cellular Lysate

Even though library **2** (**Fig. 1**) was designed for *E. coli* leader peptidase, it was found to be a useful library for screening against other proteases. In the example given, it was used to screen for napsin A substrates, a mammalian protease expressed in kidney and lung *(8,9)*. The protease inhibitors E64 and phenylmethylsulfonyl fluoride (PMSF) were from Boehringer Mannheim (Germany). HEK293 cells were from American Type Culture Collection (Manassas, VA). General reagents such as ethylene diaminoacetic acid (EDTA) were from Sigma (St. Louis).

2.5. Discovery of Surrogate Substrates for Protein Kinases Using OBOC Peptide Libraries

For putative Ser/Thr protein kinase deorphaning, we use a 7-mer peptide library of the following formula: X_3-S/T-X_3-M-TentaGel-130 kept as separate Serinyl or Threonyl pools. X positions were randomized with a proprietary set of amino acids. Methionine (M) is used as a linker to enable cleavage by cyanogen bromide to provide a soluble peptide suitable for sequencing by appropriate microanalytical methods (e.g., mass spectrometric fragmentation or Edman chemistries with the peptide still attached to the bead, whichever is more useful). The loading of the Tentagel-130 resin (Rapp Polymere GmbH, Tubingen, Germany) was 300 pmol/bead. PBS, 10X stock was from Boehringer Mannheim (cat. no. 1666-789). TM is 50 m*M* Tris, pH 7.5 and 10 m*M* $MgCl_2$. A Geiger counter was from Ludlum (Sweetwater, TX; model 3-98) equipped with a NaI probe (Ludlum, model 44–3).

2.6. Fluo-4 Method for FLIPR Assay to Detect Calcium Mobilization: Proprietary GPCR

Fluorometric Imaging Plate Reader (FLIPR) was from Molecular Devices (Sunnyvale, CA). Common tissue culture reagents were from Gibco/BRL (Rockville, MD). The 384-well plates used were from Costar (Cambridge, MA; cat. no. 3712).

1. *Compound and Wash Buffer:* 500 mL Hands' balanced salt solution (HBSS), 5 mL (1% FCS; varies with cells or surrogate ligand used in the screen; e.g., 01% FCS, BSA, or nothing), 10 mL 1 *M* Hepes, 5 mL of 250 m*M* probenecid (*see* next item—make fresh probenicid for each experiment).
2. *Probenecid:* Make fresh. Dissolve 710 mg in 5 mL of 1.0 *M* NaOH, and add 5 mL compound buffer. The concentration of this stock solution is 250 m*M* (the working concentration in the presence of the loading dye is 2.5 m*M*).
3. *Fluo-4 dye/pluronic acid solution:* Combine equal volumes of the following two stock solutions:
   a. Fluo-4 AM ester (Molecular Probes F-14201), 2 m*M* stock in anhydrous DMSO made fresh.
   b. 20% (w/v) pluronic acid (cat. no. F127, Molecular Probes, Eugene, OR) in anhydrous DMSO. Dissolve the pluronic acid at 37°C, cool to RT prior to use. The pluronic acid stock solution can be stored at RT.
4. *Loading dye solution:* Make the dye loading solution immediately prior to use.

Dilute Fluo-4 dye/pluronic acid to a working solution of 8 µ*M* dye (a dilution of 1 in 125).

5. *Agonists and compound plates:* These vary with the experiment being done. We have compounds already formatted into 384-well templates with appropriate columns free for control incubations.

3. Method

3.1 Screening for Ligands Using a Biotinylated Protein as Probe: PKA-C Subunit

3.1.1. Biotinylation of the Protein Target

1. Dissolve 1.86 mg of NHS-LC-biotin in 62 µL of dimethyl formamide (for a final stock of 30 µg/µL).
2. Make a stock of glycine at 100 mg/mL in TBS to help quench the biotinylation reaction.
3. Dissolve 3 mg of PKA-C lyophilized residue (6 µg of enzyme) in 100 µL of 6 mg/mL DTT (other sorts of protein targets need to be in a buffer lacking free amines—dialyze or desalt the protein accordingly).
4. To the PKA-C solution, add 2 µL (60 µg) of NHS-LC-biotin and react on ice for 1–2 h.
5. Quench the reaction by adding 50 µL of TBS and 1 µL (100 µg) of 100 mg/mL glycine in TBS and incubate on ice for 1–2 h.
6. Add 100 µL of 10 mg/mL BSA in TBS to provide a carrier for the desalting step.

3.2. Desalting the Biotinylated Protein

1. Equilibrate a G-25 Sephadex desalting mini-column with TBS, then with 5 mM DTT in TBS.
2. Apply the 0.25 mL PKA-C/BSA solution to the top of the desalting column. Let it pass into the matrix of the column.
3. Slowly add 5 mM DTT in TBS to the top of the column, and hand collect approx 0.35 mL fractions in premarked Eppendorf tubes (usually only 10 fractions are needed). The biotinylated protein usually comes out in the void (breakthrough) volume.
4. Identify the fractions containing protein by either UV (1 = 280 nm) measurement of aliquots or by Bradford reaction (Pierce) with aliquots of the fractions.
5. Estimate the approximate concentration of the protein target, which should now be biotinylated (see Note 2).

3.3. Screening

1. Make a stock solution of 50 mg/mL BCIP in DMF. Keep the solution dark by wrapping the container in aluminum foil. This stock solution is stable at 4°C for at least 3 mo.
2. Swell the library beads in 0.1% Triton X-100 in TBS.
3. Sonicate the swelled resin briefly to disperse the beads thoroughly.

4. Discard the solvent.
5. Block the beads using 0.1% NFDM, 0.1% Triton X-100, 0.01% gelatin, 0.1 mg/mL BSA in TBS.
6. Wash the beads with TBS.
7. Estimate the concentration of the SA-AP (or Avidin-AP) stock you are using by dividing the protein mass concentration by the summed molecular weights of SA and AP (conjugate MW = 60 kDa + 120 kDa = 180 kDa). The manufacturer's stock solutions are usually about 5 μ*M*.
8. Probe the library using 70 n*M* biotinylated PKA-C subunit and 20 n*M* SA-AP premixed (10–15 min in advance) in TBS supplemented with 0.1 mg/mL BSA and 20 m*M* $MgCl_2$. We generally try to achieve a 4:1 ratio between biotinylated target and streptavidin, which is believed to enhance sensitivity by an avidity effect.
9. Incubate for 1 h with gentle agitation.
10. Wash the beads with TBS five times.
11. Develop color by adding a developer solution to the beads that consists of 165 μg/mL BCIP in HBS. Color development usually takes about 10–20 min at RT with constant agitation. Observe the progress of the color reaction by gross inspection and by sampling aliquots of beads and checking them with a dissecting microscope. Let experience guide the determination of when the reaction has gone to useful completion, or employ control incubations (*see* **Note 3**).
12. Terminate color development by adding 0.25 vol of glacial acetic acid to inactivate the alkaline phosphatase and avoid continued staining (ultimately all of the beads can become stained, so development has to be terminated).
13. Collect stained beads by spreading them in Petri dishes with 0.01% Triton X-100 in TBS and aspirating them with hand pipets (10–200 μL capacity tips) while viewing with a dissecting microscope.
14. Specific interaction with the biotinylated target protein is established in various ways. Isolated stained beads are stripped by washing with 50 m*M* NaOH (to remove bound and aggregated protein) followed by washing with DMF (to remove deposited dye). First, if a high affinity ligand is available, reprobe the beads as before in the presence of the soluble, competing ligand ***(14)***. Second, if no ligand is known—common with genomically derived targets of pharmaceutical interest—specific interaction with the target is established by reprobing the beads in the absence of the biotinylated target protein. Third, to control for simple target-specific but nonpharmacological interactions with beads (e.g., ion exchange or nonspecific hydrophobic interactions), denature the biotinylated target protein (e.g., by heat, chemical modification, solvent effects, etc.), and reprobe the beads with the denatured target ***(6)***. If beads have ligands for a pocket present in the native protein, but not in the denatured or inactivated target, then colorless beads should be observed. In each of the three cases described above, the specifically reactive beads are the colorless ones after reaction with BCIP. Nonspecific binding beads will be stained. Collect the colorless, specific beads.

15. It is important to verify reproducible staining behavior, so expect to strip and reprobe the beads a few times. A colorless bead after a specificity test should still bind and stain in the presence of the native, biotinylated target in the absence of a competing ligand. Reproducibility is an essential key to successful performance of the bead staining approach.
16. For general precautions and advice, *see* **Note 4**.

3.4. Screening for Ligands Using a Monoclonal Antibody to the Protein Target: Xa

1. Make a stock solution of 50 mg/mL BCIP in DMF. Keep the stock dark using aluminum foil. It is stable for at least 3 mo when stored at 4°C.
2. Swell the library beads in 0.1% Triton X-100 in TBS.
3. Sonicate the swollen resin briefly to disperse beads thoroughly.
4. Discard the solvent.
5. Block the beads using 0.1% NFDM, 0.1% Triton X-100, 0.01% gelatin, 0.1 mg/mL BSA in TBS.
6. Wash the beads with TBS.
7. Estimate the concentration of the GAM-AP stock you are using by dividing the protein mass concentration by the summed MWs of GAM and AP (conjugated MW = 150 kDa + 120 kDa = 270 kDa).
8. Probe the library using 16 nM Xa, 8 n*M* mAb cat. no. 5295, and 8 n*M* GAM-AP premixed (10–15 min in advance) in TBS supplemented with 0.35 *M* NaCl (final concentration), 5 m*M* $CaCl_2$, 0.1% NFDM, 0.1% Triton X-100, 0.01% gelatin, 0.1 mg/mL BSA.
9. Incubate for 1 h with gentle agitation.
10. Wash the beads with TBS five times.
11. Develop color by adding 165 μg/mL BCIP in HBS. Color development usually takes about 10–20 min at RT with constant agitation. Observe the progress of the color reaction with a dissecting microscope and gross inspection. Let experience guide the determination of when the reaction has gone to useful completion, or employ control incubations (*see* **Note 3**).
12. Terminate color development by adding 0.25 vol of glacial acetic acid to inactivate the alkaline phosphatase and avoid continuing staining (ultimately all of the beads can become stained, so development has to be terminated).
13. Collect stained beads by spreading them in Petri dishes with 0.01% Triton X-100 in TBS and aspirating them with hand pipets (10–200 μL capacity tips) while viewing with a dissecting microscope.
14. Evaluate the specificity of staining as described for the streptavidin case using PKA-C in the previous section. For Xa, specificity was evaluated using chemically inactivated enzyme ***(6)***. Specificity also was evaluated using high affinity ligands ***(5,14)***. Finally, just as with biotinylated targets, reproducibility is critical to successful performance of histochemical screening methods, so repeat the staining and specificity steps several times to find true ligands.
15. For general precautions and advice, *see* **Note 4**.

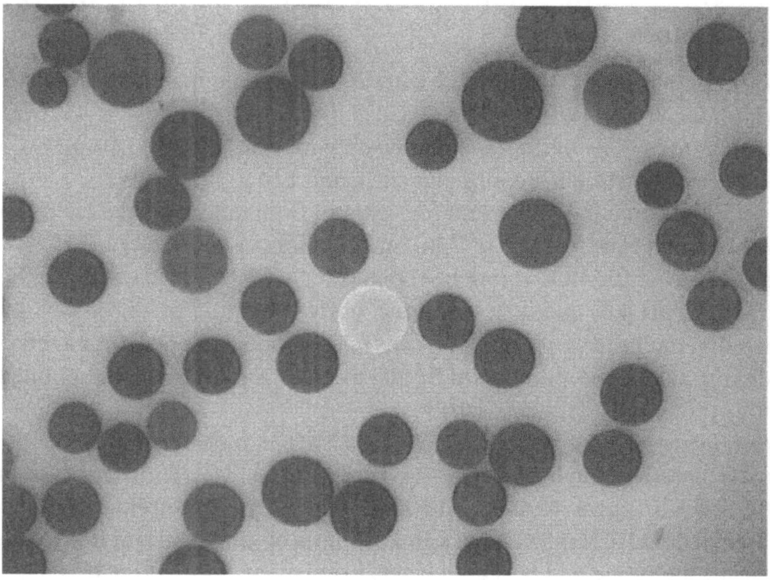

Fig. 3. Fluorescent photomicrograph of beads revealing one fluorescent bead (near the center of the field) obtained after digestion of its peptide had released the quenching Dabsyl moiety.

3.5. Substrate Library Screen Using E. coli Leader Peptidase

1. Wash 25,000 beads of library **2** (**Fig. 1**) in a fritted syringe with 100 m*M* Bis-(Tris)-propane buffer, pH 8.5.
2. Add 100 m*M* Bis-(Tris)-propane buffer, pH 8.5, 50 m*M* $MgSO_4O$ and the enzyme preparation (80 µg/mL) to the resin in a final volume of 4 mL.
3. After shaking for 18 h at RT, remove the enzymatic solution.
4. Wash the beads extensively with 100 m*M* Bis-(Tris)-propane buffer, pH 8.5 followed by 100 m*M* MES buffer, pH 4.8.
5. Examine the beads with a fluorescent microscope equipped with a red filter ($\lambda_{50\%\ \text{Transmission}}$ = 605 nm) .
6. In the case of *E. coli* leader peptidase, 45 brightly fluorescent beads were isolated and analyzed by Edman chemistry sequencing (**Fig. 3**).

   The expected sequence of AXAA was obtained. Of 45 beads, 39 were cleaved between the Ala-Ala. The remaining 6 beads were cleaved between X_3-Ala, in which case Ala was found in the X_3 position. Thus, essentially all of the isolated beads were satisfactory substrates for the leader peptidase. There was modest enrichment of certain amino acids in some of the positions, as expected. Resynthesis of 7 of the 39 sequences selected at random yielded surrogate substrates with a range of relative catalytic efficiencies from 0.3- to 19-fold compared to the known substrate **1** (**Fig. 1**). Two of the substrates were 11- and 19-fold better *(8)*.

3.6. Substrate Library Screen for Napsin A Using an Unpurified Cellular Lysate

1. Wash 25,000 beads of library **2** with 200 m*M* sodium acetate (NaOAc) buffer, pH 4.0 (*see* **Note 1**).
2. Wash with NaOAc buffer supplemented with 1 m*M* EDTA, 10 µg/mL protease inhibitors E64 (Boehringer Mannheim), and 170 µg/mL PMSF.
3. Add 500 µL of a 0.4 mg protein/mL extract from untransfected HEK293 cells to the resin in a final volume of 2 mL and add this to the library.
4. After shaking for 18 h at room temperature, wash the beads with NaOAc buffer (200 m*M*, pH 4.0) and then with MES buffer (100 m*M* pH 4.8).
5. Examine the beads with a fluorescent microscope equipped with a red filter.
6. Remove any of the fluorescent beads, as these represent the uninhibited endogenous proteolytic activity of the lysate.
7. Wash the remaining beads with 200 m*M* NaOAc buffer, pH 4.0.
8. To 200 m*M* NaOAc buffer, pH 4.0, 1 m*M* EDTA, 10 µg/mL protease inhibitors E64, and 170 µg/mL PMSF, add 500 µL of a 0.4 mg protein/mL extract of napsin A transfected HEK293 cells to a final volume of 2 mL and add this to the library.
9. Discard the lysate and wash the beads extensively with 200 m*M* NaOAc buffer, pH 4.0 (*see* **Note 5**).
10. After washing with 100 m*M* MES buffer pH 4.8, inspect the beads under the fluorescent microscope. For napsin, 21 bright beads were isolated and submitted for Edman sequencing.

    Twelve beads were cleaved at one site. Considering the fact that a cellular lysate was being used with multiple endogenous proteases, it is not surprising that some of the peptide sequences were recognized and processed several ways. Of these 12 peptides, 3 were cleaved one residue amino to the AA/AP of the *E. coli* leader peptidase sequence (**Fig. 1**). The peptides were resynthesized, and cleavage was found to occur between Leu-Met or Phe-Phe *(8)*. One of the peptides was used to monitor large-scale production of the protease, and it was used successfully in a high-throughput screening campaign *(8)*. In conclusion, this method is a rapid technique to obtain surrogate substrates for proteases in a matter of weeks.

3.7. Discovery of Surrogate Substrates for Protein Kinases Using OBOC Peptide Libraries

Washes are generally of about 10 mL volume added each time. When discharging the waste wash fluid, prevent crushing of beads by drawing a little air into the syringe to remain between the resin and the plunger. Each wash lasts about 3 min with constant agitation. All steps are done at room temperature (20–25°C) unless specified.

1. Weigh out 1 g of 7-mer Serinyl peptide library (approx 800,000 beads). Peptide libraries on TentaGel are stable for at least 2 yr when stored dry at 4°C.

2. Put the resin into a polyethylene fritted 20 mL polypropylene syringe.
3. Draw up approx 10 mL of water and swell the beads for 1 h at RT.
4. Wash the resin three times with water.
5. Draw up approx 10 mL of 0.1% Triton X-100 in water and cap the syringe.
6. Disperse the beads thoroughly by sonicating in a sonication bath for 2 min.
7. Incubate with shaking for 30 min.
8. Wash the resin three times with water.
9. Block the beads with 10 mL of 0.5 mg/mL BSA in TBS for 1 h.
10. Wash the resin three times with 50 m*M* Tris HCl, pH 7.5.
11. Wash the resin three times with 10 m*M* $MgCl_2$, 50 m*M* Tris HCl, pH 7.5.
12. Resuspend the beads in 5 mL of 10 m*M* $MgCl_2$, 50 m*M* Tris HCl, pH 7.5 (TM buffer).
13. Push out approx 1 mL of TM buffer into an Eppendorf tube. This will be used to help draw up the enzyme and radiolabel in the next step.
14. Premix 100 µL of putative protein kinase (approx 40 µg of the purified protein) with 0.5 mCi (50 µL) of γ[^{32}P]ATP with a specific activity of 6000 Ci/mmol (enzyme and label premix).
15. Attach a needle to the syringe. With a needle (**caution!** radioactive material and sharps hazards) draw up the enzyme and label premix into the syringe. Then draw up the approx 1 mL of TM buffer as a vehicle to ensure quantitative transfer of the enzyme and label from the needle and into the syringe. Note that insoluble enzyme preparations will not pass through the polyethylene frit of the syringe.
16. Mix and incubate the reaction with constant gentle agitation at RT (or 30°C) for 2 h (or longer).
17. Carefully discard the labeling mixture according to regulations.
18. Wash the resin three times with 50 m*M* Tris HCl, pH 7.5.
19. Wash the resin 11 times with 0.5 *M* NaCl, 50 m*M* Tris HCl, pH 7.5. It is useful to start monitoring unincorporated counts after the fourth wash by using different disposable 50 mL tubes and a Geiger counter equipped with a NaI probe.
20. After the fourth wash, monitor the washed counts using the hand probe (NaI probe). When the change in counts eluted tapers off, it is time for a different wash procedure. For example, the counts following the first three washes went like this (all in thousands of cpm): 39, 34, 25, 22, 13, 14, 14, 12.
21. Wash five times with 10X PBS. Counts eluted tracked like this (all in thousands of cpm): 60, 60, 24, 17, 10. At this point, the counts in the resin were 260,000 cpm, so additional washes would not be particularly helpful. Also, this extent of radioactivity approaches what is useful for finding substrates.
22. Wash three times with 0.5 *M* NaCl, 50 m*M* Tris HCl, pH 7.5.
23. Displace much of the exchangeable counts (we observed nearly half are displaced at this step) with ATP as follows: Add 8 mL of 1 m*M* ATP in 0.5 *M* NaCl, 50 m*M* Tris HCl pH 7.5 to the resin. Mix and set at 4°C overnight to several days (this is a very flexible step). The key is that truly phosphorylated beads are covalently labeled and fairly stable in this well-defined and thoroughly washed system.
24. Discard the displaceable counts.

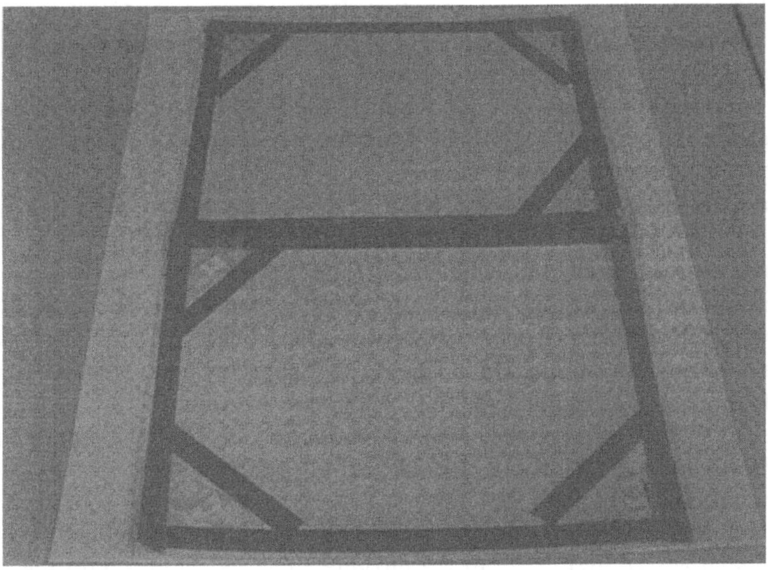

Fig. 4. Agarose and beads are spread out on acetate sheets (used for making transparencies). Borders are demarcated with masking tape, and this reduces the chance of molten agarose from running off of the sheet. Underlying absorbant pads and a level surface are important. Radioactive label (we use excess ^{32}P-ATP) diluted with ink is spotted onto Whatman paper strips and dried. These are taped onto the acetate sheets and used for orientation spots. After the agarose sets, each acetate sheet is covered with Saran wrap for autoradiography.

25. Wash the resin three times with 10X PBS. The counts in the resin should now be about 130,000–180,000 cpm.
26. Wash once with water.
27. Resuspend the resin in a small volume of water (approx 2–4 mL added) and transfer 0.5–1 mL slurries to a 50 mL tube.
28. To the aliquot of beads, add 1.5% agarose to about 20 mL total volume.
29. Seal the tube with a cap, mix with gentle inversion, and pour the molten agarose and beads onto a pre-taped acetate sheet resting on a level surface.
30. Slowly and gently manipulate the acetate sheet to spread out the molten agarose.
31. Repeat these steps ***(28–30)*** until all of the resin has been spread out in agarose on sheets of acetate.
32. Let the agarose solidify (approx 30 min).
33. Attach colored, radiolabeled markers onto the acetate sheet in distinctive patterns to facilitate localization of radiolabeled beads (**Fig. 4**).
34. Wrap the acetate sheet in Saran Wrap.
35. Perform autoradiography (1–5 h at RT; using 35 × 43 cm Biomax MS film, Kodak, Rochester, NY, cat. no. 143-5726) in film casettes fitted with intensifying screens.

36. Develop the film.
37. Using the reference spots, align the acetate sheet with the agarose embedded beads on top of the autoradiogram (**Fig. 5A**).
38. Confirm alignment of beads to spots on the film using a dissecting microscope.
39. Using a tapered spatula, excise a generous portion of the agarose and beads corresponding to each of the best autoradiographic spots (**Fig. 5B**). Combine these fragments of agarose in an Eppendorf tube with 0.2 mL of water.
40. Melt the agarose at 65°C in a heating block.
41. Combine the melted agarose and beads with 28 mL of 1.5% agarose in water and spread them on another sheet of taped acetate on a level surface.
42. Repeat autoradiography and bead excision as described above until only one bead corresponds to each autoradiographic spot (**Figs. 6A** and **B**).
43. Isolated radiolabeled beads are then individually excised and placed into separate labeled Eppendorf tubes containing 0.2 mL of water.
44. Add 0.8 mL of water to each Eppendorf with isolated beads and melt the agarose at 65°C.
45. Briefly spin the tubes in a low-speed microcentrifuge to ensure that each bead is at the bottom.
46. Carefully aspirate 0.9 mL of supernatant water.
47. To the bead in 0.1 mL, add 0.1 mL of 2 m*M* ATP in water. This will displace noncovalently bound radiolabel from the bead.
48. Incubate with ATP overnight (or for a few days, a flexible step).
49. To each Eppendorf tube, add 0.8 mL of 0.5 *M* NaCl, 50 m*M* Tris, pH 7.5, mix and incubate at RT for 30 min.
50. Spin the tubes briefly at low speed to ensure the beads are at the bottom of the tubes.
51. Carefully transfer 0.9 mL of the supernatant fluid to another labeled Eppendorf tube (this is the competed radiolabel).
52. Put the tubes into vials suitable for Cherenkov counting in a liquid scintillation counting machine (**Fig. 7**) and measure Cherenkov radiation (wide open window, no scintillant is added to the beads).
53. Compile the cpm of each bead in 0.1 mL buffer compared to the cpm of each bead's competed supernatant fluid. Express the cpm displaced as a percentage of the total radiolabel associated with the bead [i.e.,% displaced = 100 × cpm in supe/(cpm in supe + cpm in bead)]. Prioritize beads according to the extent of labeling of the bead balanced by the lowest percentage of displaceable counts. Data from an experiment using a model Arginyl-X_2-Serinyl library are shown in **Fig. 8**. Beads that have 20% of radiolabel displaced by cold ATP are likely not to be true protein kinase substrates. In the screening data shown for a proprietary protein kinase (**Figs. 4–6**), we obtained 17 covalently labeled phosphopeptides. This gave a covalent hit rate of 100×(17/800,000) = 0.002%, which is fairly selective.
54. Determine the sequence of the presumptively phosphorylated peptides by Edman sequencing or other microchemical technique.

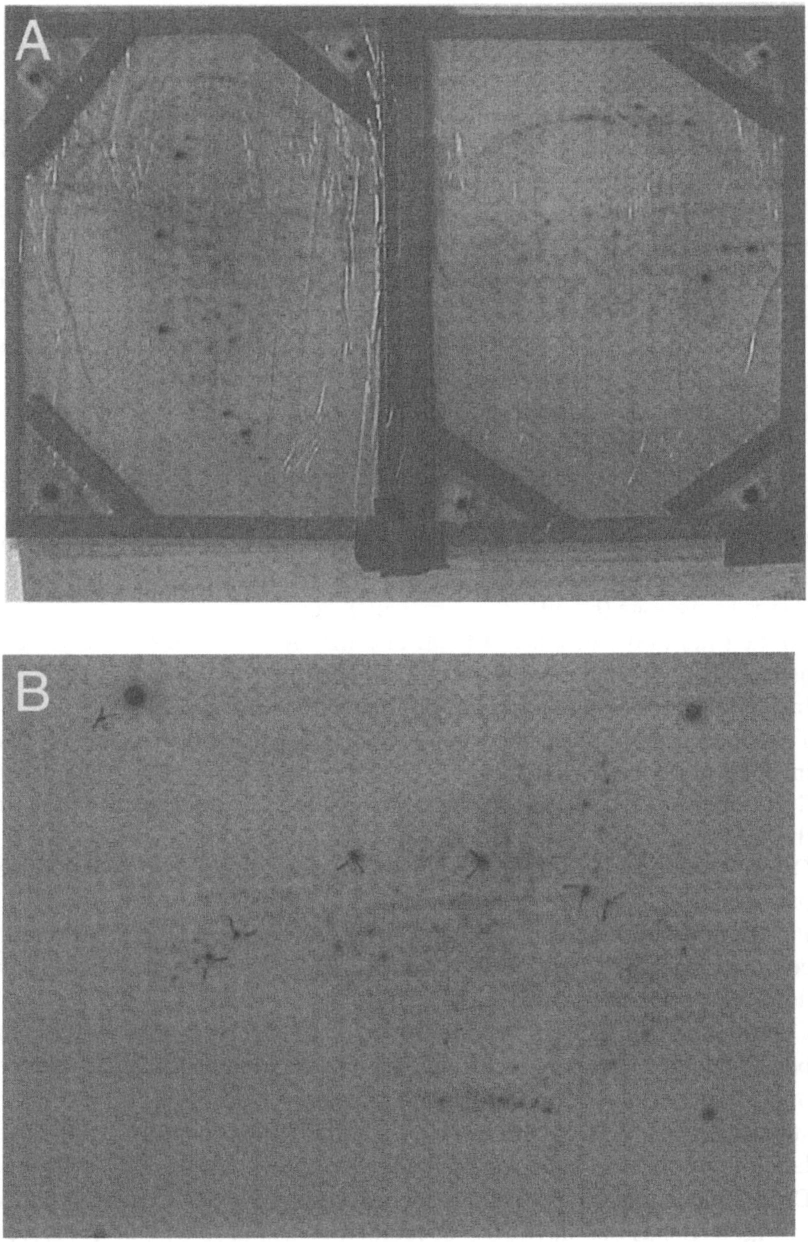

Fig. 5. Panel A, exposed autoradiogram is on top of two acetate sheets with radioactive beads embedded in agarose. After removing the Saran Wrap, the areas corresponding to the best labeled beads are cored out of the agarose for spreading again for isolation of distinct labeled beads. Panel B indicates what signals correspond to the best extent of labeling in this particular spread of library beads.

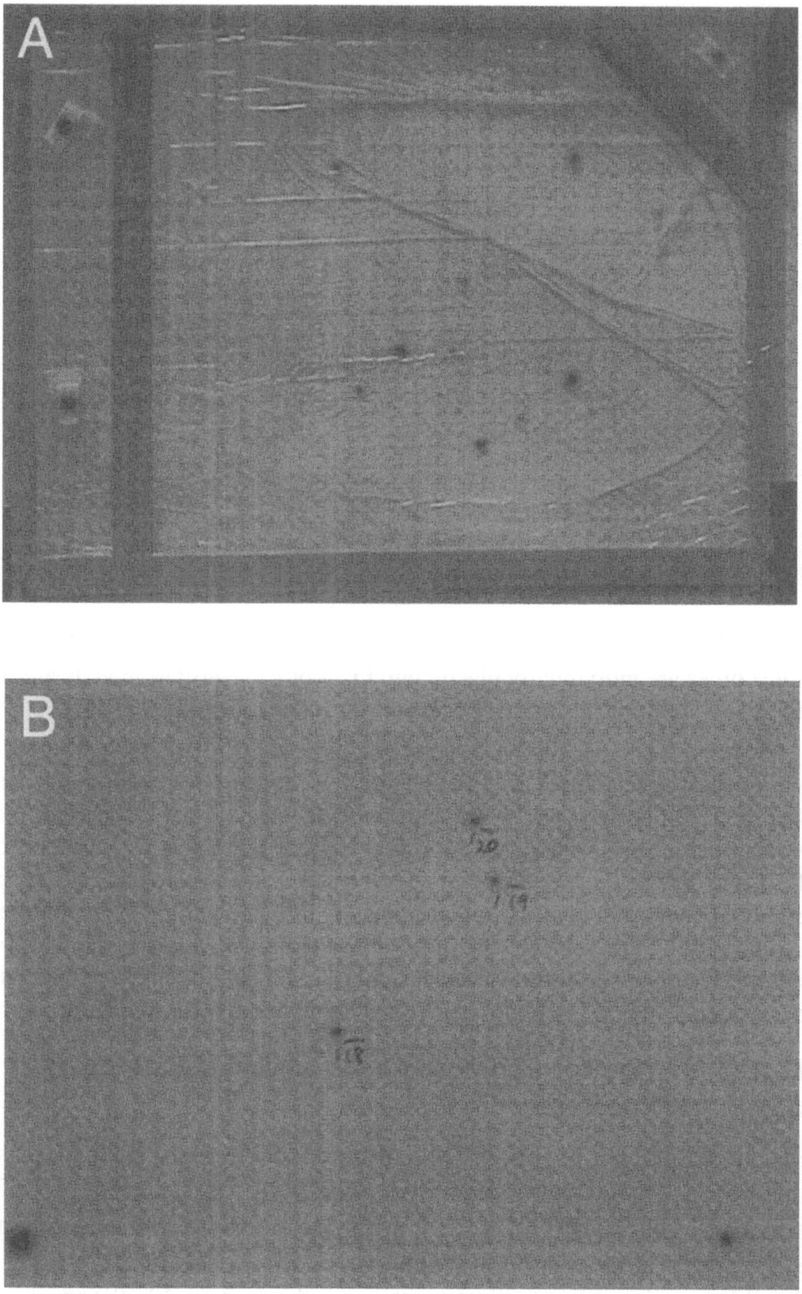

Fig. 6. Panel A, exposed autoradiogram is on top of an acetate sheet with beads spread out in agarose for the second time. Panel B, illustrates the much-reduced level of background signals on the film at this stage. The two spots along the bottom are reference marks.

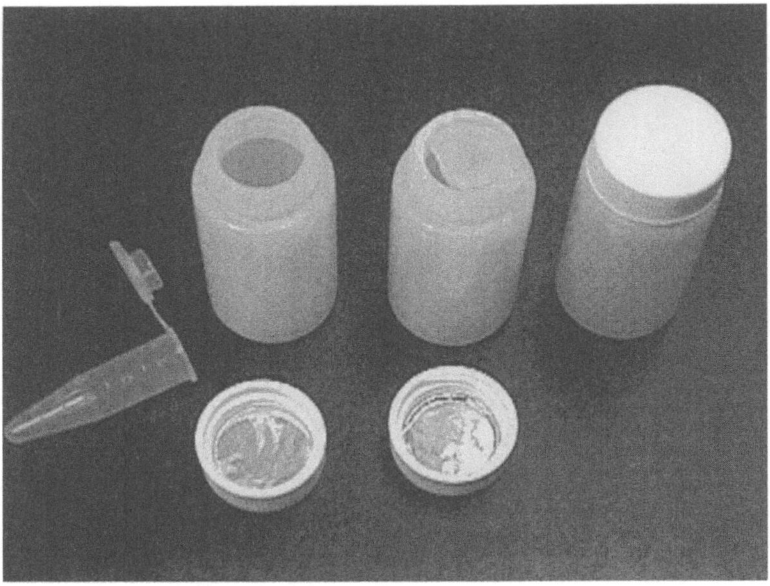

Fig. 7. Isolated radioactive beads and their competed supernatant fluids in separate Eppendorf tubes are individually placed into the tops of scintillation vials and capped for Cherenkov counting.

55. Verify that the sequences are true substrates by appropriate experiments showing that the resynthesized peptide in solution is labeled with $\gamma[^{32}P]$ATP in the presence of the protein kinase. Appropriate deletions (e.g., des-Ser, des-Thr, or des-Tyr) also should be used as controls. Detection of phosphopeptide is generally observed using thin layer chromatography (TLC) and autoradiography methods. Co-migration of radiolabeled peptide with synthetically prepared phosphopeptide should be satisfactory proof of successfully finding a new substrate and its phosphorylated product. For additional information and advice, *see* **Note 6**.

3.8. Fluo-4 Method for FLIPR Assay to Detect Calcium Mobilization: Proprietary GPCR

Cells are maintained in appropriate culture media (e.g., for CHO cells, Hams F-12 medium is often used, Gibco-BRL, Rockville, MD, cat. no. 11059-029, supplemented with 10% FCS), with appropriate antibiotic and selection agents in the case of engineered cell lines. Endogenous receptors or transiently transfected cell lines, of course, do not generally need selection. Stocks are described in detail in the case of CHO cells.

1. Plate out cells in 384-well black plates with clear bottoms (Costar cat. no. 3712), such that cells are confluent on the day of the assay (e.g., 8,000–10,000 cells/20 µL and then incubated overnight). To prevent edge effects, do not stack the plates.

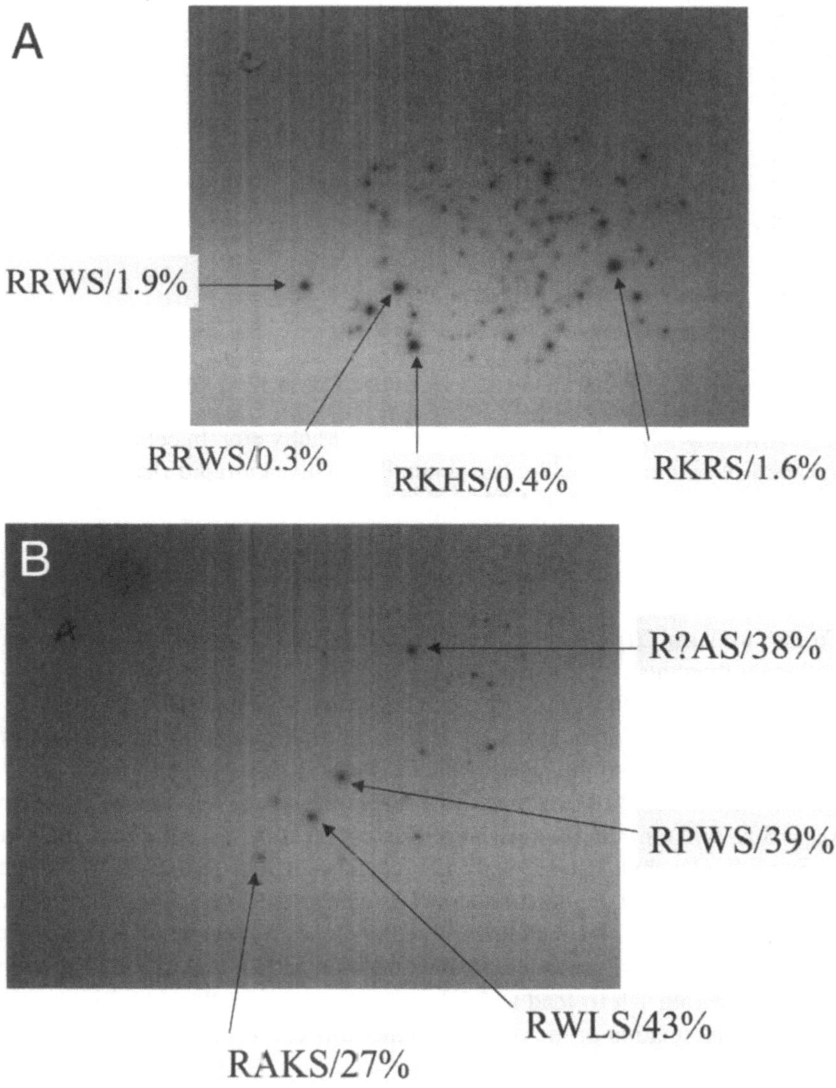

Fig. 8. Autoradiograms of a model library of structure R-X_2-S-X_3-M-TentaGel incubated with γ[^{32}P]ATP either with (Panel A) or without (Panel B) the catalytic subunit of protein kinase A (PKA-C). Only the first four Edman cycles were performed for this experiment. Side-by-side spreads of the beads and exposure on a single film for 3 h at RT are shown. The sequences of labeled beads are indicated along with the percentage of the total radiolabel competed off by nonradioactive ATP. It is clear that considerable amounts of label can associate noncovalently with diverse sequences, so competition and quantitative methods are important to successful screening. The substrate motif for PKA-C is RRXS *(13)*. "?" represents a sequencing cycle with a proprietary unnatural amino acid.

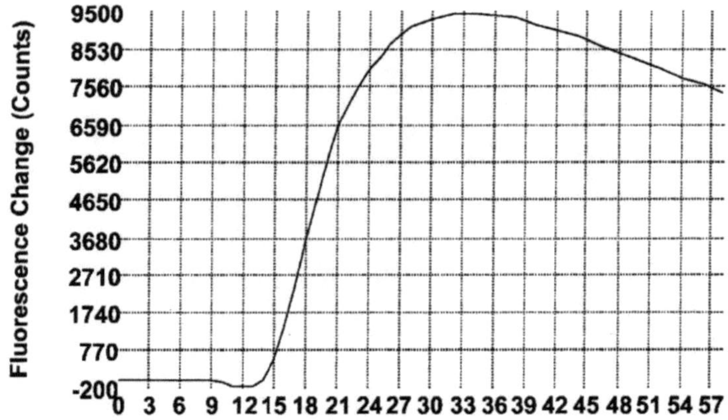

Fig. 9. A tracing of a calcium response to GPCR activation in cells is shown. The curve shape can be quite variable, but generally involves a rapid rise followed by a gradual decay in fluorescent signal. The lag is rather short and is mixing and diffusion limited.

2. To each well, add 20 µL loading medium, incubate at 37°C for 30–60 min (critical—no longer than this).
3. Wash four times with wash buffer, leaving a final volume of 20 µL (*see* **Note 7**).
4. If antagonist compounds are being tested, pin tool them into the wells prior to loading the plates into the rack for the FLIPR.
5. On the FLIPR, add 20 µL agonist (2X concentration) and measure (λ_{ex} = 488 nm, λ_{em} = 500–560 nm) the fluorescent response over time—usually read for 1 min.

Some final technical aspects of FLIPR screens will be shared. First, the specificity of agonist or antagonist hits must be confirmed with appropriate controls. These include using other cell lines, receptor clones, stereochemical analogs, etc. Second, important logistical constraints include the fact that this FLIPR method must stay within 1 h to load and 1 h to read, otherwise the signal is impaired. Complete automation of the screen is rather difficult because of these logistical constraints.

Qualitative aspects of the FLIPR screen merit attention. When an agonist is found, it should produce a classical response as illustrated in **Fig. 9**. The rise in signal should be rapid (diffusion limited after addition of agonist) with a gradual decline. All three phases—lag, rise, and decay—should be observed within 1 min. The signal may not return to baseline in this timeframe. There should be a satisfactory dose-dependence to the response as shown in **Fig. 10**. The concentration-response curve should not be too steep or shallow. Optimally, a surrogate agonist should be submicromolar to be both economical for screening and less likely to be confounded by signaling through unintended receptor pathways. An example of actual screening data for a proprietary genomically derived GPCR is shown in

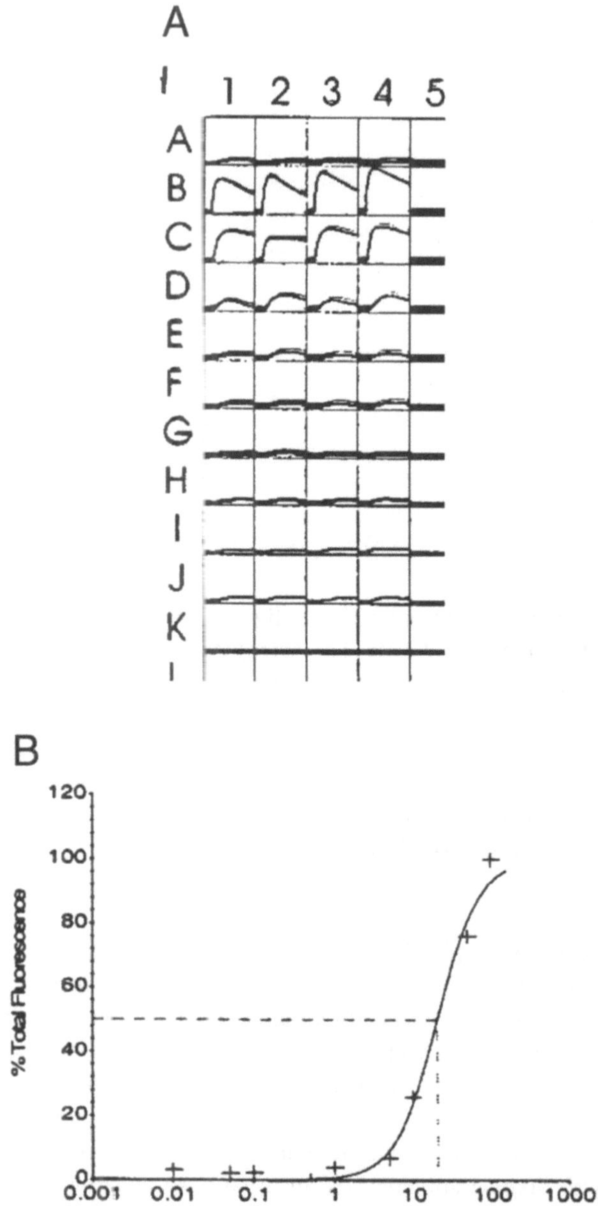

Fig. 10. Dose-dependent agonist stimulation of a GPCR. Panel A, the raw data showing graded responses of the fluorescent signal in response to increasing concentrations of the agonist are illustrated. Panel B, plot of the resulting concentration-response curve. A range from 0, 0.1–100 n*M* of agonist is illustrated. The EC_{50} is 20 n*M* and the slope is nominal. Most surrogate agonists should be nanomolar agents in order to be useful for high-throughput screening. This potency is important for fidelity as well as economy.

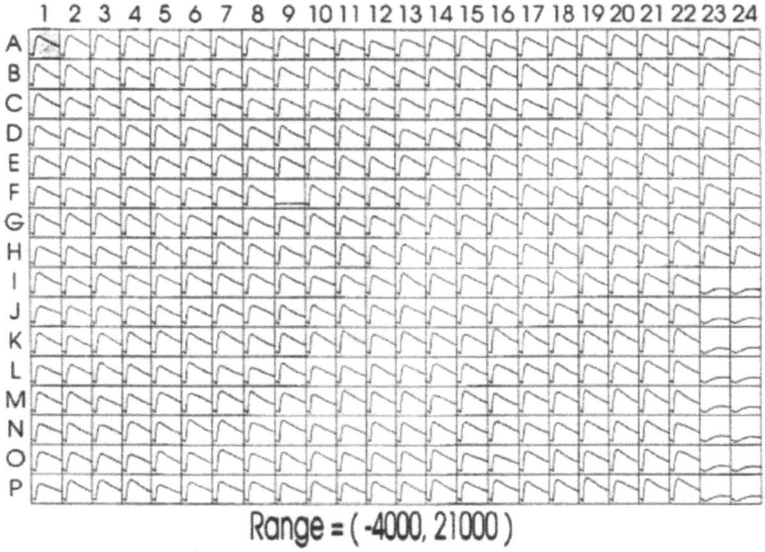

Fig. 11. Raw data curves from a 384-well plate. Data from a screen of a proprietary oGPCR for antagonists are shown. The surrogate agonist generates reliable fluorescent signals across the plate. A potential antagonist is seen in well F9.

Fig. 11. Narrow inter-assay and intra-assay variations are critical parameters for successful high-throughput screens using FLIPR technology. Some final advice is provided when you *see* **Note 8**.

4. Notes

1. Polyethylene glycol (PEG) length was found to be critical. Low loading resin with bead diameters of 300–500 μm and $PEGA_{1900}$ worked well. Beads with PEG_{900} did not work. It is also important to note that these PEGylated resins have extremely fragile beads when in a dry state. They also stick to various surfaces such as glass and plastics when dry, complicating manipulation. Therefore, when handling the resin, it is best to work with it after swelling in methanol or some other solvent. Other resins might be used, but remember that they must be compatible with diverse solvents—from organic to aqueous—and should be penetrable by proteases to generate sufficient fluorescence to be seen. Besides the resin, the selection of fluorescent and quenching pairs is very important. In the case of the *E. coli* leader peptidase, Dabsyl and EDANS did not work. Dabsyl was still selected as a quencher partly because it gives a red color to the beads, and this simplifies handling because the beads are easier to see. Dabsyl also increases the contrast between reactive yellow beads and beads that have intact peptides. Nitrotyrosine and 2-aminobenzoic acid (ABZ) provided a good fluorescent signal after cleavage, however, the fluorescing blue color was rather hard to

discern among colorless beads.

2. Successful biotinylation can be confirmed by performing a dot-blot experiment with SA-AP comparing the binding of SA-AP to the target protein before and after biotinylation.
3. There are several ways to have control incubations. For example, if a ligand is known, positive control beads with the ligand attached can be processed in parallel with the library beads. Once the positive control beads have stained well, then the library staining reaction is done. Alternatively, if no ligand is available, then two aliquots of a library can be processed in parallel in which one sample lacks the genomically derived target. In this case, let the staining reaction progress until there is a good range of staining of beads that have been exposed to the target compared to fairly colorless beads in the sham treated library.
4. Sometimes the beads are not well blocked, which was commonly seen with the earlier versions of polysytrene-based beads. In this case, the target protein or detection enzymes (SA-AP or antibody-AP conjugates) are rapidly adsorbed to nonspecifically binding beads, reducing the final effective concentration of the probe molecules. We call this probe depletion. It is useful to check for depletion by saving an aliquot of the probe cocktail prior to probing the library and comparing its enzymatic activity and amount of target protein against an aliquot obtained after incubation with the library beads. Depletion greater than 20% probably should be addressed. Some solutions to probe depletion that we have used include using different resins for synthesis of the library, increasing the concentration of the probe or the total volume of the probe solution, or changing the blocking buffer. Clearly, changing resin is the last choice. Buffer components found to be strongly useful to minimize probe depletion and spurious staining behavior include higher ionic strengths, additional BSA, gelatin, NFDM, and nonionic detergents. Sometimes, library beads need to be sonicated in the presence of the blocking buffer to get satisfactory results. As for other resins, see the section in this chapter on protease screening. In addition, we have used Sepharose beads with some degree of success; however, stripping the beads is done with 6 *M* guanidine hydrochloride because NaOH hydrolyzes the resin. Furthermore, libraries made on sepharose tended to degrade during storage despite various storage conditions, so it was never quite as robust as TentaGel.
5. Sometimes, especially in the case of highly labile exogenous proteases, a second incubation with lysate is needed in order to get satisfactorily fluorescent beads.
6. Previous versions of this method ***(12)*** used lower specific activity $\gamma[^{32}P]$ATP. The higher specific activity we prefer increases sensitivity considerably, so that autoradiography takes much less time. In earlier methods, after spreading the beads in agarose on a glass plate, the agarose was allowed to dry overnight leaving a thin, somewhat brittle film. We use acetate sheets, which are flexible ensuring easy manipulation on a microscope stage unlike a glass plate. Furthermore, we do not let the agarose dry out, thereby avoiding brittleness. Also, drying the sample can make it more difficult to remove agarose from the beads by melting. Finally, earlier methods relied on extensive and even harsh washing conditions to

exclude noncovalent labeling of the beads. We believe that the gentler, quantitative competition step with unlabeled ATP is a significant improvement for identifying covalently labeled beads and for prioritizing beads for peptide sequencing.

7. In **step 3** we use an automated plate washer (Skatron EMBLA). This washer must be well maintained to ensure uniform and reproducible volumes of media remaining in the wells prior to addition of compounds or agonists. To minimize clogging of the cannulas, we purge the system with water and leave the cannulas immersed in water between uses.
8. Some assays are sensitive to the media used. For example, serum may have endogenous agonists, or may cause adsorptive or other loss of added surrogate agonist or antagonist test compounds. Therefore, you may need to change the composition of the assay media while maintaining satisfactory cellular viability. We have not observed problems from phenol red, which was widely believed to quench fluorescent signals. Also, for a wide variety of cell types, we have found that seeding 7,500–10,000 cells in 20 µL gives optimal responses. Seeding at 12,000 cells per well in 20 µL results in nutrient depletion and poor cellular responses. Finally, optimal adherence is achieved after letting the seeded cells settle for 18–24 h prior to assay.

References

1. Singh, S., Lowe, D. G., Thorpe, D. S., et al. (1988) Membrane guanylate cyclase is a cell-surface receptor with homology to protein kinases. *Nature* **334,** 708–712.
2. Furka, A. and Bennett, W. D. (1999) Combinatorial libraries by portioning and mixing. *Comb. Chem. High Throughput Screen.* **2,** 105–122.
3. Lam, K. S., Salmon, S. E., Hersh, E. M., Hruby, V. J., Kazmierski, W. M., and Knapp, R. J. (1991) A new type of synthetic peptide library for identifying ligand-binding activity. *Nature* **358,** 82–84.
4. Ohlmeyer, M. H., Swanson, R. N., Dillar, L. W., et al. (1993) Complex synthetic chemical libraries indexed with molecular tags. *Proc. Natl. Acad. Sci. USA* **90,** 1099–10926.
5. Thorpe, D. S. and Walle, S. (2000) Combinatorial chemistry defines general properties of linkers for the optimal display of peptide ligands for binding soluble protein targets to tentagel microscopic beads. *Biochem. Biophys. Res. Commun.* **269,** 591–595.
6. Ostrem, JA, Al-Obeidi, F., Safar, P., et al. (1998) Discovery of a novel, potent, and specific family of factor Xa inhibitors via combinatorial chemistry. *Biochemistry* **37,** 1053–1059.
7. Meldal, M., Svendsen, I., Breddam, K., and Auzanneau, F. I. (1994) Portion-mixing peptide libraries of quenched fluorogenic substrates for complete subsite mapping of endoprotease specificity. *Proc. Natl. Acad. Sci. USA* **91,** 3314–3318.
8. Rosse, G., Kueng, E., Page, M. G. P., et al. (2000) Rapid identification of substrates for novel proteases using a combinatorial peptide library. *J. Comb. Chem.* **2,** 461–466.

9. Schauer-Vukasinovic, V., Bur, D., Kitas, E., et al. (2000) Purification and characterization of active recombinant human napsin A. *Eur. J. Biochem.* **267,** 2573–2580.
10. Gray, N. S., Wodicka, L., Thunnissen, A. M., et al. (1998) Exploiting chemical libraries, structure, and genomics in the search for kinase inhibitors. *Science* **281,** 533–538.
11. Thorpe, D. S. (2000) Forecasting roles of combinatorial chemistry in the age of genomically derived drug discovery targets. *Comb. Chem. High Throughput Screen* **3,** 421–436.
12. Wu, J., Ma, Q. N., and Lam, K. S. (1994) Identifying substrate motifs of protein kinases by a random library approach. *Biochemistry* **33,** 14825–14833.
13. Kemp, B. E. and Pearson, R. B. (1990) Protein kinase recognition sequence motifs. *TIBS* **15,** 342–346.
14. Thorpe, D. S., Yeoman, H., Chan, A. W. E., Krchnak, V., Lebl, M., and Felder, S. (1999) Combinatorial chemistry reveals a new motif that binds the platelet fibrinogen receptor, gpIIbIIIa. *Biochem. Biophys. Res. Commun.* **256,** 537–541.

III

Computational Library Design

17

Design of Virtual Combinatorial Libraries

Alex M. Aronov

1. Introduction

Last year marked the completion of the sequencing effort on the human genome. Advances in the fields of genomics and bioinformatics are widely expected to bring forth a large number of new biochemical targets for drug design by linking specific diseases to single genes or collections thereof. The pharmaceutical and biotechnology sector is now likely to become one of the most active industrial fields in the new century. A dramatic increase in the number of pharmaceutical targets in the near future would create a bottleneck of sorts at the stage of pharmaceutical drug discovery. Historically, this field has used a serial process, screening and optimizing compounds one at a time. However, the advent of combinatorial chemistry and high-throughput screening has significantly altered the face of drug discovery. Rapid generation and screening of combinatorial libraries has become commonplace in both industry and academia; with the ever-growing offering of organic reagents along with the vast palette of organic reactions, the chemical space accessible to combinatorial chemists has dramatically expanded. The application of combinatorial technology, however, is not without a potential caveat—even as the cost of synthesis and testing of a single chemical entity has fallen, it skyrockets when multiplied by the thousands or millions of combinatorial library members. Some of the potential solutions include (i) use of the three-dimensional target structure information in the design process; (ii) application of pharmacophore and quantitative structure-activity relationship (QSAR) methods to focus on libraries that most closely resemble screening hits; and (iii) reduction in the scale on which both medicinal chemistry and screening are currently performed. These solutions imply a need for a rational approach to reducing library size, custom-

From: *Methods in Molecular Biology, Combinatorial Library Methods and Protocols*
Edited by: L. B. English © Humana Press Inc., Totowa, NJ

izing the library for a given target, and more efficient *in silico* screening of a library (or, potentially, multiple libraries) of compounds against a large number of targets of interest. The challenge in accomplishing these goals is in being able to reduce the library while increasing the potential hit rate of individual library members.

2. Directions in Library Design

Combinatorial library design attempts to choose the best set of substituents for a combinatorial synthetic scheme, from all available candidates, to maximize the chances of finding a useful compound such as a drug lead *(1)*. Library design involves operations on virtual combinatorial libraries, which, in turn, are collections of compounds that can be synthesized by a reaction scheme from a set of available reagents *(2)*. By focusing on well-characterized reactions, this approach circumvents a bottleneck in current computational approaches to *de novo* ligand design: the assessment of synthetic accessibility. Possible types of virtual libraries include (i) diverse, (ii) focused, and (iii) complementary libraries. An ideal diverse library is a collection of chemical entities without gaps and redundancies, with an even distribution with regard to a given chemical space. Design of focused libraries involves incorporating a bias into the library, usually based on a similarity metric, docking scores, or QSAR predictions. A complementary library would fill in the voids discovered in another library. Initial efforts in the field focused on maximizing diversity in virtual libraries. This would typically be achieved by calculating a property space and specifying any fixed substituents, followed by sampling the remaining candidates with D-optimal design *(3)*. However, experience indicates that many of the libraries designed in such a manner yielded molecules that were too large, flexible, insoluble, or lipophilic to qualify as attractive pharmaceutical leads. Maximum diversity too often meant moving away from preferred ranges in these and other properties. This setback gave birth to a conceptually different approach to library design, what could be termed "knowledge-based library analysis."

Knowledge-based analysis is best exemplified by the approach taken by Lipinski et al. *(4)*, who studied the United States Adopted Names compound list and produced a set of rules, now commonly referred to in the literature as the "Lipinski rule of 5." According to this set of rules, one would expect a molecule to demonstrate poor permeation or absorption properties when it has greater than 5 hydrogen bond donor groups, greater than 10 hydrogen bond acceptors, molecular weight above 500, and the Clog *P* above 5. Ghose et al. *(5)* went further and examined the Comprehensive Medicinal Chemistry (CMC) database and computed "qualifying ranges" (defined as covering more than 80% of the compounds) of several computed physicochemical properties such as Clog *P*, molecular weight, molar refractivity, and number of atoms, as well as a qualitative characterization based on the occurrence of functional groups and certain important substructures. The computed qualifying range of Clog *P* values is between –0.4 and 5.6 with an average of 2.52, for molecular weight it is between 160 and 480 with an average value of 357, and for molar refractivity it is between 40 and 130 with an average of 97. The molecules contained between 20 and 70 atoms (average of 48), with tertiary amines, alcoholic hydroxyls, and carboxamides as the most abundant functional groups in the CMC database. Molecules generally conforming to this set of empirical rules are often referred to as "drug-like." Not surprisingly, benzene was by far the most abundant substructure in the database. Another study of common molecular frameworks within the CMC was published by Bemis and Murcko *(6)*. They reported, for example, that the shapes of 50% of the drugs in CMC can be described by only 32 of the most frequently occurring two-dimensional frameworks (these frameworks did not account for variations in atom type, hybridization, or bond order). This observation of rather limited topological diversity within the database of known drugs can be used to create screening libraries incorporating the spectrum of the most frequently encountered "medicinal" shapes, as illustrated by Fejzo et al. *(7)* in creating a diverse set of drug-like molecules for NMR screening.

2.1. Ligand Structure-Based Design

In the field of computer-aided drug design, two broad applications can be identified: (i) design of focused libraries where the main activity is prediction of binding to a particular protein target and (ii) design of exploratory primary libraries to be screened against a panel of protein targets that might be either structurally similar or unrelated *(8)*. In cases when there is little or no structural information available about the target, a starting point for library design is typically the calculation of descriptors to characterize molecular structures, followed by QSAR to optimize for the desired library characteristics. Potential descriptors of use in this process include molecular weight, Clog *P*, descriptors containing information about chemical functionality (e.g., hydrogen bond donors or acceptors), as well as molecular similarity between known ligands and library members. Both rule of 5 *(4)* and scaffold similarity to reported structures *(6)*, described in **Subheading 2.**, can serve as a possible optimization tool for macro properties such as absorption, distribution, metabolism, excretion, and toxicology (ADMET) rather early on in the design process. Ajay et al. *(9)* have demonstrated the utility of Bayesian neural networks in distinguishing between drug-like and nondrug-like molecules. Available Chemicals Directory (ACD) was used in the study as a surrogate for non-drug-like molecules, and structures from Comprehensive Medicinal Chemistry (CMC) database were properly classified by the program as being drug-like for over 90% of the CMC. Both one-dimensional (1D) parameters containing information about the entire molecule and two-dimensional (2D) parameters encoding information about specific functional groups were used to train a learning system.

Although traditional whole-molecule 1D/2D descriptors have been widely used to address a variety of diversity-related tasks, they tend to make rather poor chemistry-space metrics ***(10)***. Many of these "traditional" descriptors, such as Clog P, surface area, pK_a, etc., are highly correlated; the axes of a vector space should ideally be orthogonal (uncorrelated). Some of the descriptors are strongly related to pharmacokinetics but are very weakly related to compound activity. The main issue is the fact that aforementioned descriptors describe molecules as a whole, conveying little information about the detailed substructural differences that are the basis for structural diversity ***(10)***. Pearlman and co-workers proposed using cell-based algorithms that are based on partitioning chemistry space into hypercubic cells as a potential way of selecting a diverse subset of compounds from a much larger population. Implementation of these algorithms led to the low-dimensional chemistry space based on BCUT metrics, descriptors that incorporate connectivity information in addition to atomic properties relevant to intermolecular interaction, such as atomic charge, polarizability, hydrogen bonding, etc. The metrics were then validated using activity-seeded, structure-based clustering ***(10)***.

In addition to 1D/2D descriptors, three-dimensional (3D) metrics are sometimes used in library design. Calculation of 3D properties involves generating an energetically relevant structure or ensemble of structures. One would expect the descriptors containing 3D information to be more predictive of biological activity than the 1D/2D descriptors. Several studies have documented that the opposite can sometimes be the case; however, that more likely testifies to the incompleteness of the 3D metric rather than its inferiority. McGregor and Muskal utilized the pharmacophore concept widely used in drug design to develop their 3D metric ***(8)***. A pharmacophore is a set of functional group types present in a particular geometrical arrangement that represent interactions made in common by a set of known ligands with the receptor, such as hydrogen bonding, charge, and hydrophobic interactions. The pharmacophore library constructed arbitrarily based on enumerating distances and group types was matched onto the full conformationally expanded combinatorial library and resulting pharmacophore fingerprints were reduced to a low-dimensional representation, which can then be used in the design and profiling of novel virtual libraries.

2.2. Protein Structure-Based Design

A more common approach to library design involves the use of the 3D structure of the target. Molecular docking, which utilizes rapid docking and scoring routines to screen a list of virtual ligands, has become an integral part of structure-based drug discovery. Combinatorial docking of enumerated libraries or of a common scaffold as a skeleton to which substituents are added in a combinatorial manner has produced potent ligands for a variety of targets *(11)*. DOCK *(12)* is one of the more widely distributed docking programs; it systematically attempts to find a match between each ligand and a set of site points (spheres), then uses a precalculated grid to estimate binding energy. Other examples of docking software include AUTODOCK *(13)*, which employs simulated annealing and is thus best suited for ligands with a small number of rotatable bonds; FLEXX *(14)*, which initially docks base fragments then incrementally builds up substituents; and FLO98 *(15)*, which performs Monte Carlo perturbation followed by energy minimization in Cartesian space.

While virtual screening is capable of rapidly producing collections of potential ligands, accurate scoring of the docked configuration remains problematic. The major shortcomings of the scoring schemes that are currently in common use include: (i) they do not take into account flexibility of the bonding site, (ii) the entropic effects in binding are treated in a crude, empirically-based manner, and (iii) they score on the basis of the bound complex alone instead of considering the difference between the bound and free solvated forms. Most of the existing scoring functions fall into one of two categories: *(1)* empirical scoring functions, such as Ludi function *(16)*, SCORE *(17)*, ChemScore *(18)*, and PLP *(19)* or *(2)* molecular mechanics functions, such as MMFF *(20)*, FLOG *(21)*, DOCK energy score, and DOCK chemical score. Results of a study of the major scoring functions tested on the known complexes of p38 mitogen-activated protein (MAP) kinase, human immunodeficiency virus-1 (HIV-1) protease, and inosine monophosphate dehydrogenase (IMPDH) indicate that DOCK energy score, PLP, and ChemScore tend to perform best across a range of targets *(22)*. The study also suggests that using a consensus scoring approach with several scoring functions significantly enriches the library in true positives. An interesting new scoring method was recently published by Pearlman and Charifson *(23)*. Termed OWFEG (one-window free energy grid), this method attempts to bridge the gap between high-accuracy low-throughput free energy simulations and empirical approaches by using a single molecular dynamics (MD) simulation to generate a grid surrounding a molecule of interest that represents free energy for insertion of a probe group at any point on that grid. Together with rapid scoring potential, the method performed as good as or better than most existing empirical scoring schemes.

2.3. Reducing the Library to the Reagent Matrix

Methods for library design discussed in **Subheadings 2.1.–2.2.** deal primarily with ways to reduce the library size in terms of products. However, the reality of combinatorial synthesis, which is most often done in 96-well plates, necessitates the design not of the optimal product collection itself, but the best reagent library that would most closely approximate the optimal product library. A number of techniques have been used toward the identification of library subsets; most can be classified as either reagent- or product-based models. While product-based approaches would generally be expected to yield better results, reagent-based methods have the advantage of being additive with respect to the size of each reagent list. Not unexpectedly, the benefits from product-based combinatorial optimization depend on the descriptor used. Jamois et al. ***(24)*** showed that high-order connectivity indices such as Daylight fingerprints span extended paths, possibly reaching across different substituents and therefore increasing the benefit of product-based methods.

Another challenging computational problem is determining which subset of the larger library is both optimally dense in "model-matching" compounds and amenable to the experimental matrix constraints. A clear limitation of operating within these matrix constraints is the fact that matrix synthesis may greatly sacrifice the diversity of the resultant library. An algorithm developed by Stanton et al. ***(25)*** detects the maximally dense submatrix of 1's within the larger matrix, i.e., the best combination of rows and columns. A branch and bound technique developed is more rigorous, but tends to become intractable with larger compound sets, while a "cut-down" solution presented in the study yields results closely approximating that of a true solution.

3. New Trends in Virtual Library Design

Many different algorithms for structure-based design can be divided roughly into two classes: *de novo* design, which builds ligands tailored to fit the target, and docking, which searches for existing compounds with good complementarity to the target. Both of these paradigms have traditionally treated the enzyme or receptor as a rigid body, considering only one active site conformation. Rather than exploring the receptor conformational space by theoretical means, one could instead make use of the increasingly available X-ray and nuclear magnetic resonance (NMR) structural data on protein structure and flexibility. A set of related crystal structures can be treated as snapshots of a dominant conformation perturbed by different ligands, crystallization conditions, point mutations, etc. ***(26)***. The most straightforward method would then evaluate ligand-receptor binding energies for each structure and use a Boltzmann-weighted energy average at each point as the score. Faster methods of adding up scores from different structures were also proposed, such as geometry-weighted and energy-weighted averaging.

Another trend that is rapidly gaining popularity is biological fingerprinting of proteins and ligands. Originally proposed as an experimental approach to ligand fingerprinting by Terrapin *(27)*, it involves the use of experimentally determined in vitro binding potency of compounds against a diverse reference panel of proteins in generating the so-called "affinity fingerprints." The underlying assumption is that compounds that have similar affinity fingerprints against the targets in the reference panel would have similar affinity for another receptor as well. This approach takes the binding interaction as the basis for the similarity metric. Briem and Kuntz *(28)* replaced Terrapin's experimental approach with a computer docking simulation. Their results indicated that the method is suitable for finding significant similarities of compounds of the same biological activity; however, it was outperformed by a traditional 2D similarity metric. The virtual reference panel consisted of only eight proteins; one would expect the system to perform significantly better once a large number of diverse targets is incorporated into the set. A potential advantage of this approach is the facilitated "lead-hopping," i.e., selecting from the vastness of structurally dissimilar molecules *M* structures most likely to have the same critical biological property as the compound of interest *(29)*. The traditional way to create lead-hops involves including compounds from entirely different chemical classes into high-diversity combinatorial libraries using methods discussed in **Subheading 2.1.** *(30)*. Since lead-hopping is potentially advantageous for reasons of intellectual novelty for patenting rights, discovery of novel scaffolds with interesting biological activity is one of the priorities in drug design today. A reciprocal process of profiling proteins or other biological targets based on their binding to a set of different ligands has also been described *(31)*. Virtual biological fingerprinting methods allow for the libraries to be placed on a continuum ranging from universal libraries that contain hits against every target to completely directed libraries that contain hits against one specific target. While, initially, libraries with greater target breadth may contain more condensed information about binding preferences, subsequent optimization would make libraries with narrow specificity more desirable. The ultimate Webster's dictionary of combinatorial drug design would be the database of interactions of all potential ligands with all potential targets *(32)*. The protocol described by Lamb et al. scales linearly in the number of libraries and linearly in the number of targets. A useful variation on the multiple ligands versus multiple targets approach is gene family-specific ligand profiling *(33)*, which allows for rapid customization of ligands within target families. By including target family-specific information to direct docking, one could easily restrict sampling with a potential associated gain in speed and accuracy. Another potential time-saving solution would involve aligning targets based on known family relationships or active-site geometries to avoid evaluating libraries against similar targets in early stages of the discovery process. Since the virtual activity profile for a ligand family (a scaffold) across the target family is often conserved from one protein to the next *(32,33)*, we could reduce the full matrix to evaluation of columns of clusterheads, or, as they are now referred to in the literature, the "gatekeeper targets."

4. Conclusions

The past several years have witnessed the evolution of structure-based design into an indispensable tool in drug discovery. At the same time, advances in combinatorial chemistry have necessitated the use of new computational approaches to design novel libraries. While currently available scoring functions as well as similarity and physicochemical filters enable considerable enrichments of active molecules, further improvements are necessary. The coming years promise more exciting developments in the field.

References

1. Martin, E. and Wong, A. (2000) Sensitivity analysis and other improvements to tailored combinatorial library design. *J. Chem. Inf. Comput. Sci.* **40,** 215–220.
2. Gorse, D. and Lahana, R. (2000) Functional diversity of compound libraries. *Current Opinion in Chemical Biology* **4,** 287–294
3. Martin, E. J. and Critchlow, R. E. (1999) Beyond mere diversity: tailoring combinatorial libraries for drug discovery. *J. Comb. Chem.* **1,** 32–45.
4. Lipinski, C. A., Lombardo, F., Dominy, B. W., and Feeney, P. J. (1997) Experimental and computational approaches to estimate solubility and permeability in drug discovery and development settings. *Adv. Drug Delivery Rev.* **23,** 3–25.
5. Ghose, A. K., Viswanadhan, V. N., and Wendoloski, J. J. (1999) A knowledge-based approach in designing combinatorial or medicinal chemistry libraries for drug discovery. 1. A qualitative and quantitative characterization of known drug databases. *J. Comb. Chem.* **1,** 55–68.
6. Bemis, G. W. and Murcko, M. A. (1996) The properties of known drugs. 1. Molecular frameworks. *J. Med. Chem.* **39,** 2887–2893.
7. Fejzo, J., Lepre, C. A., Peng, J. W., et al. (1999) The SHAPES strategy: an NMR-based approach for lead generation in drug discovery. *Chem. Biol.* **6,** 755–769.
8. MacGregor, M. J. and Muskal, S. M. (2000) Pharmacophore fingerprinting. 2. Application to primary library design. *J. Chem. Inf. Comput. Sci.* **40,** 117–125.
9. Ajay, Walters, W. P. and Murcko, M. A. (1998) Can we learn to distinguish between "drug-like" and "nondrug-like" molecules? *J. Med. Chem.* **41,** 3314–3324.
10. Pearlman, R. S. and Smith, K. M. (1999) Metric validation and the receptor-relevant subspace concept. *J. Chem. Inf. Comput. Sci.* **39,** 28–35.

11. Boehm, H.-J. and Stahl, M. (2000) Structure-based library design: molecular modeling merges with combinatorial chemistry. *Current Opinion in Chemical Biology* **4,** 283–286.
12. Ewing, T. J. A. and Kuntz, I. D. (1997) Critical evaluation of search algorithms for automated molecular docking and database screening. *J. Comput. Chem.* **18,** 1175–1189.
13. Goodsell, D. S., Morris, G. M., and Olson, A. J. (1996) Automated docking of flexible ligands: application of AUTODOCK. *J. Mol. Recognit.* **9,** 1–5.
14. Rarey, M., Wefing, S., and Lengauer, T. (1996) Placement of medium-sized molecular fragments into active sites of proteins. *J. Comput.-Aided Mol. Des.* **10,** 41–54.
15. McMartin, C. and Bohacek, R. S. (1997) QXP: powerful, rapid computer algorithms for structure-based drug design. *J. Comput.-Aided Mol. Des.* **11,** 333–344.
16. Boehm, H.-J. (1992) The computer program LUDI: a new method for the de novo design of enzyme inhibitors. *J. Comput.-Aided Mol. Des.* **6,** 61–78.
17. Wang, R., Liu, L., Lai, L., and Tang, Y. (1998) SCORE: a new empirical method for estimating the binding affinity of a protein-ligand complex. *J. Mol. Model.* **4,** 379–394.
18. Eldridge, M. D., Murray, C. W., Auton, T. R., Paolini, G. V., and Mee, R. P. (1997) Empirical scoring functions: I. The development of a fast empirical scoring function to estimate the binding affinity of ligands in receptor complexes. *J. Comput.-Aided Mol. Des.* **11,** 425–445.
19. Gehlhaar, D. K., Verkhivker, G. M., Rejto, P. A., Sherman, C. J., Fogel, D. B., and Freer, S. T. (1995) Molecular recognition of the inhibitor AG-1343 by HIV-1 protease—Conformationally flexible docking by evolutionary programming. *Chem. Biol.* **2,** 317–324.
20. Halgren, T. A. (1996) Merck molecular force field. I. Basis, form, scope, parametrization, and performance of MMFF94. *J. Comput. Chem.* **17,** 490–519.
21. Miller, M. D., Kearsley, S. K., Underwood, D. J., and Sheridan, R. P. (1994) FLOG—A system to select quasi-flexible ligands complementary to a receptor of known three-dimensional structure. *J. Comput.-Aided Mol. Des.* **8,** 153–174.
22. Charifson, P. S., Corkery, J. J., Murcko, M. A., and Walters, W. P. (1999) Consensus scoring: a method for obtaining improved hit rates from docking databases of three-dimensional structures into proteins. *J. Med. Chem.* **42,** 5100–5109.
23. Pearlman, D. A. and Charifson, P. S. (2001) Improved scoring of ligand-protein interaction using OWFEG free energy grids. *J. Med. Chem.* **44,** 502–511.
24. Jamois, E. A., Hassan, M., and Waldman, M. (2000) Evaluation of reagent-based and product-based strategies in the design of combinatorial library subsets. *J. Chem. Inf. Comput. Sci.* **40,** 63–70.
25. Stanton, R. V., Mount, J., and Miller, J. L. (2000) Combinatorial library design: maximizing model-fitting compound within matrix synthesis constraints. *J. Chem. Inf. Comput. Sci.* **40,** 701–705.
26. Knegtel, R. M. A., Kuntz, I. D., and Oshiro, C. M. (1997) Molecular docking to ensembles of protein structures. *J. Mol. Biol.* **266,** 424–440.

27. Kauvar, L M., Villar, H. O., Sportsman, J. R., Higgins, D. L., and Schmidt, D. E. Jr. (1998) Protein affinity map of chemical space. *J. Chromatogr. B* **715,** 93–102.
28. Briem, H. and Kuntz, I. D. (1996) Molecular similarity based on DOCK-generated fingerprints. *J. Med. Chem.* **39,** 3401–3408.
29. Andrews, K. H. and Cramer, R. D. (2000) Toward general methods of targeted library design: topomer shape similarity searching with diverse structures as queries. *J. Med. Chem.* **43,** 1723–1740.
30. Weber, L. (2000) High-diversity combinatorial libraries. *Current Opinion in Chemical Biology* **4,** 295–302.
31. Kauvar, L M., Higgins, D. L., Villar, H. O., et al. (1995) Predicting ligand binding to proteins by affinity fingerprinting. *Chem. Biol.* **2,** 107–118.
32. Lamb, M. L., Burdick, K. W., Toba, S., et al. (2001) Design, docking, and evaluation of multiple libraries against multiple targets. *PROTEINS: structure, function, and genetics* **42,** 296–318.
33. Aronov, A. M., Munagala, N. R., Kuntz, I. D., and Wang, C. C. (2001) Virtual screening of combinatorial libraries across a gene family. In search of inhibitors of *Giardia lamblia* guanine phosphoribosyltransferase. *Antimicrob. Agents Chemother.*, **45,** 2571–2576.

18

Approaches to Library Design for Combinatorial Chemistry

Stefan Güssregen, Bernd Wendt, and Mark Warne

1. Introduction

With the advent of high-throughput technologies in drug discovery, combinatorial chemistry has superseded natural products as the prime source of compounds to be tested in biological screening *(1)*. The reason being that compared with classical synthesis, high-throughput chemistry can produce a much larger number of compounds within a much shorter period of time. However, owing to the combinatorial nature of the problem and the large number of reagents available, many more compounds are theoretically accessible via a given synthetic route than it is actually feasible to synthesize. Therefore, library design techniques are employed to identify those reagents that yield a library enriched with the desired properties. Early on, most applications of these techniques concentrated on designing very diverse libraries, with the idea that these libraries would be tested against a variety of biological assays. The ideal library in this case would be with no voids, no redundancy, and an even distribution with regard to a given chemical space. Nowadays, it is equally common to design targeted libraries to have a maximal activity in a selected assay. Here, all compounds would be designed to be similar to given hits or leads. Other applications of library design techniques include the computational evaluation of templates, i.e., prioritization of candidate libraries for synthesis and screening (*see* **Note 1**).

In this chapter we will discuss the principles of library design for combinatorial chemistry by focusing on two major applications: design of diverse libraries for general screening and design of nondiverse targeted libraries for hit/lead follow-up. Both will be exemplified using the homopiperazine chemistry (**Fig. 1**) as an example (*see* **Note 3**).

From: *Methods in Molecular Biology, Combinatorial Library Methods and Protocols*
Edited by: L. B. English © Humana Press Inc., Totowa, NJ

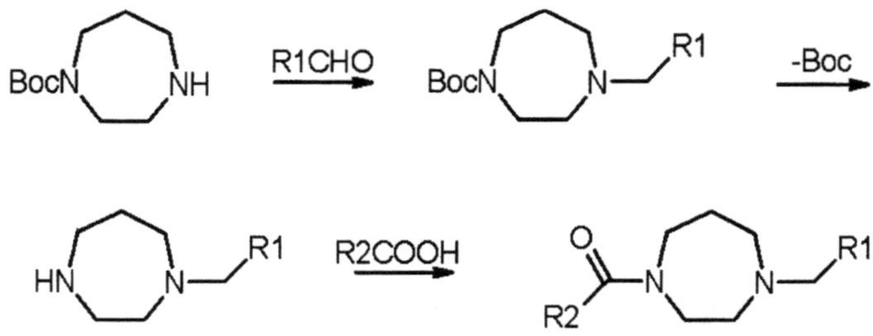

Fig. 1. The synthetic route to the homopiperazine compounds. Assuming that the actual synthesis would be carried out in solution phase, one would start from monoprotected homopiperazine introducing a first set of building blocks (R1 building blocks). These could come, for instance, from a set of alkyl aldehydes added via reductive amination. In a deprotection step the intermediates would be prepared for the addition of the second set of building block (R2 building blocks) to form the final products. Acylation of the secondary amine of the homopiperazine intermediate with aromatic carboxylic acids is quite common and will be used in this example (*see* **Note 2**).

1.1. Strategies for Library Design

Library design comprises a series of tasks: a search for suitable reagents, conversion into synthons (building blocks), removal of unsuitable synthons, calculation of molecular properties and descriptors, selection of a final set of products, and the generation of product structures and corresponding reagent lists (reaction matrix). The various design strategies described in the literature *(2–8)* typically differ by the order in which the tasks are executed and whether they are carried out on the building block or at the product stage (*see* **Fig. 2**). The purpose for which the library is being assembled and the nature of the chemistries required will determine the strategy to implement.

1.1.1. Reagent-Based Design

Reagent-based design, where the entire selection process is carried out on the building blocks, is the simplest to perform because the size of the problem is reduced from $m \times n$ to $m + n$ building blocks. Product structures are generated in the very last step for registration purposes only. However, many desirable descriptors for diversity and similarity (*see* **Subheading 1.1.2.**) used in library design are only meaningful on full structures and thus cannot be applied to building blocks.

1.1.2. Product-Based Design

Product-based design, on the other hand, involves generating $m \times n$ product structures within virtual libraries very early on in the design process, and properties as well as molecular descriptors are to be calculated for those. An addi-

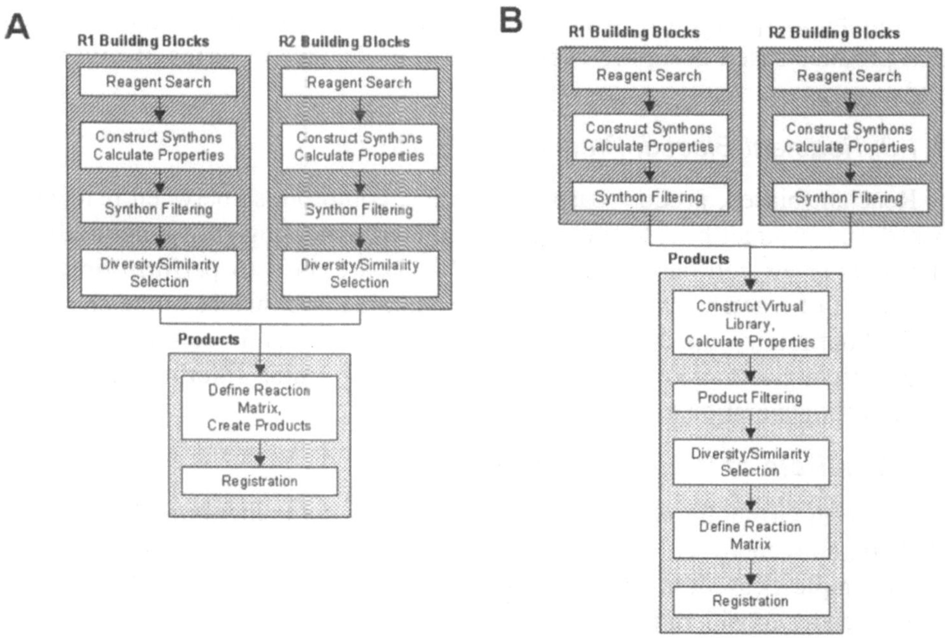

Fig. 2. Various approaches for library design. For simplicity only two points of diversity are considered. (A) Reagent-based library design. Suitable reagents are being searched and converted into the corresponding synthons. For each synthon the desired properties and descriptors are calculated. Filtering and selection of reagents for the library is carried out on a synthon basis exclusively. The product structures are only generated in the very last step for the registration of the library. (B) Product-based approach. In contrast to reagent-based design, a virtual library of product structures is constructed before or after some crude filtering on synthons has been carried out. Molecular properties and descriptors are being calculated for each individual product. As filtering and selection is carried out on a product basis, a sparse reaction matrix is obtained at first. It has to be condensed to a full matrix in an additional step identifying the optimal reagents for the library.

tional complication of the design process in this case results from the fact that a set of individual products are being selected in the end (sparse reaction matrix) whereas in combinatorial chemistry, one synthesizes one or more full matrices. Therefore an additional step deconvoluting the selected products into their corresponding reagents and identifying those reagents that would generate the most products originally selected is necessary. It should be noted that by doing this, additional products outside of the original selection may be synthesized. The performance of product-based design vs reagent-based design has been extensively discussed in the literature ***(3,9,10)***.

Despite being questioned in a recent publication *(8)*, it is generally believed that product-based design is generating better designs in terms of diversity and coverage.

1.1.3. Reactant-Biased, Product-Based Design

Reactant-biased, product-based design is a compromise between reactant-based and product-based strategies. Simulated annealing *(6)*, genetic algorithms *(11)*, or cell-based techniques *(12,13)* are used to perform optimal selections of reactants while retaining the combinatorial nature of the chemistry and optimizing the diversity of the resulting products. In addition, the use of genetic algorithms offers the advantage of performing filtering, selection, and matrix generation at the same time *(11)*.

1.2. The Role of Virtual Libraries

The term virtual library is used quite frequently in the literature these days. Unfortunately there is no common definition on what a virtual library really is. In the simplest case it is a database of fully enumerated structures of products that are under consideration to be made. The generation of such a virtual library involves software that maps the reaction sequence and the corresponding sets of building blocks onto a combinatorial representation *(7)* (*see* **Note 4**). This combinatorial representation can be partially or fully enumerated to generate product structures, which are then stored in structural databases (*see* **Note 5**). Subsequently, properties and descriptors to be used in the selection process are to be calculated. As there are limitations in terms of file size etc., such virtual libraries cannot hold more than a few million products. This makes it necessary to filter out reagents (*see* **Subheading 1.3.**) prior to construction of the virtual library.

More advanced technologies *(14)* enable the use of a wider definition: the virtual library comprises all products that are potentially synthetically accessible via a given synthetic scheme and pool of available reagents. As in the case described above, the reaction sequence and building blocks are mapped onto a combinatorial representation. In addition, those technologies can represent filtered and selected sets of individual products without needing to enumerate their full structures. Thus, reusable virtual libraries can be constructed independent of the specific design goal at hand, making the virtual library a true representation of the corresponding chemistry. The size of such virtual libraries can reach up 10^{13} products (*see* **Note 1**).

1.3. Filtering of Synthons and Products

The purpose of one or more "filtering" steps in library design is to remove unwanted synthons or products from the selection pool (virtual libraries). The various filters that are currently being applied can be grouped roughly into the following areas: drug-like character, chemical compatibility, and availability. Additional filters may be necessary depending on the specific goals for the design.

The drug-like character is very important in order not to waste time and money when trying to identify compounds that could possibly be developed into drugs ***(2)***. Therefore, many companies routinely use two-dimensional (2D) substructural filters to remove compounds containing chemical moieties that are insoluble, chemically reactive, or correlated with toxicity ***(15–17)***. Pharmacokinetic properties like log *P*, molecular weight, and the number of hydrogen bond donors and acceptors are equally important ***(18)***. Lipinski's "Rule of five" is most popular here ***(19)***. The entropic penalty that flexible molecules experience on binding to the target can be taken into account by filtering on the number of rotatable bonds of a molecule.

Much validation information is gathered during development of the chemistry. It is essential for successful library production that this information is included in the design process and that all building blocks are removed that contain functional groups, which are incompatible with the given chemistry. As library design and chemistry validation often take place at the same time, this list of functional groups changes over time. Therefore the quality of the library design can be improved by using the augmented list of functional groups. Sometimes quality assurance (QA) issues make it necessary to remove products from the selection pool with more than one chiral center or without chromophores.

The best library design is useless if the required reagents are not available. Most pharmaceutical companies have an extensive set of proprietary in-house reagents with very high availability that are used exclusively for library design. In our laboratories, we use a system that scores each commercially available reagent by price, quantity on offer, and supplier/catalog reliability (*see* **Note 6**). All reagents scoring less than a predefined threshold are removed from the selection pool. It should be noted that the reagent scores need to be updated on a daily basis, as information on reagent availability changes frequently.

1.4. Descriptors for Molecular Diversity

Molecular descriptors are calculated for each synthon or product to be considered in order to guide the final selection. The molecular descriptors and selection algorithms chosen for this purpose should be able to maximize the chances of finding more active compounds and to improve their potency because this is the objective of synthesizing and testing libraries.

Diversity and similarity are not inherent properties; they depend on the specific descriptors used. Therefore, the natural choice is to use such descriptors that are based on properties relevant to drug-receptor interactions. In addition, useful descriptors follow the neighborhood behavior *(20)*, which is based on the similarity property principle *(21)*: Two compounds that are very similar to each other based on a given descriptor should exhibit similar biological activity as well. This concept relates to the basic principle in medicinal chemistry, where small changes are introduced into compounds in order to obtain a small change (to the better hopefully) in potency. Many different descriptors for different applications like BCUT (Burden-CAS-University of Texas) values *(13)*, topological and graph theory indices, 3-/4-point pharmacophore fingerprints *(22,23)*, and various molecular properties (log *P*, volume, dipole moment) have been reported in the literature, and computational methods for measuring diversity have already been extensively reviewed *(7,24–26)*. Not all of them are valid descriptors inasmuch as they do not show neighborhood behavior. One example is the octanol/water distribution coefficient (log *P*) that is very important as a pharmacokinetic property and is relevant to drug-receptor interactions. However, nobody would expect that two compounds with a similar log *P* value would also necessarily exhibit a similar biological activity.

Two classes of descriptors that have been proven *(20)* to display favorable neighborhood behavior are substructural fingerprints *(27,28)* and, even more so, steric fields based on single topomeric conformation of sidechains *(29)*. The former (substructural fingerprints) represents molecules as strings of bits, where each bit accounts for the presence or absence of a given substructural moiety. The latter one describes the 3D shape of molecules. Both are being used extensively in our own library design activities. "Topomeric fields" have proven to be very useful descriptor in order to design focused libraries and to perform lead-hopping, where one starts with an active structure from one chemical series and finds shape-similar compounds in a different chemical series *(30,31)*.

1.5. Design of Homopiperazine Libraries for General Screening and Lead Follow-up

To exemplify the techniques described above, we have used commercially available software tools to undertake the following design work (*see* **Note 1**). A three-component virtual library (core and two reagents) was created using the homopiperazine chemistry shown in **Fig. 1**. Various physical properties and substructural fingerprints, such as diversity descriptors, were calculated for the virtual compounds, after which a reduced subset was generated using reagent filtering techniques. The general screening library was created by selecting compounds from the filtered subset employing a molecular diversity algorithm in a reagent-based design. From the general screening library, a single compound was arbitrarily defined as biologically active. Using a prod-

Table 1
Results from General Screening and Focussed Library Design[a]

Step	Number of aldehydes	Number of acids	Number of virtual products
Search in ACD	2525	4298	10852450
Druglike	2515	4292	10794380
Compatible	2272	3663	8322336
Nitro-filter	2124	3259	6922116
MW-filter	1117	1085	1211945
Clog *P* filter	437	1078	471086
Diversity-selection	50	50	2500
Similarity search	Product	Product	1231
Focus-selection	10	10	100

[a] See text for discussion.

uct-based approach a focused library of compounds that fulfilled specified similarity criteria was generated.

The results of our library designs are summarized in **Table 1**. The effect of the different filter steps on the number of reagents and the number of virtual products reflects the real case where the highest cut comes from molecular weight and log *P* filtering. For brevity, in this example we employ a limited number of filters relative to those implemented in a true design. Within a true design, there are many more queries for the notlist-, the druglike-, and the incompatibility filters. Starting from a virtual library of more than 10 million products, the diversity selection is carried out on a subset of 470,000 compounds (5%) and yielded a final set of 2500 compounds (0.03%), which broadly covers the chemistry of this virtual library. From the 2500 diversely selected compounds, one was defined as active. A similarity search of the 470,000 property filtered compounds yielded 1231 (0.015%) compounds, of which 100 were extracted into a 10 reagent × 10 reagent combinatorial matrix. The coverage of chemical space for both the general screening library and the focus library is exemplified in **Fig. 3** by visualizing the substructural fingerprints of corresponding compounds using principle component analysis followed by a modified nonlinear mapping technique *(32)*. The general screening library is compared to the focused library around an individual compound of the screening library. Here the distances between the compounds of the screening library are, as required, very large and distributed all over the graph. The focused library represents the reverse case, distances of compounds from the focused library to the lead compound are rather short so that all members are bundled around the lead compound.

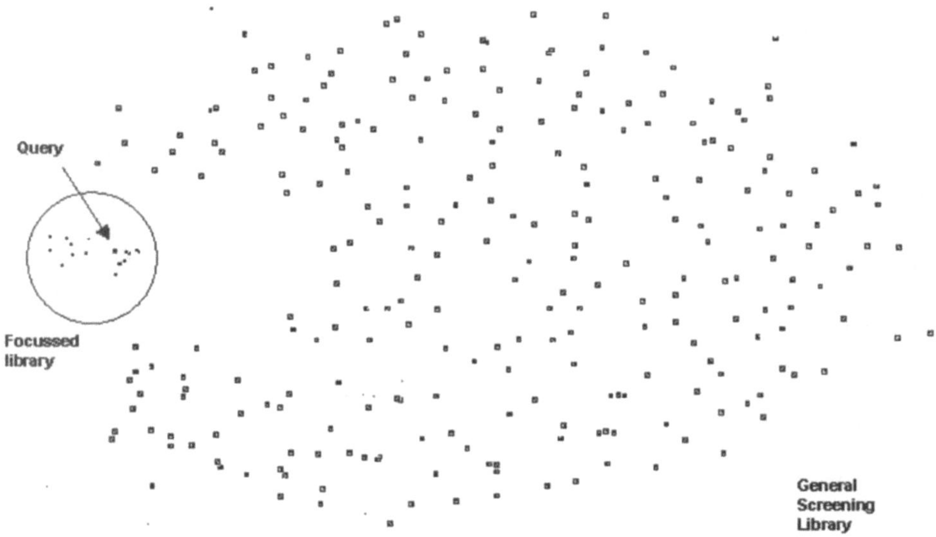

Fig. 3. Nonlinear map *(32)* for the two subsets of the general screening and focused library design (*see* text for discussion). The subsets of 250 (general screening library) and 22 (focused library) representative compounds have been obtained by applying the OptiSim method *(33)*.

2. Materials

2.1. Hardware, Software, and Data to Prepare a Reagents-Database from ACD

1. A computer system from Silicon Graphics (Silicon Graphics, Inc., Mountain View, CA) using the IRIX operating system (version 6.5 or higher).
2. ACD—Available Chemicals Directory (MDL Information Systems, Inc., San Leandro, CA) (*see* **Note 7**) *(34)*.
3. Software packages Sybyl version 6.7, Unity version 4.2, and ChemEnlighten version 2.8 including modules Legion, Selector, and CLOGP (Tripos, Inc., St. Louis, MO) (*see* **Note 8**).
4. Perl programming software (www.perl.com) (*see* **Note 9**).

2.2. Input Data for Design of General Screening Library

1. The query substructure for reagent searching (*see* **Note 10**): Enter the query into a text file using an editor like "vi":
   a. For aliphatic aldehydes:
   `CCH=O`
   b. For aromatic acids:
   `Hev:CC(=O)OH`

2. A text file containing the notlist for reagent searching: Enter line notations for unwanted fragments, e.g., to remove bifunctional acids (line 1 of the file) as well as a list of atom types (elements) (line 2) you do not want to see in the reagent structures (*see* **Note 11**):
   a. For aliphatic aldehydes
      ```
      CH=O.CH=O
      ```
   b. For aromatic acids:
      ```
      CC(=O)OH.CC(=O)OH
      Any[not=H,C,N,O,F,P,S,Cl,Br]
      ```
3. The Shell-Script (*see* **Note 12**) sybyl_wrapper.sh, also a text file:
   ```
   #!/bin/sh
   . $TA_ROOT/lib/profile
   sybyl @<
   setvar INPUT $2
   setvar OUTPUT $3
   setvar ARG $4
   take $1
   @
   ```
4. The SPL-script "reagent_convert.spl" for converting reagent structures into synthon structures (*see* **Note 13**):
   ```
   setvar fh1 %open($INPUT "r")
   setvar fh2 %open($OUTPUT "w")
   while %not(%eof($fh1))
     setvar ct %sln_ct_read($fh1)
     setvar synthon %sln_mutate_sln($ct $ARG)
     setvar regid %sln_get_sln_attr($ct "regid")
     if %pos("<" $synthon)
       setvar len %math(%strlen($synthon) - 1)
       setvar synthon %cat(%substr($synthon 1 $len)
         ";name=" $regid ">")
     else
       setvar synthon %cat($synthon "<na me=" $regid ">")
       endif
     setvar out %write($fh2 $synthon )
     % sln_ct_delete($ct)
   endwhile
   % close()
   ```
5. The SPL-Script "csln_create.spl" for generating the combinatorial sybyl line notation (CSLN):
   ```
   echo %sln_generate_csln($INPUT $ARG $INPUT) >$OUTPUT
   ```
6. The text file "non_druglike.sln" for creating a druglike subset (*see* **Note 10**):
   ```
   N=[!r]N<na me=diazo>
   CC(=O)C(=O)C<na me=alpha-diketone>
   ```

7. The file "incompatible.sln" for creating a compatible subset (*see* **Note 10**):
```
Hev:COH<na me=phenols>
C[not=C*~Het]C(=O)C[not=C*~Het]<na me=ketone>
```
8. Input code, programmed in Perl, to generate a combinatorial matrix of products from compounds selected in a focused library (*see* **Note 14**).

3. Methods

3.1. Building of Reagent Database

1. Convert RDF-files from ACD-CD into Unity Database (*see* **Note 15**): For each RDF-file as supplied on CD run program dbimport using command line:
```
dbimport -database ACD.tdb -type maccs -property_data
-coords -maccs_regname 'MDLNUMBER' <na me-of-rdfile>
```
2. To speed up structural searches generate molecular fingerprints (*see* **Note 16**): Run program dbmkscreen on database created in first step using command line:
```
dbmkscreen -class 2d -database ACD.tdb
```

3.2. Design of a Homopiperazine Library for General Screening

1. Search for proper reagents running program dbsearch using command line:
```
dbsearch -database ACD.tdb -qtype 2d -qfile aldehydes.query
-hitlist aldehydes.hits -notlist aldehydes.notlist
```
2. Convert reagents into synthons (*see* **Notes 2** and **4**) using sybyl_wrapper script in combination with SPL-script slnconvert.spl:
```
/bin/sh sybyl_wrapper.sh slnconvert.spl aldehydes.hits
homopips.X1 'CH=O R1=1,1 CH2R1 R1=1,1'
```
3. Repeat **steps 1** and **2** for acids:
```
dbsearch -database ACD.tdb -qtype 2d -qfile acids.query
-hitlist acids.hits -notlist acids.notlist
/bin/sh sybyl_wrapper.sh slnconvert.spl acids.hits
homopips.X2 'C(=O)OH R1=1,1 C(=O)R1 R1=1,1'
```
4. Build combinatorial SLN (*see* **Note 17**) using sybyl-wrapper-script in combination with SPL-script csln_create.spl:
```
/bin/sh sybyl_wrapper.sh csln_create.spl homopips
homopips.csln 'X1N[2]CH2CH2N(X2)CH2CH2CH2@2 1,X1R1=2;
10,X2R1=9'
```
5. Partially enumerate CSLN (*see* **Note 5**) on variation site X1 running program dbexplode using command line:
```
dbexplode -format "&1N" -enum "Y_02=1" -output
homopips.X1.chom homopips.csln
```
6. Create Unity database from enumerated structure list using program dbimport:
```
dbimport -database homopips.X1.tdb -type sln
homopips.X1.chom
```
7. Calculate fingerprints for all structures using command line:
```
dbmkscreen -class 2d -database homopips.X1.tdb
```

8. Calculate molecular weights for all structures using program dbcomputecolumn:
   ```
   dbcomputecolumn -database homopips.X1.tdb -colname
   molwt -calctype molwt
   ```
9. Calculate Clog *P*-values for all structures:
   ```
   dbcomputecolumn -database homopips.X1.tdb -colname
   clogp -calctype clogp
   ```
10. Create druglike subset (*see* **Note 18**):
    ```
    Echo 'Any' _ dbsearch -database homopips.X1.tdb
    -notlist non_druglike.sln -report regkey _ dbset -da-
    tabase homopips.X1.tdb -create -setname druglike
    ```
11. Create subset of druglike structures having no incompatible groups:
    ```
    echo 'Any' _ dbsearch -database homopips.X1.tdb
    -notlist incompatible.sln -use_subset druglike -report
    regkey _ dbset -database homopips.X1.tdb -create
    -setname compatible +buffer
    ```
12. Calculate number of occurences of nitro groups (*see* **Note 19**):
    ```
    dbcomputecolumn -database homopips.X1.tdb -colname
    nitro -calctype patterncount -query 'N[+1](=O)O[-1]'
    ```
13. Filter for all reagents that are druglike, compatible, and have no nitro group
    ```
    dbfilter -database homopips.X1.tdb -in_selset
    compatible -out_selset nitro -query data nitro=0
    ```
14. Filter for reagents having a molecular weight below 217 (*see* **Note 20**):
    ```
    dbfilter -database homopips.X1.tdb -in_selset
    compatible.dir/nitro -out_selset molwt -query data
    molwt=0-605
    ```
15. Filter for reagents having a Clog *P* between -5 and 6 (*see* **Note 21**):
    ```
    dbfilter -database homopips.X1.tdb -in_selset
    compatible.dir/nitro.dir/molwt -out_selset clogp
    -query data clogp=-5-6
    ```
16. Select diverse set of reagents (*see* **Note 22**) using OptiSim algorithm in ChemEnlighthen:
    ```
    dbdiverse -database homopips.X1.tdb -in_selset
    compatible.dir/nitro.dir/molwt.dir/clogp -out_selset
    design_1 -column standard_2FPRINT:0.15 -subsamplesize
    5 -max_sel 50
    ```
17. Write out the registration id's of the selected reagents:
    ```
    echo 'Any' | dbsearch -database homopips.X1.tdb
    -report regkey -use_subset compatible.dir/nitro.dir/
    molwt.dir/clogp/design_1 >X1_1.mfcd
    ```
18. For the next step create a list of registration ids which are supposed to come from unavailable reagents:
    ```
    head -18 X1.mfcd > unavailable.X1_1.mfcd
    ```
19. Replacement selection for unavailable reagents (*see* **Note 6**), first remove unavailable reagents from the previous design:

```
cat unavailable.X1.txt _ dbset -database
homopips.X1.tdb -remove -setname design_1
```

20. Next remove unavailable reagents from filtered set:
```
cat unavailable.X1.txt _ dbset -database
homopips.X1.tdb -remove -setname compatible.dir/
nitro.dir/molwt.dir/clogp
```
21. Finally select another set of reagents to fill up the gaps:
```
dbdiverse -database homopips.X1.tdb -in_selset
compatible.dir/nitro.dir/molwt.dir/clogp -out_selset
design_2 -starter_set design_1 -column
standard_2FPRINT:0.15 -subsamplesize 5 -max_sel 50
+restore_excluded
```
22. Repeat **steps 5–21** for variation site X2.
23. For registration purposes enumerate structures using selected reagents from both variation sites. First extract registration ids of reagents:
```
echo 'Any' | dbsearch-database homopips.X1.tdb -report
regkey -use_subset compatible.dir/nitro.dir/molwt.dir/
clogp.dir/design_2 >X1_2.mfcd
echo 'Any' | dbsearch-database homopips.X2.tdb -report
regkey -use_subset compatible.dir/nitro.dir/molwt.dir/
clogp.dir/design_2 >X2_2.mfcd
```
24. Then enumerate those products made up by the selected reagents in X1 and X2, respectively:
```
/bin/sh sybyl_wrapper.sh enumerate.spl 'homopips.csln
X1_2.mfcd X2_2.mfcd' homopips.design.chom
X1N[2]CH2CH2N(X2)CH2CH2CH2@2
```

3.3. Design of an Analog Library Based on a Biologically Active Homopiperazine Compound

1. Export reagent id's from the general screening library which fulfil the required molecular weight, Clog *P*, druglikeness, and chemistry compatibility criteria for druglike compounds (*see* **Note 23**):
```
echo 'Any' | dbsearch-database homopips.X1.tdb -report
regkey -use_subset compatible.dir/nitro.dir/molwt.dir/
>X1_2.list
```
2. Repeat for X2:
```
echo 'Any' | dbsearch-database homopips.X2.tdb -report
regkey -use_subset compatible.dir/nitro.dir/molwt.dir/
clogp >X2.list
```
3. To generate products that may be used for similarity searching, enumerate all the potential products from the reagents and core with the command:
```
/bin/sh sybyl_wrapper.sh enumerate.spl 'homopips.csln
X1.list X2.list' focus.hits X1N[2]CH2CH2N(X2)CH2CH2CH2@2'
```

4. Create Unity database from enumerated structure list using the unity program dbimport:
```
dbimport -database focus.tdb -type sln focus.hits
```
5. Calculate 2D structural fingerprints for all structures using command line:
```
dbmkscreen -class 2d -database focus.tdb
```
6. The compound identified as a lead must be entered in a text file such that it can be used for searching for similar compounds in a unity database. Create a query file (named, for example, active.query) containing the SLN of the structure to be searched with. Use a text editor such as vi, or export the SLN of a named structure from a unity hitlist or database. The contents of the query file should be three lines long. The first line is the SLN of the query. The second line should contain a single dot. The third line contains the similarity cut off value (a double precision value between 0.00% and 100.00%), which will be applied for the similarity search. For example:
```
C[1](:C(:CH:CH:C(:CH:@1)OH)C(=O)N[14]CH2CH2N
(CH2CH2CH2@14)CH2CH2C(CH3)(CH3)CH3)OH
.
cutoff=85.00
```
7. Search using the query to find compounds in the unity database focus.tdb that have a structural similarity of 85% or greater using the command (*see* **Note 24**):
```
dbsearch -database focus.tdb -dbtype unitydb -qtype
sim2d -qfile active.query -report reg -hitlist
focus_results.hits -qfmt sln -fprintcol
standard_2FPRINT
```
8. Import hit compounds into a new Unity database and extract a list of registration id's (*see* **Note 25**):
```
dbimport -database focus_results.tdb focus_results.hits
dbexport -database focus_results.tdb -type tsv -query
regid -output product_list.txt -property_data '*'
```
9. Generate an optimum 10 × 10 combinatorial matrix for similarity search products. If using the Perl script described in the **Subheading 2.2.**, use the command (*see* **Note 26**):
```
gen_matrix.pl -nR110 -nR2 10 -nRuns 100 product_list.txt
```

4. Notes

1. At Tripos we are using a proprietary library-design technology, ChemSpace™, to assist in the selection of novel compounds. Chemspace™, which is tightly integrated with the library production process operated at our synthetic laboratories in Bude, UK, is used to create and search large virtual libraries at a rate of two trillion compounds per hour. This computational approach may be used to optimize the diversity of combinatorial libraries, or, conversely, to seek out similar molecules for focused libraries. In addition, the topomeric field descriptor incorporated into the Chemspace™ technology is a powerful tool in the field of patent navigation. It is possible to "lead hop" out of a patented series of

biologically active compounds by searching Tripos' virtual libraries and identifying topomeric neighbors with different chemical scaffolds.

2. These reagent classes are picked arbitrarily. In practice one might want to add additional reagent classes like acid chlorides, sulfonylclorides, or isothiocyanates as building blocks.
3. This chemistry has recently been developed in our laboratories and compounds of this type are part of our LeadQuest® screening library.
4. There are at least two alternative ways in how virtual combinatorial libraries can be built: reaction-scheme entries or product-only transformations. The first method requires a description of a reaction, a number of rules for how to convert the reagent structures into its synthon forms; the combination of synthons will then result in product structures. The product-only method is based on a core structure with defined variation sites on which different substituents can be selected. In the example given in this chapter, a combination of both methods is used. First reagents are converted into synthons and in a subsequent step a core molecule is given with two variation sites where the synthons formed in the first step are selected as substituents to form the product structures.
5. Enumeration is the process where the core structure and a given set of r-groups are combined to build complete product structures, given a set of four substituents on variation site 1 (X1) and another four substituents on variation site 2 (X2) enumeration will generate $4 \times 4 = 16$ individual product structures.
6. One of the most critical factors in combinatorial chemistry is the availability of reagents. There are several reasons for unavailability: high prices, small quantities, bad qualities, outdated catalog information, unreliable suppliers, to mention just a few. Within our company's chemical repository system ChemCore™, we have introduced a scoring system to determine reagent availability. Depending on the library size and the amount of reagent material required, a certain range of score is applied for filtering. Each reagent in ChemCore™ has an availability score, so that filtering on these scores will pass only available reagents. In an analogous way, a reactivity-based score is being developed such that reagent success rates (yields and purities based on analysis of products) are being accumulated, given a certain reactivity score range filtering will only pass reagents with previously specified reactivities.
7. Apart from the ACD there are many other sources of catalogs of reagents available. Many chemical suppliers offer their compound collections on CD in a similar form to the ACD.
8. Apart from Tripos (St. Louis, MO) there are other software vendors offering analogous software packages like "Diversity Explorer" from Accelrys (San Diego, CA) or "Project Library" from MDL (San Leandro, CA).
9. Perl is an interpreted language optimized for scanning arbitrary text files, extracting information from those text files, and printing reports based on that information. It is also a good language for many system management tasks. The standard release of Perl is distributed only in source code form. You can find this at http://www.perl.com/CPAN/src/latest.tar.gz, which is in standard Internet format (a gzipped archive in POSIX tar format).

10. Within the Tripos molecular modeling package 'Sybyl' a language called Sybyl-Line-Notation (SLN) *(35)* is often used to define structures, fragments, Markushes, and as in this case substructural queries. SLN is derived from Daylight's (Santa Fe, NM) SMILES *(36)* for string presentations of structures and is mainly used by computational chemists.
11. The reagent search normally yields a large pool of monomers. On the basis of functional groups, fragments, or element types that should not be in the synthon, the list can be reduced. The notlist is the primary tool to get rid of unwanted monomers. Each structural hit resulting from the search is checked for any occurrence of a structural query and, if there is, the hit is rejected. The reagent structures from ACD often contain mixtures or counterions, which need to be stripped off from the reagent structure. In this example, we did not apply this step of stripping salts and mixtures, therefore product structures resulting from this design could potentially carry counterions or mixture parts. The support group at Tripos offers highly efficient tools for this and other purposes (http://www.tripos.com/services).
12. The main intention was to give a very simple example that even the non-expert can reproduce, therefore, every step that is normally carried out by a computational chemist within Sybyl is converted into a single clear command. To circumvent a graphical user interface, a shell script is needed that handles input, output and, command arguments for use in Sybyl.
13. Sybyl has a rich macro programming language (SPL) that allows constructing complex batch operations within short scripts. This is mainly used by computational chemists. For detailed information there is an SPL-Manual. Tripos also offers an archive of SPL-scripts and trainings courses for SPL-programming (*see* **Note 11**).
14. The application of combinatorial chemistry in drug design is not solely limited to general screening libraries, it may also be used when designing a lead follow up library. Having determined a selection of reagents, which, when bonded to the appropriate scaffold, mimic the lead compound in structure, it may be necessary to enumerate these reagents into a combinatorial matrix. Various matrix selection methods may be employed, from manual selection to complex algorithms *(37)*. In this case, the use of a simple script programmed in Perl (available from the authors) allows the rapid selection of the most frequently occurring reagents that fit the structural query.
15. The compound database on the ACD-CD is supplied in RDF (Reaction Data File) format *(38)*. For usage of this database within the Tripos software package a conversion into the Unity format is required.
16. The molecular fingerprints contain information about the presence of molecular fragments encoded in binary format. They were originally introduced to speed up large searches in chemical databases and were also used for similarity *(39)*. Through validation studies *(20,27,28)* their ability for the diversity selections has been revealed.
17. The combinatorial SLN is a very compact and dense format of a combinatorial library where the core structure with its variation sites is defined together with the corresponding r-groups (substituents).

18. In the example, given substructural searches are applied to remove groups that generally react with biological molecules like proteins or nucleic acids.
19. The occurrence of a nitro group in a drug molecule is controversially discussed—some people consider it inappropriate, some people tolerate it. Depending on the purpose or target of the combinatorial library, there will always be functional groups that are either required or inappropriate for the design of a library. Substructural searches are the most convenient way to deal with such conditions.
20. In this example, we make substituent selections, therefore we need to filter for substituent properties. Molecular weight is an additive property of a structure. To filter for substituents of a given size range, you need to calculate the molecular weight of the core structure plus the molecular weight of the constant substituent on the other variation site and add this to the maximum value of molecular weight range ***(40)***. In our example, the molecular weight of the core was 166. If we want to limit the molecular weight of products up to a value of 600, we subtract 166 from 550 and divide by 2, which gives 217. The molecular weight of our constant substituent on the other variation site was 222. When we now add 222 + 166 + 217 we get 605, this is the maximum value used for molecular weight filtering on variation site X1.
21. Clog *P*-values of structures derived from the sum of Clog *P*-values of its corresponding substituents with its open valences filled by methyl groups correlate well with Clog *P*-values calculated on the whole structure ***(40)***. As mentioned in **Note 20**, to filter substituents you need to calculate the contribution of the core structure plus the constant substituent on the other variation site and add it to the maximum value of the allowed Clog *P* range. Analogous to the molecular weight we have added the Clog *P* of the core structure (0.31), the Clog *P* of the constant substituent (3.44) and the allowed range on either variation site (2.35) to use a range of –5 to +6 for Clog *P* filtering.
22. Several selection methods ***(41)*** are known. In our example, we have used the OptiSim-algorithm ***(33)*** to perform a representative selection of compounds.
23. In this example, we filter compounds for properties such as molecular weight and Clog *P* based on reagents. To accelerate to speed of the process, we use filtered data generated from the general screening library. However, generation of all potential products with property data, followed by product based filtering is possible if time and resources allow.
24. Molecular similarity is calculated from 2D fingerprints for all product structures. This is achieved by the utilization of the Tanimoto coefficient ***(42)***, which measures similarity of structures on a pairwise basis. The Tanimoto coefficient is the ratio of the number of common bitsets in two molecules, divided by the number of bit sets in either. Tanimoto coefficients are expressed as ratios between 0 and 1, or percentages between 0 and 100. A ratio of 0.85 or greater has been shown as a valid approach for quantifying molecular similarity ***(20)*** using 2D Unity fingerprints. Apart from using Tanimoto coefficients to determine the difference between of structures quantified by 2D fingerprints, other validated descriptors include connectivity rules, log *P*/molar refractivity, "3D" fingerprints ***(43)***, and topomeric fields ***(29)***.

25. It is not possible to extract a list of product registration id's from a unity hitlist, hence the list must be imported into a new database, after which lists of id's may then be extracted.
26. The advantages of using a combinatorial matrix when synthesizing a focus library is that, as with a general screening library, you can make many products with relative ease. In addition, if your focus library is a continuation of work from a general screening library, then it is likely that you will have synthetic protocols in place already. One disadvantage of building a combinatorial library, rather than "cherry picking" the compounds, is that you may choose reagents based on the frequency that they appear in product compounds, rather than their novelty.

References

1. Gibbon, J. A., Taylor, E. W., and Braeckman, R. A. (1998) in *Combinatorial chemistry and molecular diversity in drug discovery*, (Gordon, E. M. and Kerwin, J. F., eds.), Wiley-Liss, New York, NY, pp. 453–474.
2. Martin, E. J. and Critchlow, R. E. (1999) Beyond mere diversity: Tailoring combinatorial libraries for drug discovery. *J. Combinat. Chem.* **1,** 32–45.
3. Gorse, D. and Lahana, R. (2000) Functional diversity of compound libraries. *Curr. Opin. Chem. Biol.* **4,** 287–294.
4. Walters, W.P, Stahl, M. T., and Murcko, M. A. (1998) Virtual screening—an overview. *Drug Discov. Today* **3,** 160–178.
5. Leach, A. R. and Hann, M. M. (2000) The in silico world of virtual libraries. *Drug Discov. Today* **5,** 326–336.
6. Good, A. C. and Lewis, R. A. (1997) New methodology for profiling combinatorial libraries and screening sets cleaning up the design process with HARPick. *J. Med.Chem.* **40,** 3926–3936.
7. Van Drie, J. H. and Lajiness, M. S. (1998) Approaches to virtual library design. *Drug Discov. Today* **3,** 274–283.
8. Linusson, A., Gottfries, J., Lindgren, F., and Wold, S. (2000) Statistical molecular design of building blocks for combinatorial chemistry. *J. Med. Chem.* **43,** 1320–1328.
9. Gillet, V. J., Willet, P., and Bradshaw, J. (1997) The effectiveness of reactant pools for generation of structurally diverse combinatorial libraries. *J. Chem. Inf. Comput. Sci.* **37,** 731–740.
10. Jamois, E. A., Hassan, M., and Waldman, M. (2000) Evaluation of reagent-based and product-based strategies in the design of combinatorial library subsets. *J. Chem. Inf. Comput. Sci.* **40,** 63–70.
11. Gillet, V. J., Willet, P., Bradshaw, J., and Green, D. V. S. (1999) Selecting combinatorial libraries to optimize diversity and physical properties. *J. Chem. Inf. Comput.* Sci. **39,** 169–177.
12. Pearlman, R. S. and Smith, K. M. (1998) Novel software tools for chemical diversity. *Perspect. Drug Discov. Design.* **9,** 339–353.
13. Pearlman, R. S. and Smith, K. M. (1999) Metric validation and the receptor-relevant subspace concept. *J. Chem. Inf. Comput. Sci.* **39,** 28–35.

14. Cramer, R. D., Patterson, D. E., Clark, R. D., Soltanshahi, F., and Lawless, M. S. (1998) Virtual compound libraries: A new approach to decision making in molecular discovery research. *J. Chem. Inf. Comput. Sci.*, **38,** 1010–1023.
15. Rishton, G. M. (1997) Reactive compounds and *in vitro* false positives in HTS. *Drug Discov. Today* **2,** 382–384.
16. Lewis, R. A., Mason, J. S., and McLay, I. M. (1997) Similarity measures for rational set selection and analysis of compbinatorial libraries: the diverse property-derived (DPD) approach. *J. Chem. Inf. Comput. Sci.* **37,** 599–614.
17. Hann, M., Hudson, B., Lewell, X., Lifely, R., Miller, L., and Ramsden, N. (1999) Strategic pooling of compounds for high-throughput screening. *J. Chem. Inf. Comput. Sci.* **39,** 897–902.
18. Clark, D. E. and Pickett, S. D. (2000) Computational methods for the prediction of "drug-likeness." *Drug Discov. Today* **5,** 49–58.
19. Lipinski, C. A., Lombardo, F., Dominy, B. W., and Feeney, P. J. (1997) Experimental and computational approaches to estimate solubility and permeability in drug discovery and development settings. *Adv. Drug Delivery Rev.* **23,** 3–25.
20. Patterson, D. E., Cramer, R. D., Ferguson, A. M., Clark, R. D., and Weinberger, L. E. (1996) Neighborhood behavior: A useful concept for validation of "molecular diversity" descriptors. *J. Med. Chem.* **39,** 3049–3059.
21. Johnson, M. A. and Maggiora, G. M. (1990) *Concepts and applications of molecular similarity,* Wiley, New York.
22. Mason, J. S., Morize, I., Menard, P. R., Cheney, D. L., Hulme, C., and Labaudiniere, R. F. (1999) New 4-point pharmacophore method for molecular similarity and diversity applications: overview of the method and applications, including a novel approach to the design of combinatorial libraries containing priviledged substructures. *J. Med. Chem.* **42,** 3251–3264.
23. Picket, S. D., Mason, J. S., and McLay, I. M. (1996) Diversity profiling and design using 3D pharmacophores: pharmacophore-derived queries (PDQ). *J. Chem. Inf. Comput. Sci.* **36,** 1214–1223.
24. Drewry, D. H. and Young, S. S. (1999) Approaches to the design of combinatorial libraries. *Chemometr. Intell. Lab. Syst.* **48,** 1–20.
25. Gorse, D., Rees, A., Kaczorek, M., and Lahana, R. (1999) Molecular diversity and its analysis. *Drug Discov. Today* **4,** 257–264.
26. Mason, J. S. and Hermsmeier, M. A. (1999) Diversity assessment. *Curr. Opin. Chem. Biol.* **3,** 342–349.
27. Brown, R. D. and Martin, Y. C. (1996) Use of structure-activity data to compare structure-based clustering methods and descriptors for use in compound selection. *J. Chem. Inf. Comput. Sci.* **36,** 572–584
28. Matter, H. (1997) Selecting optimally diverse compounds from structure databases: a validation study of two-dimensional and three-dimensional molecular descriptors. *J. Med. Chem.* **40,** 1219–1229.
29. Cramer, R. D., Clark, R. D., Patterson, D. E., and Ferguson, A. M. (1996) Bioisosterism as a molecular diversity descriptor: steric fields of single topomeric conformers. *J. Med. Chem.* **39,** 3060–3069.

30. Andrews, K. M. and Cramer, R. D. (2000) Toward general methods of targeted library design: topomer shape similarity searching with diverse structures as queries. *J. Med. Chem.* **43,** 1723–1740.
31. Cramer, R. D., Poss M. A., Hermsmeier, M. A., Caulfield, T. J., Kowala, M. C., and Valentine, M. T. (1999) Prospective identification of biologically active structures by topomer shape similarity searching. *J. Med. Chem.* **42,** 3919–3933.
32. Clark, R. D., Patterson, D. E., Soltanshahi, F., Blake, J. F., and Matthew, J. B. (2000) Visualizing substructural fingerprints. *J. Mol. Graphics Mod.* **18,** 404–411.
33. Clark, R. D. (1997) OptiSim: An extended dissimilarity selection method for finding diverse representative subsets. *J. Chem. Inf. Comput. Sci.* **37,** 1181–1188.
34. This compendium of all commercially offered compounds may be obtained from MDL Information Systems, Inc., 140 Catalina Street, San Leandro, CA 94577, USA.
35. Ash, S., Cline, M. A., Homer, R. W., Hurst, T., and Smith, G. B. (1997) SYBYL line notation (SLN): a versatile language for chemical structure representation. *J. Chem. Inf. Comput. Sci.* **37,** 71–79.
36. Weininger, D. J. (1988) SMILES, a chemical language and information system. 1. Introduction of methodology and encoding rules. *J. Chem. Inf. Comput. Sci.* **28,** 31–36.
37. Agrafiotis, K. and Lobanov,V. S. (2000) Ultrafast algorithm for designing focused combinatorial arrays. *J. Chem. Inf. Comput. Sci.* **40,** 1030–1038.
38. Dalby, A., Nourse, J. G., Hounshell, W. D., et al. (1992) Description of several chemical structure file formats used by computer programs developed at Molecular Design Limited. *J. Chem. Inf. Comput. Sci.* **32,** 244–255.
39. Willett, P. (1986) *Similarity and clustering in chemical information systems,* Research Studies Press, Letchwork, U. K.
40. Shi, S., Peng, Z., Kostrowicki, J., Paderes, G., and Kuki, A. (2000) Efficient combinatorial filtering for desired molecular properties of reaction products. *J. Mol. Graphics. Mod.* **18,** 478–496.
41. Holliday, J. D. and Willett, P. (1996) Definitions of "dissimilarity" for dissimilarity-based compound selection. *J. Biomol. Screening* **1,** 145–151.
42. Willett, P. and Winterman, V. (1986) A comparison of some measures for the determination of intermolecular structural similarity. *Quant. Struct. Activ. Relat.* **5,** 18–25.
43. Kubinyi, H. (ed.) (1993) *3D QSAR in drug design,* ESCOM, Leiden, The Netherlands.

19

Reagent-Based and Product-Based Computational Strategies in Library Design

Eric A. Jamois

1. Introduction

With the current and ever-growing offering of reagents along with the vast palette of organic reactions, virtual libraries accessible to combinatorial chemists have dramatically increased in size. Yet, extracting suitable subsets for experimentation is an essential step in the design of combinatorial libraries. Several approaches to this problem can be envisaged, involving either reagent-based or product-based considerations. Reagent-based designs tend to be popular with chemists, as they provide highly practical means of weeding extensive reagent lists based on property and substructure considerations. They can also provide suitable reagent selections for model studies. However, reagent-based designs also overlook the extent of chemical transformations involved in generating products. Several studies have demonstrated the superiority of product-based designs in yielding diverse and representative subsets. Although more computationally intensive, product-based approaches present significant advantages, and they are also more amenable to the incorporation of drug-like restraints.

In the last few years, the efficient design of combinatorial libraries has become increasingly important for both lead identification and lead follow-up programs *(1–4)*. The vast palette of available reagents has had a large impact on the size of synthetically accessible libraries for both general screening and targeted applications. Extremely large libraries are often inaccessible and must be reduced to smaller subsets in order to accommodate current limitations of synthesis and screening equipment. Indeed, multistep reactions often involve common reactants for which large selections are available from commercial sources. The Available Chemicals Directory provides a large inventory of

From: *Methods in Molecular Biology, Combinatorial Library Methods and Protocols*
Edited by: L. B. English © Humana Press Inc., Totowa, NJ

reagents from which an appropriate selection can be made. The problem therefore becomes the selection of the reagents that best suits our needs in terms of reactivity, selectivity, cost, product diversity, product drug-likeness, and possibly other criteria, taken either individually or in combination.

A number of techniques have been used toward the identification of library subsets ***(5–11)***. A first category of techniques involves reagent-based selections, that is, selections involving reagent properties only. Selection criteria may involve such diverse factors as reactivity, selectivity, cost, substructure, and also three-dimensional (3D) pharmacophore information ***(12,13)***. A second class of techniques involves the selection of diverse sets of products. Again, a range of clustering ***(11)*** and dissimilarity-based methods ***(6–10)*** has been used for selections at the product level. One of the limitations of dissimilarity-based methods applied to products is the lack of combinatorial constraints, thus the selections produced are synthetically inefficient. A third type of technique involves combinatorially constrained product selections. In this case, the combinatorial array may be maintained by the selection of reagents, but the evaluation of diversity of the resulting subset is performed at the product level. Such a procedure using genetic algorithms or Monte Carlo optimization has previously been described ***(14,15)***. Recently, techniques combining multiple optimization criteria such as diversity, cost efficiency, and drug-like character have appeared ***(16)***. It is therefore possible to combine the practical aspects of reagent selection or reagent bias with the more rigorous product-based approaches.

One major distinguishing feature between the techniques is the computational resource required to identify library subsets. Most reagent-based selection techniques require little computational resource owing to the limited size of reagent lists. As a matter of fact, many selections have been performed through visual inspection of the reagent lists. Also, the size of the problem is only additive with respect to the size of each list. On the other hand, product-based techniques often require complex and time-consuming procedures due to the multiplicative nature of the problem. A reagent array of $50 \times 150 \times 200 \times 350$ for a four-substituent system $R1 \times R2 \times R3 \times R4$ would generate 525 million products. While the reagent lists can be handled separately by common analysis techniques, handling the full set of products would pose a serious challenge.

2. Reagent-Based Design

Reagent-based design has been in practice among chemists for many years. Its practical appeal and efficiency cannot be denied. In recent years, chemists have been selecting building blocks that incorporate drug-like fragments, with the hope that these would result in drug-like molecules. Chemists have also

A

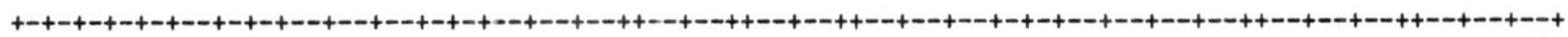

B

Fig. 1. (A) Selection by maximum dissimilarity (sample size = 40). (B) Selection by D-optimal design (sample size = 40).

learned to avoid building blocks with ambident reactivity yielding mixtures of products. Both of these procedures can be performed on the reagent lists using simple substructure considerations. Additional considerations involve cost and availability where reagents that are inexpensive and/or already in inventory should be selected preferentially. A number of reagent-based design strategies have been reported and illustrate the popularity of this approach ***(12,13,17)***.

As a typical problem involving reagent selection, we may be interested in investigating the scope and selectivity of an organic reaction. This may be a valuable model study as we are considering a very large virtual library. Knowing that some reactions will go to completion quickly but provide mixtures while others will simply not proceed, already narrows down the set of reagents to be considered. Somewhere in between these extreme behaviors, we can find reagents for which the reaction goes to completion and also provides the desired selectivity. If we consider reactivity as the parameter governing the scope and selectivity or our reaction, we can provide a diverse sampling of reagents based on these criteria. Many schemes have been reported for the selection of diverse subsets involving a variety of diversity metrics ***(6–10)***. As these metrics vary in their behavior, that is, the sampling of property space that they provide, we may be tempted to ask which one(s) should be selected to solve a particular problem. In this example, it can be argued that we are more interested in the extremes of our reactivity scale rather than with the middle where we do not expect any problem. As an illustration, if we could choose between a relatively uniform sampling provided by Maximum Dissimilarity (**Fig. 1A**) and a skewed sampling from D-Optimal design (**Fig. 1B**) favoring the extremes of our reactivity scale, the latter would give more information on the usable scope of our reaction.

Reagent-based design can also been applied to generate diverse or, in some cases, focused subsets based on bioisosteric replacement ***(18)***. Several methods have been used in the selection of monomers, including maximum dissimilarity ***(6)***, D-Optimal design ***(18)***, and clustering ***(15)***. Hierarchical cluster analysis has been well validated in compound selection *(11)*; it has also provided acceptable results in our earlier work *(15)*. Although it does not provide an optimal solution, the method is fast and easy to implement—reagents are clipped prior to analysis and only retain the fragment portion attached to the scaffold. At this point several sets of descriptors can be considered for analysis, ranging from MDL ISIS and Daylight fingerprints to physicochemical descriptors *(19)*. In the case of physicochemical descriptors, principal component analysis (PCA) should be performed prior to clustering and retain 85–90% of the original variance. This procedure allows for weighting of the principal components in subsequent analyses. In this fashion, we can compensate for possible correlations between descriptors that could bias the analysis. Hierarchical cluster analysis is then performed on the principal components. Complete or average linkage should be preferred over single linkage, since the latter tends to produce clusters of uneven sizes. Once clustering is complete, selections of reagents can be made by simply choosing the desired clustering level, a representative reagent is drawn for each cluster (usually near the cluster center).

The limitations of reagent-based selections have been pointed out, since considerations at the reagent level do not directly reflect the nature of the corresponding products *(14,15,20)*. Although better than random, selections performed at the reagent level are suboptimal in terms of diversity due to the nonadditive nature of the descriptors involved. It has been shown that better selections can be obtained using product-based considerations *(14,15)* The incorporation of drug-like properties follows a similar logic. Although it is possible to select building blocks with drug-like considerations, the extent of synthetic transformations makes this also approach suboptimal. It is possible that a building block be considered unsuitable when taken individually, but that, in combination with other reagents, results in acceptable drug-like products. Each reagent should therefore be examined in its possible combinations with all other reagents and evaluated according to the drug-likeness of the corresponding products. Consequently, it is advisable that even simple drug-like rules such as the Lipinski rule of five *(21)* be applied at the product level. More complex models involving computational ADME (absorption, distribution, metabolism, excretion) do not usually lend themselves to reagent-based considerations and should realistically be applied at the product level *(22)*.

3. Product-Based Design

Product-based library design involves a more complex optimization procedure, which we term "combinatorial optimization," where the reagent selection is optimized against the properties of the corresponding products. In this scheme, the combinatorial nature of the sublibrary is maintained through combinatorial constraints, while evaluation of diversity, focusing, or other criteria

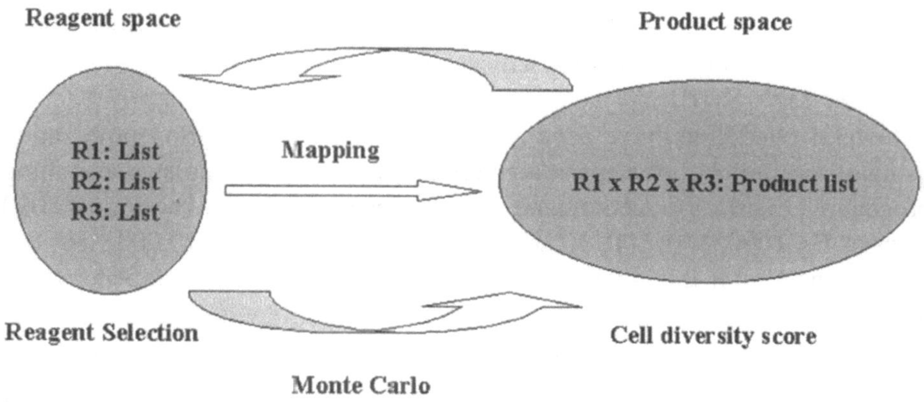

Fig. 2. Combinatorial optimization of reagent selection based on product properties.

is performed on the products. Procedures using either a Monte Carlo optimization or genetic algorithms have been reported ***(14,15)***. These procedures are more computationally intensive than simpler reagent-based considerations, since conventional techniques require full enumeration of the products, descriptor calculation for the entire library, and optimization of the subset. Dealing with problems of this complexity should be envisaged only in the perspective of substantially higher quality sublibraries than what could be obtained via simple reagent-based analysis.

The Monte Carlo optimization we describe starts from a random selection of reagents, which is then mapped to the corresponding selection of products (**Fig. 2**). This sublibrary is evaluated against a selected target function ***(23)***. A single change is then made in one of the reagent lists where one reagent is traded for another one. This change is then mapped to the modified sublibrary, which is again evaluated against the target function. If the target function is improved, the reagent swap is accepted as "beneficial" to the desired property; if the target function is worsened, the reagent swap is either accepted or rejected depending on a temperature parameter. The lower the temperature, the less likely a detrimental reagent swap will be accepted. The optimization ends when no reagent swaps can be proposed that would improve the target function.

The combinatorial optimization process described above attempts to identify a selection of reagents, which provides the desired product properties. If we take the example of a virtual library consisting of a $100 \times 100 \times 100$ array (1,000,000 possible products) for $R1 \times R2 \times R3$ and seek to isolate subsets of $10 \times 10 \times 10$ (1000 compounds), the total number of possible solutions is ${C_{100}}^{10} \times {C_{100}}^{10} \times {C_{100}}^{10} = 5 \times 10^{39}$. This is a formidable number, which makes it generally impossible to systematically investigate every possible subset. So we rely on the optimization procedure to provide a near optimal solution. Related studies using genetic algorithms or the Monte Carlo procedure described above suggest that the subsets obtained with such procedures are only slightly suboptimal ***(14,15)***.

Several conditions have been reported for the Monte Carlo optimization using a large number of steps at a given temperature or a simulated annealing procedure ***(15,16)***. We recommend an annealing procedure involving 100,000 steps at $T = 1000$, 300, 100, 30 and 10 K with a minimum of 10,000 steps (no improvement after this given number of cycles). We also recommend trying different seed values (i.e., different random starting points) in the optimization. We have previously ascertained the consistent quality of the subsets returned although the solutions themselves may be different ***(15)***.

Several diversity and space coverage measurements have also been considered in the combinatorial optimization process ***(15)***. Diversity and space coverage can be evaluated using a number of cell-based methods. These methods, implemented as diversity metrics ***(23)***, evaluate how much of the space occupied by the complete library is filled by the subset. For example, the cell-based fraction metric attempts to select one compound from each cell in order to cover as many cells as possible. However, owing to the combinatorial constraint, the objective to cover all occupied cells can seldom be achieved. The cell-based Chi^2 metric attempts to level out the distribution so as to provide an even allocation of compounds to cells. Cell-based entropy and cell-based density metrics attempt to select more than one compound from the most populated cells, in order to respect the level of occupancy of each cell. The following metrics can be used as target functions in the combinatorial optimization process:

Cell-based fraction: F = Cells occupied by subset/number of occupied cells

Cell-based Chi^2: $X^2 = \Sigma (N_i - N_{\text{ave}})^2$

Cell-based entropy: $S = -\Sigma (N_i \times \log(N_i))$

Cell-based density: $D = -\Sigma [N_i \times \log(N_i/M_i)]$

where: N_i = Number of compounds in cell i for subset
M_i = Number of compounds in cell i for complete library
N_{ave} = Average number of compounds per cell expected for subset
Σ = Sum over cell occupied by subset

Cell-based fraction and cell-based density have consistently provided the most satisfactory results in our earlier work ***(15)***. The main difference between these two cell-based metrics relates to diversity and representativeness. Whereas cell-based fraction is designed to sample the complete library with diversity as a primary objective, cell-based density draws more compounds from more densely populated regions of the property space, hence yielding potentially more representative subsets.

A number of molecular descriptors have been used in library design and in the characterization of molecular diversity ***(18,19)***. Similar descriptors to those described earlier in the reagent-based design section are typically used in the analysis of products. The cell-based approaches described above provide fast measurements of space coverage. However, a low-dimensionality space is required. A number of techniques are routinely used for dealing with the high-dimensional problems. In the case of physicochemical descriptors, PCA can routinely be applied. The general guideline is to extract 85–90% of the original variance contained in the set of descriptors. This is usually achieved in three to five components. In the case of fingerprint descriptors (MDL ISIS or Daylight), an analogous procedure is known as multidimensional scaling (MDS) ***(24)***. However, the procedure often fails to extract more than 40–50% of the distance information contained in the native fingerprints in the first three to five MDS dimensions. We found clustering a viable alternative to provide a 1D representation of the fingerprint space. Especially since the desired number of clusters is known at the outset (equal to the size of the desired sublibrary), a fast clustering method such as relocation clustering can be used. Following on this idea, each cluster may be described as the equivalent of an occupied cell.

Whereas early computational methods for library subsetting concentrated on either diversity or similarity to known leads, other aspects such as drug-likeness are also part of the library design process. In this fashion, we can produce libraries whose hits can be more easily optimized into successful drug candidates. This approach requires that we pursue several objectives simultaneously—diversity (or similarity) and drug-likeness, for example. It has been shown that drug-likeness can be achieved with minimal impact on the diversity of the compounds selected ***(16)***, so we can obtain sets that are almost equally diverse but different in the drug-like character. Such outcome is not entirely surprising when we look at the vast number of possible sublibraries that can be generated. The large ensemble of solutions, first perceived as a liability, may be turned into an asset where we now have the flexibility to provide libraries that are both drug-like and diverse.

Ideally, we would like to apply product-based design criteria for properties that are not adequately represented by the reagents (such as diversity and drug-like character) and reagent-based design criteria for practical considerations such as availability, cost, or ease of handling. For example, in addition to diversity and drug-like character, we would like to introduce a bias so that reagent selection is directed toward those preferred by the chemists. A reagent deemed "desirable" would receive no penalty and could easily be incorporated, whereas one deemed "less desirable" would receive a penalty and have less chance to be involved in the resulting sublibrary. Such a process involves simultaneous optimization against a number of criteria that need to be balanced appropriately. If the criteria involved in the optimization are normalized, then we can ensure the proper balance between them and reach an acceptable compromise solution. In this fashion, we can provide sublibraries that are combinatorial, reasonably diverse, drug-like, and use mostly "desirable" reagents. A procedure involving such conditions has been reported recently ***(16)***.

As we mentioned earlier, product-based library design introduces considerable computational complexity, involving library enumeration, descriptor calculation, and the optimization procedure itself. The latter can be performed rather efficiently so only the former two steps remain as bottlenecks. Novel approaches have appeared in the literature involving partial enumeration as well as Markush-based descriptor calculation ***(25,26)*** These breakthroughs make product-based library design more efficient, especially when dealing with very large virtual libraries. In partial enumeration, the selection of reagents for a given position (R1) is performed by considering a fixed selection of reagents for the other substituent positions (R2, R3, . . . , etc). Markush-based descriptor calculation bypasses the library enumeration step all together. In this approach, product properties are derived directly from assembly of the individual reagent or R-group properties. The method takes advantage of the semi-additive behavior of descriptors in going from reagents to products. It produces the same values for the descriptors as would be obtained with a conventional approach, but with several orders of magnitude better performance ***(26)***.

4. Conclusion

One of the challenges in modern library design is to involve both theoretically sound methods and practical considerations. Whereas the theoretical bases rely primarily on the computational chemists, the practical aspects are dictated by the experimental chemists. The implementation of a successful library design strategy draws expertise from these two groups and conciliates the appeal of theory with the reality of experiments. Computational design as we know it has seldom produced the final answer to a library design problem, as touch ups are often required. For example, it is common to provide replacements for reagents that were initially selected but, for some reason (commercial availability or other), cannot be obtained or used. Thanks to computational methods, we can make educated suggestions for such replacements. We can also evaluate alternate propositions, such as those proposed by experimental chemists, and ensure that the resulting library retains most or all the characteristics that were originally intended.

Acknowledgment

The author would like to thank Dr. Robert Brown at Accelrys for helpful scientific discussions.

References

1. Gallop, M. A., Barrett, R. W., Dower, W. J., Fodor, S. P. A., and Gordon, E. M. (1994) Applications of combinatorial technologies to drug discovery. 1. Background and peptide combinatorial libraries. *J. Med. Chem.* **37,** 1233–1251.
2. Gallop, M. A., Barrett, R. W., Dower, W. J., Fodor, S. P. A., and Gordon, E. M. (1994) Applications of combinatorial technologies to drug discovery. 2. Combinatorial organic synthesis, library screening strategies, and future directions. *J. Med. Chem.* **37,** 1385–1401.
3. Kick, E. K., Roe, D. C., Skillman, A. G., et al. (1997) Structure-based design and combinatorial chemistry yield low nanomolar inhibitor of Cathepsin D. *Chemistry and Biology* **4,** 297–307.
4. Czarnik, A. W. and DeWitt, S. H. (eds.) (1997) *Combinatorial chemistry.* American Chemical Society, Washington, DC.
5. Johnson, M. and Maggiora, G. M. (eds.) *Concepts and applications of molecular similarity.* Wiley, New York, NY.
6. Snarey, M., Terrett, N. K., Willett, P., and Wilton, D. J. (1997) Comparison of algorithms for dissimilarity-based compound selection. *J. Mol. Graphics Mod.* **15,** 372–385.
7. Lajiness, M. S. (1997) Dissimilarity-based compound selection techniques. *Perspect. Drug Discovery Design* **7/8,** 65–84.
8. Marengo, E. and Todeschini, R. (1992) A new algorithm for optimal distance-based experimental design. *Chemo-metrics Intell. Lab. Syst.* **16,** 37–44.
9. Holliday, J. D., Ranade, S. S., and Willett, P. (1995) A fast algorithm for selecting sets of dissimilar structures from large chemical databases. *Quant. Struct.-Activity Relationships* **14,** 501–506.
10. Hassan, M., Bielawski, J. P., Hempel, J. C., and Waldman, M. (1996) Optimization and visualization of molecular diversity of combinatorial libraries. *Molecular Diversity* **2,** 64–74.
11. Brown, R. D. and Martin, Y. C. (1996) Use of structure-activity data to compare structure-based clustering methods and descriptors for use in compound selection. *J. Chem. Inf. Comput. Sci.* **36,** 572–584.
12. Zheng, W., Cho, S. J., and Tropsha, A. (1998) Rational combinatorial library design. 1. Focus-2D: A new approach to targeted combinatorial chemical libraries. *J. Chem. Inf. Comput. Sci.* **38,** 251–258.
13. Leach, A. R., Green, D. V. S., Hann M. M., Judd, D. B., and Good, A. C. (2000) Where are the GaPs? A rational approach to monomer acquisition and selection. *J. Chem. Inf. Comput. Sci.* **40,** 1262–1269.
14. Gillet, V. J., Willett, P., and Bradshaw, J. (1997) The effectiveness of reactant pools for generating structurally-diverse combinatorial libraries. *J. Chem. Inf. Comput. Sci.* **37,** 731–740.

15. Jamois, E. J., Hassan, M., and Waldman, M. (2000) Evaluation of reagent-based and product-based strategies in the design of combinatorial library subsets. *J. Chem. Inf. Comput. Sci.* **40,** 63–70.
16. Brown, R. D., Hassan, M., and Waldman, M. (2000) Combinatorial library design for diversity, cost efficiency, and drug-like character. *J. Mol. Graphics Mod.* **18,** 427–437.
17. Lewell, X. Q., Judd, D. B., Watson, S. P., and Hann, M. M. (1998) RECAP-retrosynthetic combinatorial analysis procedure: a powerful new technique for identifying privileged molecular fragments with useful applications in combinatorial chemistry. *J. Chem. Inf. Comput. Sci.* **38,** 511–522.
18. Langer, T. and Hoffmann, R. D. (1998) New principal components derived parameters describing molecular diversity of heteroaromatic residues. *Quant. Struct.-Act. Relat.* **17,** 211–223.
19. Brown, R. D. (1997) Descriptors for diversity analysis. *Perspect. Drug Discovery Design* **7/8,** 31–49.
20. Leach, A. R., Bradshaw, J., Green, D. V. S., and Hann, M. M. (1999) Implementation of a system for reagent selection and library enumeration, profiling, and design. *J. Chem. Inf. Comput. Sci.* **39,** 1161–1172.
21. Lipinski, C. A., Lombardo, F., Dominy, B. W., and Feeney, P. J. (1997) Experimental and computational approaches to estimate solubility and permeability in drug discovery and development settings. *Adv. Drug Delivery Rev.* **23,** 3–25.
22. Eagan, W. J., Merz, K. M., and Baldwin, J. J. (2000) Prediction of drug absorption using multivariate statistics. *J. Med. Chem.* **43,** 3867–3877.
23. Cerius^2, Version 4.5, Accelrys Inc., 9685 Scranton Rd., San Diego, CA 92121, USA.
24. Everitt, B. S. and Dunn, G. (eds.) (1992) *Applied multivariate data analysis.* Oxford University Press, New York, NY.
25. Agrafiotis, K. A. and Lobanov, S. L. (2000) Ultrafast algorithm for designing focused combinatorial arrays. *J. Chem. Inf. Comput. Sci.* **39,**1161–1172.
26. Barnard, J. M., Downs, G. M., Scholley-Pfab, A. and Brown, R. D. (2000) Use of Markush structure analysis techniques for descriptor generation and clustering of large combinatorial libraries. *J. Mol. Graphics Mod.* **18,** 452–463.

20

Designing Combinatorial Libraries for Efficient Screening

John I. Manchester and David S. Hartsough

1. Introduction

Half a century ago, the term "drug discovery" conjured images of adventures into the jungle, beneath the sea, and atop mountains in search of frogs, sponges, lichens, or any unstudied life form that, ground up, might exhibit inhibitory effects toward a major human disease. More romantic and exciting science cannot be, to those of us too young to have participated in the "old" drug discovery paradigm, and perhaps also not more laborious, unpredictable, and frightening when included in a business plan. Combinatorial chemistry and high-throughput screening evolved to fill the need for a more systematic approach to discovery, in which miniaturization and automation were applied, as in traditional manufacturing processes, to reduce costs and cycle times. But to the contrary, the cost associated with producing clinical candidates seems to have actually risen with the application of these technologies ***(1–3)***.

One reason for the apparent lack of efficiency in the newer discovery paradigm is that, at first, generally little attention was paid to the overall information content of combinatorial libraries. The emphasis instead was on developing novel chemistries from which the maximum number of compounds could be produced; as the thinking goes, screen enough compounds and finding new drugs is a certainty. Around the mid-1990s, a flaw in this logic began to be generally recognized. The flaw is generally stated along these lines: something like 10^{180} different chemical compounds are possible, yet only 10^{17} seconds have elapsed since the beginning of the universe; thus, screening even a million compounds per second since the beginning of time is by itself an insufficient strategy for finding a drug (*see*, for example, **ref. *4***). Practically, libraries are not screened in search of the perfect drug, but for a lead compound, which

From: *Methods in Molecular Biology, Combinatorial Library Methods and Protocols*
Edited by: L. B. English © Humana Press Inc., Totowa, NJ

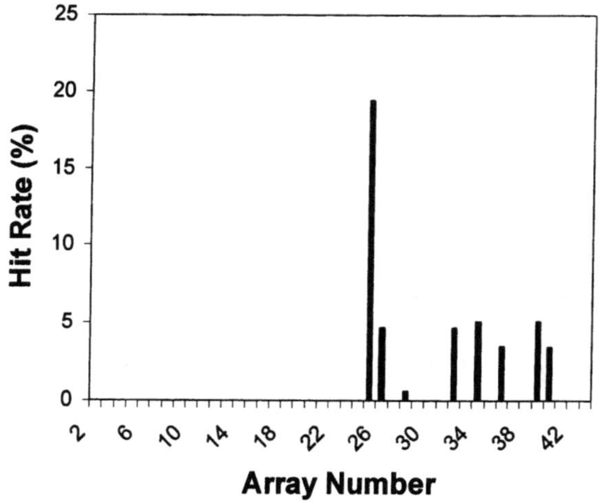

Fig. 1. Combinatorial libraries are usually prioritized for further investigation based on their observed hit rate in a high-throughput screen. Hit rate is the percentage of compounds in a given library that bind to a receptor at or above a pre-defined threshold. For libraries containing thousands of diverse compounds, one might expect to observe a more or less uniform hit rate. This plot shows that, for a representative high-throughput screen, hit-rates are far from uniform for a number of typical, early combinatorial libraries, indicating that combinatorial libraries are inherently biased by the presence of a common scaffold. This observation further indicates that similar information can be obtained from screening a diverse subset of a combinatorial library compared to screening the full array.

can be optimized into a drug via successive cycles of structural modification, synthesis, and testing. In addition, only a small portion of the chemical space represented by the 10^{180} possible compounds are "drug-like" or pharmacologically meaningful molecules. It is also true that libraries are often designed to mimic a natural or otherwise known ligand for the target receptor, further reducing the search space. So the paradigm does work, but it is still inefficient.

A successful approach to reducing this inefficiency is designing chemical diversity into combinatorial libraries, by selecting diverse subsets of the reagents available for a library based on a given chemical reaction, by selecting diverse subsets of synthesized libraries, or by performing constrained optimizations considering the chemical diversity of the resulting library *in silico*—in other words, designing libraries for maximal information content. This last approach is particularly useful when screening is being conducted to detect a lead compound (a compound close to the optimum activity, but not necessarily optimal itself). The approach works because combinatorial libraries exhibit much less internal diversity than generally thought. **Figure 1** shows the hit rates (the fraction of compounds in a library that bind to a receptor above a desired threshold) for a number of representative combinatorial libraries, containing several thousand compounds each, for high-thoughput screening against a particular receptor. If each library were "truly" diverse, activity would be distributed more or less evenly among them. However, most libraries exhibit no activity at all, whereas libraries showing any appreciable activity do so with hit rates above 5%, and sometimes much higher.

Another reason for the apparent lack of efficiency is the high attrition rate of compounds entering the discovery pipeline. About 40% of compounds with promising activity and selectivity for a given target fail in clinical trials due to poor pharmacokinetic or "ADME" (Absorption, Distribution, Metabolism, and Elimination) profiles *(5)*. A poor pharmacokinetic profile for an orally administered medication results in low bioavailability and can be broadly attributed to poor absorption from the gut, distribution to tissue other than the site of action, rapid metabolism or elimination, or confounding interactions of two or more of these properties. In addition, inhibition of metabolic pathways can lead to drug–drug interactions. Absorption has received a great deal of recognition as perhaps the most important of the ADME properties, and the most difficult to correct via structural modifications that do not also obliterate activity. A significant amount of work, best represented by the "rules of 5" paper *(6)*, has shown that absorption is related primarily to the physicochemical characteristics of a given compound. In that work, the authors also observed that combinatorial chemistry and high-throughput screening favor detection of compounds that are, on average, more lipophilic and higher in molecular weight than those from more traditional drug discovery paradigms. Heavy, lipophilic compounds generally exhibit poor aqueous solubilities, a property that is important for bioavailability. Compounds with poor solubilities are likely to remain poorly soluble during efforts to optimize activity because improving solubility involves addition of hydrophilic groups or removal of hydrophobic groups, either of which is likely to interfere with activity. Moreover, efforts to optimize activity generally involve increases in molecular weight and hydrophobicity. To achieve large numbers of "interesting" compounds, combinatorial libraries are generally high in molecular weight and often quite hydrophobic, taking advantage of commercially available reagents with small numbers of functional groups that will give high synthetic yields and acceptable stability during storage. Thus, combinatorial libraries must be designed to contain compounds with reasonable aqueous solubilities and physicochemical properties to improve their chances of finding potential drugs or lead compounds when coupled with high-throughput screens.

Metabolism is a more difficult property to assess because it relies primarily on interaction with specific enzymes (mostly cytochromes P450 or CYPs) in the gut and liver, and a simple consideration of physicochemical properties of the compounds in a library usually does not provide any insight into their probable metabolic characteristics. Recently, focused effort has been applied to develop computational models for drug metabolism, including stability and inhibition of the major human CYP isoforms *(7–14)*. These models put computational assessment of metabolic properties within reach of the design of combinatorial libraries.

Targeting combinatorial libraries using structural information about the receptor or its known ligands, inferred pharmacophores, or any other useful model for ligand binding can also reduce the search space and improve the efficiency of screening. Most companies are taking advantage of structural data to design libraries specifically targeting G-protein coupled receptors (GPCRs) and ion channels for the treatment of diseases ranging from hypertension to obesity.

In this chapter, we illustrate one approach to designing combinatorial libraries in which diversity, ADME, and activity are simultaneously optimized. The approach described goes one step further to help guard against screening compounds with redundant properties. The resulting libraries are diverse, unique with respect to a reference set of compounds (such as a corporate compound repository), and "drug-like" as defined by a number of physicochemical properties. Additional properties, such as high-resolution ADME, toxicity, or receptor-affinity models are easy to include because the approach incorporates a consensus scoring method to lump all these properties into an overall per-compound "desirability" variable. The goal of the approach is to pick a subset of a combinatorial library that, when screened, will provide as much useful information as if the entire virtual library had been screened. This "optimal" screening library is shown schematically in **Fig. 2**.

2. Overview of Library Optimization

2.1. Reagent-Based vs Product-Based Optimization

The idea of screening subsets of combinatorial libraries according to maximal information content is not new. A number of techniques have been developed over the past decade, each with a particular strength. Fundamentally, the methods divide into either reagent-based or product-based selections. Reagent-based methods have the advantage that they avoid enumeration, or performing the full combinatorial cross, of all members in a combinatorial library. Consider the following reaction, which describes the combination of a ketone, an aldehyde and a hydrazine to form a chalcone:

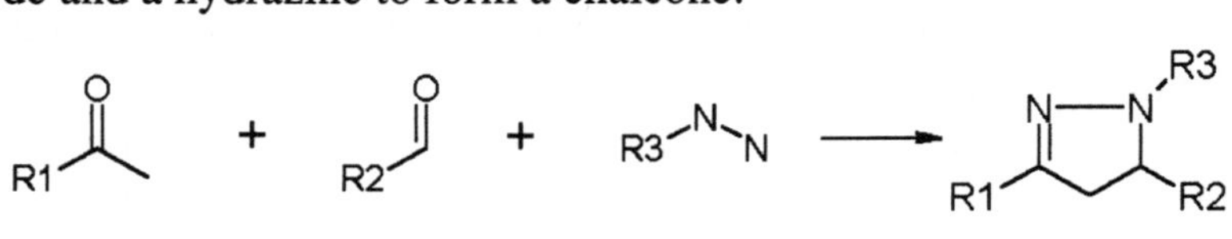

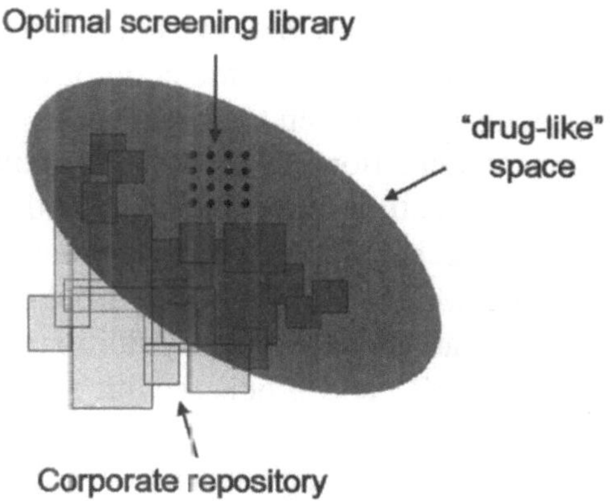

Fig. 2. Within the accessible chemical diversity space, it is generally accepted that certain regions contain compounds that, when synthesized, will exhibit acceptable physicochemical and pharmacokinetic (so-called "drug-like") properties. A combinatorial library designed for efficiency, then, should provide adequate but nonredundant coverage of a part of this drug-like region, while avoiding space occupied by a historical compound collection that may have already been screened against a given target. Thus, "optimal" libraries should be diverse, unique with respect to existing corporate repositories, and exhibit favorable or "drug-like" characteristics.

In the Available Chemicals Directory [a compilation of small molecules available from commercial sources *(15)*], there are approximately 1500 aromatic aldehydes, 1800 acetophenones, and 650 hydrazines that will yield products with reasonable purity *(16)*. Thus, considering just commercially available reagents, a library of about 1.8 billion compounds is accessible to this chemotype. Explicitly storing and manipulating a billion virtual compounds is nontrivial, even with good software and a fast computer. Reagent-based library optimization methods avoid these problems by, for example, selecting a diverse subset of reagents for each dimension in the array. In this case, only a few thousand structures need to be explicitly dealt with, and the optimization can be performed roughly a million times faster than a product-based optimization. The problem with reagent-based approaches is that they do not adequately represent chemical diversity when comparing two or more libraries *(17)*. So reagent-based optimizations are cheap and certainly better than doing nothing at all, but they are of limited utility when searching for the most diverse subset of a given combinatorial library.

Approaches have been developed for circumventing the need to fully enumerate large virtual libraries *(18,19)*; however, not all of these methods provide an adequate description of diversity within the final library, or require that the library not be fully combinatorial, losing the benefit of chemical efficiency gained from combinatorial chemistry. We have not found the size of the virtual library to be a problem—not because our methodology is necessarily revolutionary, but because chemists prescreen their reagent lists prior to performing an optimization. Many combinations of the several thousand commercially available reagents for the reaction in **Scheme 1** would produce compounds with obviously poor physicochemical or pharmacokinetic properties. Simply excluding high-molecular-weight and highly hydrophobic reagents significantly reduces the pool of candidate reagents. Additional filtering steps, such as excluding reagents that will yield products with exposed reactive groups, typically reduces the virtual library to a very manageable size. Practically, during optimization we usually deal with libraries with a few million members or less.

2.2. Chemical Diversity: Descriptors and Chemical Space

Chemical diversity remains a subjective concept. The problem is that the most useful way of representing compounds, by chemical structure, is qualitative in nature, and there is no agreed-upon standard for representing an arbitrary set of structures on a quantitative scale where computational techniques can be readily applied. The most common approach is to calculate or predict various properties from the chemical structure, such as molecular weight and polarizability, metrics that summarize molecular topology, or molecular "fingerprints," which are usually a set of a thousand or so binary variables, corresponding to specific structural motifs, that are each set to "true" for compounds possessing those motifs. These properties and metrics together are usually called "descriptors." Again, the problem is that there is no agreed-upon set of descriptors that give a uniformly adequate representation of chemical structure. A common approach is to calculate a large number of descriptors for a set of molecules, then to apply a statistical technique to reduce the effective dimensionality of the resulting chemical space by extracting a small number of latent variables. The idea here is that we know that variations in chemical structure occur, but can only reliably capture this variation mathematically by calculating a large number of descriptors, each capturing a particular aspect of the variation. Techniques such as principal components analysis (PCA) and multidimensional scaling (MDS) effectively lump descriptors together into new variables that more concisely and reliably capture the essence of the variation. The result is a low-dimensional (fewer than 10) space, in which similar compounds reside near each other.

Another approach is to calculate more abstract descriptors, which themselves can be treated as latent variables. Examples of these are Breneman's transferable atom equivalents (TAEs) ***(20)*** and Pearlman's BCUT descriptors ***(21)***. TAEs are the result of various mathematical simplifications of a molecule's *ab initio* electronic distribution, mapped to its surface; BCUTs are computed as the eigenvalues of matrices constructed using various atom-centered and atom-pair properties. Both examples are significant abstractions of chemical structure, and are affected by multiple but related structural factors. Chemical spaces can be constructed by choosing a small number of BCUTs that capture most of the variation within a set of compounds. A number of software tools to calculate descriptors, construct chemical spaces, and perform diversity comparisons are commercially available *(22)*.

2.3. Chemical Diversity: Distance-Based or Cell-Based?

Once the structural features of a set of compounds have been numerically captured with a suitable set of computed descriptors, the question of assessing diversity arises. Chemical diversity is somewhat subjective, but a number of approaches for assessing diversity have been developed, generally either distance-based or cell-based. Distance-based diversity methods are based on pair-wise calculation of some distance metric for all compounds in a library. Thus, for a library with N compounds, distance-based methods require that on the order of N^2 separate calculations be performed. As the size of a virtual library increases, distance-based methods quickly become impractical and finally impossible at even a modest size. The alternative approach is to divide the chemical space accessible to a library into discrete bins, based on each of the descriptors that define that space. The result is a grid that resembles, for the three-dimensional case, a Rubik's Cube in which there are 27 cells (3 principal axes with 3 bins along each axis). Instead of considering relationships between all pairs of compounds, only the relationship between each compound and the grid (i.e., which cell it belongs to) is examined. Thus, for computing diversity for N compounds, on the order of N calculations are required. Although they can be used with very large numbers of compounds, cell-based methods are practical only for chemical spaces with a small number of dimensions and so some form of dimensionality reduction or highly abstracted descriptors are required. However, cell-based methods have the benefit that a cell-based chemical space can be re-used as a common reference frame for comparing diversity not only within libraries but *between* libraries without any significant computational burden. By contrast, distance-based methods require the computation of a common reference frame formed by all pair-wise distances for the superset of two libraries prior to computing a diversity metric, which is often impossible in practice unless only very small libraries are being considered.

The tool described in this chapter uses cell-based diversity, in a six-dimensional chemical space formed by the BCUT descriptors that describe most of the abstract structural variation present in the ArQule corporate repository. For each virtual library considered, we divide the space into a number of cells that is at least as large as the sum of the number of compounds in the full virtual library and the corporate collection, so that each compound can theoretically occupy a single cell.

2.4. Subset Selection

Once the range of target properties and a suitable method to assess the diversity of a combinatorial library have been determined, some method to extract the combinatorial subset of compounds with the best trade-off among diversity and the target properties is necessary. Experimental design techniques have been elegantly applied to the problem ***(23–27)***, but these require assumptions about the shape of the response surface and/or the nature of variations in chemical structure present in the library of interest that are not always justified. Several other approaches, with various strengths, have been proposed ***(28–30)***. Probably the most generally applicable and efficient approaches are based on stochastic optimization. These methods search for and modify candidate subsets of reagents that give rise to better and better libraries, until after some time no better library can be found. Simulated annealing (SA) and genetic algorithms (GAs) are the stochastic methods that have been most often applied to library optimization ***(31–36)***.

Genetic algorithms are modeled after natural selection and search for an "optimal" solution to a problem by allowing a population of individual solutions to evolve, subject to a "fitness" constraint. In the present case, the candidate solutions are individual combinatorial libraries that are subsets of the full virtual library, and the fitness constraint is calculated using a suitable "fitness function." Individual libraries are specified by encoding the candidate reagents for each dimension of a library as a gene on a chromosome; each gene is a list of binary variables that indicate the presence or absence of the corresponding reagent in the individual library. By specifying which reagents are present along each gene, the resulting library can be readily deduced. An initial population of libraries is generated by randomly setting the desired number of reagents in each gene as "present," and then point-mutations and cross-over events among compatible genes are allowed in the population to simulate mating. As each generation is created, only a certain number of children (and optionally some parents) are allowed to survive according to their fitness. Because the search space is confined (new reagents cannot evolve into the population), the algorithm eventually converges to an optimal solution. Convergence is generally considered to occur when either the value of the fitness no longer improves for any of the individuals in a population from generation to generation, or when all individuals in the population have identical genetic makeup through several generations.

Stochastic optimization procedures suffer from two primary drawbacks: first, that they depend on the starting point; and second, that they are not guaranteed to find the so-called "global optimum," or the truly best solution. In fact, there is no optimization procedure except an exhaustive search that is guaranteed to find the global optimum; thus, this is a limitation inherent to all complex systems. However, stochastic optimizations are guaranteed upon convergence to find an optimal solution—not necessarily the best, but one of the best solutions to the problem at hand. We have addressed the first point by performing the optimization starting from numerous initial populations; so although there is still dependence on the starting point, this dependence is effectively averaged out and is thus less pronounced from run to run.

3. MapMaker™, Library Optimization at ArQule

MapMaker™ is a web-based tool developed at ArQule for assisting chemists in designing libraries prior to synthesis. Thus, the tool operates on *virtual* libraries, specified by the reaction scheme and a number of candidate reagents for each R-group or dimension in the library. Information flow in MapMaker™ is shown in **Fig. 3**. The virtual library is fully enumerated, followed by property and chemical-space-coordinate calculations. A genetic algorithm is then applied to select a subset of the candidate reagents. These reagents are selected to produce an "optimum" subset of the full virtual array, that is, a relatively small combinatorial library that is diverse, unique, and "drug-like." Schematically, the system functions as a black box, hiding the details of database query, library enumeration, property calculation and the optimization algorithm from the user. Chemists simply provide a reaction scheme and a set of candidate reagents, and MapMaker™ produces a set of optimized reagent lists that can be sent directly to production.

Not shown in **Fig. 3** is an optional consensus scoring step following property calculation. Currently, MapMaker™ uses a voting scheme known as Borda counting[1] for determining the consensus score. Voting is one of several methods for computing a consensus score *(39)*, which we chose here for its simplicity and ability to accommodate an arbitrary number of candidate properties. Borda counting differs from the plural vote, commonly used to select elected officials in the United States, in that it allows voters to express their preferences for all candidates in an election. Thus, it gives rise to fewer paradoxes than the plural vote *(40)*. Borda counting is implemented in MapMaker™ by

[1] Borda counting *(37)* has a colorful history. Developed in response to dissatisfaction with the often paradoxical outcome of plural votes used to elect members to the French Academy of Sciences, it was eventually adopted by the *Académie* as a superior system, until that decision was reversed by Napolean in the early 19th century. Incidentally, Kenneth Arrow received the 1972 Nobel Prize in Economics in part for his theorem, a corollary of which is that the only voting system without paradoxes is a dictatorship *(38)*.

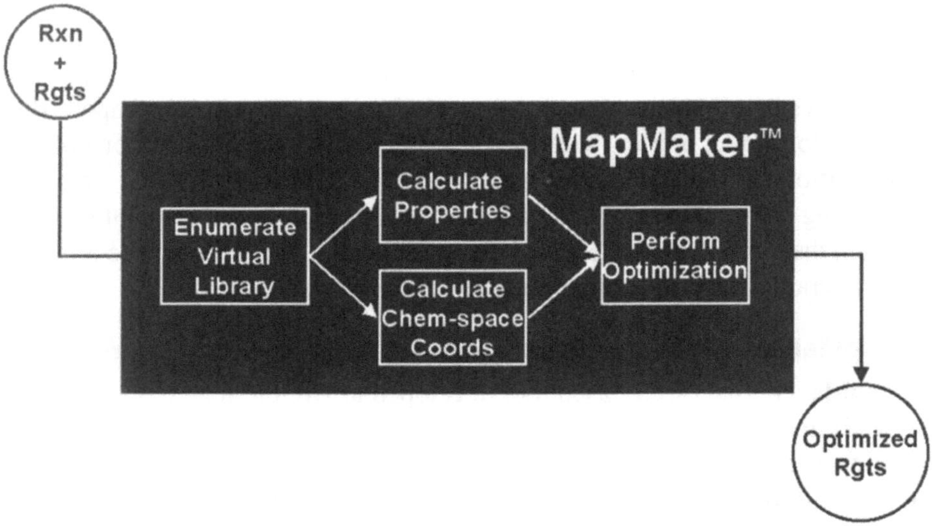

Fig. 3. MapMaker™ is a combinatorial library optimization tool developed at ArQule, currently used in the design of screening libraries. The system is represented as a black box, since the details of operation are hidden by a web interface, to which chemists provide a reaction scheme, lists of candidate reagents, and the number of reagents desired for each dimension of a library, and from which they retrieve lists of reagents that encode for the optimized library. Internally, MapMaker™ enumerates the full virtual array, calculates the desired properties and coordinates in chemical space for the virtual compounds, and performs the optimization using a genetic algorithm. Consensus scoring following property calculation allows the system to optimize around an arbitrary number of computed properties, which are determined at run-time.

treating each compound in a library as a candidate, and each property as a voter. Operationally, for each property, compounds are ranked by their deviation from an ideal value. In the "election," all the ranks for each compound are added up; the compound with lowest score, or overall deviation from ideality, wins. This overall deviation from ideality score (or "Desirability") is used in assessing the fitness of a library via the following function, which describes the value of selecting a given set of n compounds from a virtual library:

$$f(n) = \frac{w_1 \text{Diversity} + w_2 \sum_i^n \text{Desirability}_i}{n(w_1 + w_2)} - w_3(\Delta n) \quad (1)$$

Thus, the overall "Desirability" of a library is simply the sum of the individual Desirability scores for all of its compounds. The Diversity term in the fitness function is computed simply as the number of cells uniquely occupied by the candidate library in the BCUT chemical space formed by both the full virtual library and ArQule's corporate repository. Note that the Diversity term calculated in this way implicitly accounts for uniqueness because the candidate libraries that also avoid overlap with the reference collection are favored over those that only scatter compounds within the space accessible to the virtual library. Δn is a term that allows the candidate library to deviate in size from that requested. Finally, the coefficients w_1, w_2, and w_3 are weights that allow the user to influence the trade-offs among diversity, property-ideality, and size that are made during optimization.

4. An Example Combinatorial Library Optimization

As an example, consider the reaction in **Scheme 1** for formation of a chalcone library *(16)*. Prescreening of commercially available reagents from the ACD resulted in 236 reagents for R1, 338 for R2, and 75 for R3, for a virtual library containing about six million compounds. We used MapMaker™ to select 10 reagents from R1, 20 from R2, and 10 from R3. The search space is quite large (there are on the order of 10^{61} possible solutions); thus, the use of a search algorithm such as a genetic algorithm is justified in this case. For the optimization, we chose equal weights for Diversity and Desirability in eq. 1, and an arbitrarily high weight for w_3 to ensure a library of exactly 2000 compounds following optimization. Individual compounds were scored according to a computational model for aqueous solubility *(41)* and the "rule of 5" *(6)*. The "rule of 5" was developed at Pfizer, based on a historical analysis of physicochemical properties that compounds successful in early clinical trials held in common; these compounds were assumed to exhibit good solubility and absorption into the bloodstream across the gut wall. The analysis revealed that compounds have a better chance of being orally available (excluding active transport and first-pass effects) if they have no more than 5 H-bond donors, 10 H-bond acceptors, molecular weight below 500, and predicted log *P* below 5.0. In our implementation, a compound is scored based on the number of these rules it violates (perfect score = 0; worst score = 4). Borda counting is used to assign a consensus score to each compound.

If everything works, MapMaker™ should select a $10 \times 20 \times 10$ library that is more diverse, distinct from the corporate repository, and on average exhibits more desirable characteristics than the full virtual library. **Figure 4** shows the diversity of the optimized library selected by MapMaker™, relative to the average diversity for three $10 \times 20 \times 10$ libraries formed by selecting reagents at random. Diversity was computed as the fraction of compounds in the library that occupy cells all by themselves in the chemistry space; a perfect diversity by this definition is 1. The optimized library has a nearly perfect score, whereas the random libraries possess about half the diversity. Uniqueness of the optimized

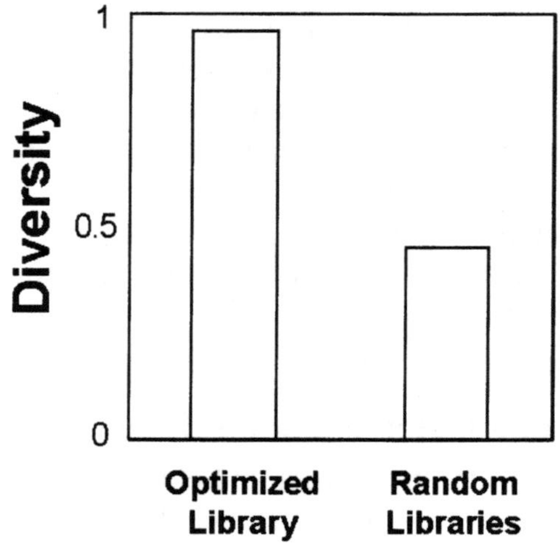

Fig. 4. Chemical diversity is calculated as the fraction of compounds in a library that singly occupy cells in chemical space, and is shown here for an optimized library of 2000 compounds (a 10 × 20 × 10 combinatorial library) selected by MapMaker™ (left bar), and for the average of three 2000-member combinatorial libraries formed by selecting reagents at random. The optimized library has a nearly optimal diversity score of 1, whereas picking reagents without optimization would result in about half the screening efficiency, or about half the information (about 50% of the compounds in the random libraries occupy redundant cells).

library relative to the corporate repository can be calculated in an analogous way, by considering the fraction of compounds in the candidate library that lie in cells also occupied by compounds from the reference collection. This metric is an indicator of the degree of overlap between the two sets of compounds, the worst sore being 0 and the best score 1. This overlap score is shown for the optimized library and the average random library in **Fig. 5**. Also shown is the average overlap for three 10 × 20 × 10 libraries formed by a random selection of reagents from each dimension. The full virtual and randomly selected libraries both overlap the corporate repository by about 0.18, whereas in the optimized library the overlap drops to about 0.12. Thus, the chances of screening compounds with redundant structural characteristics both within the library and with respect to any screening performed on a historical compound collection are significantly reduced by library optimization.

Figure 6 shows the predicted aqueous solubility profile of the full virtual library and the library selected by MapMaker™, and **Fig. 7** shows similar profiles for the number of violations of the "rule of 5." In both cases, the optimized library is shifted significantly in the direction of more favorable properties. This shift can be made more pronounced, at the expense of some

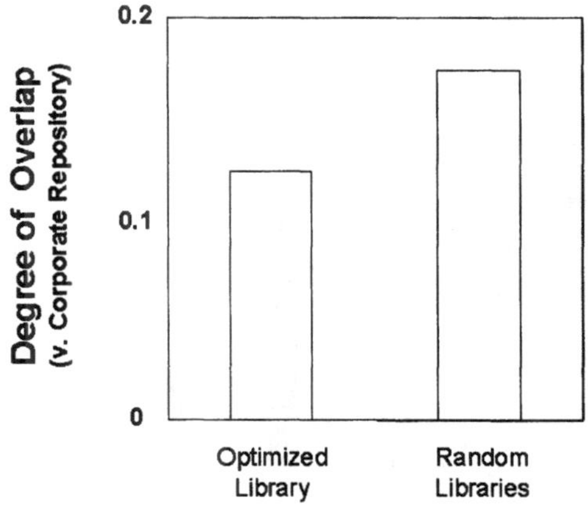

Fig. 5. The degree of overlap between a combinatorial library and a reference collection can be calculated as the fraction of compounds in the candidate library that lie in cells also occupied by one or more compounds from the reference set. Here we show degree of overlap with the ArQule compound repository for the optimized library of 2000 compounds vs the average of three libraries formed by randomly selecting reagents (as in **Fig. 4**). In this case, optimization reduces the redundancy relative to the reference collection by about 1/3.

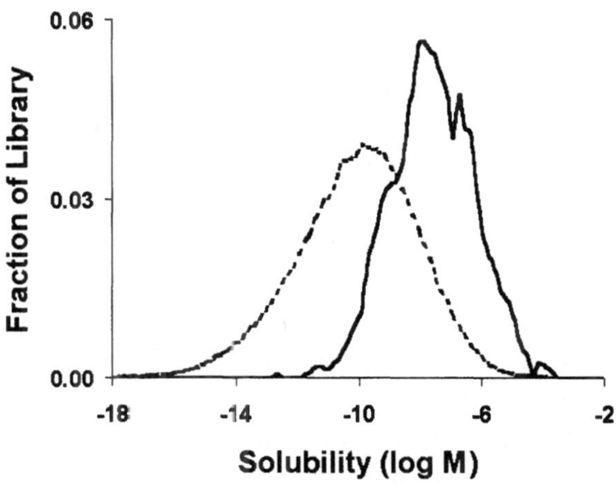

Fig. 6. Aqueous solubility for virtual compounds can be assessed using a suitable computational model, here one that estimates the $\log_{10}$ of the thermodynamic solubility of a compound in neat water at 25°C and 1 atm ***(41)***. The solubility profile of an optimized combinatorial library (solid line) is shifted to higher solubility by about two orders of magnitude, relative to the full virtual library (dashed line), as a result of optimization by MapMaker™ in this example. Owing to the use of a fitness function (eq. 1), a more pronounced shift could be obtained at the expense of some chemical diversity in the resulting library.

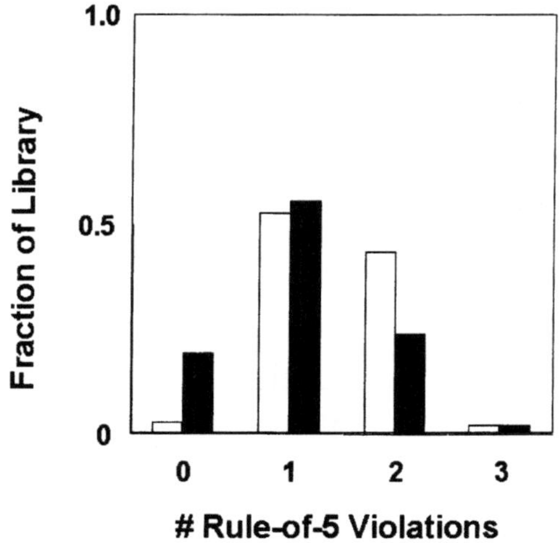

Fig. 7. A popular indicator of oral bioavailability, excluding effects due to active transport and first-pass metabolism, is the number of "rule of 5" violations *(6)*. In this case, the profile of the number of violations for an optimized library (solid bars) is shifted to lower values, relative to the full virtual array (unfilled bars), indicating that optimization selected a library with improved "drug-like" characteristics. This shift could be made more pronounced by more heavily weighting the "Desirability" term in eq. 1 at the expense of some diversity.

chemical diversity in the resulting library, by more heavily weighting the "Desirability" term in eq. 1 relative to the Diversity term. The current library is most suitable for a primary screen, where the importance of information about the structural characteristics relevant to activity outweighs finding compounds with optimal physicochemical and/or pharmacokinetic properties. When screening to uncover preliminary structure-activity relationships, some "bad" compounds are actually desirable because they provide some insight about what structural characteristics within a chemotype confer poor properties.

The current library can be easily targeted for a specific receptor by including a computational model for receptor binding, as often is developed as a lead optimization program proceeds, and of course by including more specific computational models for ADME characteristics.

5. Conclusion

Combinatorial chemistry and high-throughput screening are powerful techniques for exploring how structural diversity relates to activity toward a target receptor. However, owing to the enormous chemical diversity accessible using even simple chemistries and commercially available reagents, these techniques alone cannot in practical use provide a complete map of structure-activity relationships between compounds and a receptor. They must be combined with intelligent designs, where the composition of the combinatorial library is tailored to the goals of screening. For exploratory or primary screening, maximal chemical diversity is sought from the minimum number of compounds; thus small, diverse libraries with at least minimally acceptable physicochemical properties that may impinge on bioavailability are appropriate. For follow-up screening or during lead optimization, where particular regions of chemical space are to be more thoroughly searched (e.g., constrained by a pharmacophore), physicochemical properties and complementarity to the receptor are more important, and some loss of diversity can be tolerated. We have described one tool that helps chemists design combinatorial libraries tailored to specific screening objectives. Chemists can weight diversity more or less heavily against an arbitrary number of properties, which can be retrieved either from computational models or a database. Tools such as this one are helping us to unleash the full potential of automated synthesis and screening technologies, allowing scientists to design better experiments to test hypotheses, and more time to formulate those hypotheses and think about the results.

References

1. Prentis, R. A., Lis, Y. and Walker, S. R. (1988) Pharmaceutical innovation by the seven UK-owned pharmaceutical companies (1964–1985). *Br. J. Clin. Pharmac.* **25,** 387–396.
2. DiMasi, J. A. (1995) Success rates for new drugs entering clinical testing in the United States. *Clinical Pharmacology and Therapeutics* **58,** 1–14.
3. Venkatesh, S. and Lipper, R. A. (2000) Role of the development scientist in compound lead selection and optimization. *J. Pharm. Sci.* **89,** 145–154.
4. Martin, Y. C. (1997) Challenges and prospects for computational aids to molecular diversity. *Perspectives in Drug Discovery and Design* **7/8,** 159–172.
5. Kennedy, T. (1997) Managing the drug discovery/development interface. *Drug Discovery Today* **2,** 436–444.
6. Lipinski, C. A., Lombardo, F., Dominy, B. W., and Feeney, P. J. (1997) Experimental and computational approaches to estimate solubility and permeability in drug discovery and development settings. *Adv. Drug Deliv. Rev.* **23,** 3–25.
7. Ekins, S., Bravi, G., Binkley, S., et al. (2000) Three- and four-dimensional-quantitative structure activity relationship (3D/4D-QSAR) analyses of CYP2C9 inhibitors. *Drug Metabolism and Disposition* **28,** 994–1002.
8. Ekins, S., Bravi, G., Wikel, J. H., and Wrigthon, S. A. (1999) Three-dimensional-duantitative structure activity relationship analysis of cytochrome P450 3A4 substrates. *J. Pharm. Exp. Therap.* **291,** 424–433.
9. Ekins, S., Bravi, G., Ring, B. J., et al. (1999) Three-dimensional quantitative structure activity relationship analyses of substrates for CYP2B6. *J. Pharm. Exp. Therap.* **288,** 21–29.
10. deGroot, M. J., Ackland, M. J., Horne, V. A., Alex, A. A., and Jones, B. C. (1999) A

novel approach to predicting P450 mediated drug metabolism. CYP2D6 catalyzed N-dealkylation reactions and qualitative metabolite predictions using a combined protein and pharmacophore model for CYP2D6. *J. Med. Chem.* **42,** 4062–4070.

11. Hardman, J. G., Limbird, L. E., Molinoff, P. B., Ruddon, R. W., and Gilman, A. G. (eds.) (1996) Goodman & Gilman's The Pharmacological Basis of Therapeutics, 9th ed. McGraw-Hill, New York
12. Korzekwa, K. R., Jones, J. P., and Gillette, J. R. (1990) Theoretical studies on cytochrome P450 mediated hydroxylation: a predictive model for hydrogen atom abstractions. *J. Am. Chem. Soc.* **112,** 7042–7046.
13. Korzekwa, K. R. and Jones, J. P. (1993) Predicting the cytochrome P450 mediated metabolism of xenobiotics. *Pharmacogenetics* **3,** 1–18.
14. Korzekwa, K. R., Grogan, J., DeVito, S., and Jones, J. P. (1996) Electronic models for cytochrome P450 oxidations. *Advances in Experimental Medicine & Biology* **387,** 361–396.
15. MDL Information Systems, I. (1999) Available Chemicals Directory, 99.2 ed., San Leandro, CA
16. Powers, D. G., Casebier, D. S., Fokas, D., Ryan, W. J., Troth, J. R., and Coffen, D. L. (1998) Automated parallel synthesis of chalcone-based screening libraries. *Tetrahedron Lett.* **54,** 4085.
17. Gillet, V. J., Willett, P., and Bradshaw, J. (1997) The effectiveness of reactant pools for generating structurally-diverse combinatorial libraries. *J. Chem. Inf. Comput. Sci.* **37,** 731–740.
18. Agrafiotis, D. K. and Lobanov, V. S. (2000) Ultrafast algorithm for designing focused combinatorial arrays. *J. Chem. Inf. Comput. Sci.* **40,** 1030–1038.
19. Bravi, G., Green, D. V.S., Hann, M. M., and Leach, A. R. (2000) PLUMS: a program for the rapid optimization of focused libraries. *J. Chem. Inf. Comput. Sci.* **40,** 1441–1448.
20. Breneman, C. M. and Rhem, M. (1997) QSPR analysis of HPLC column capacity factors for a set of high-energy materials using electronic van der Waals surface property descriptors computed by transferable atom equivalent method. *J. Comp. Chem.* **18,** 182–197.
21. Pearlman, R. S. and Smith, K. M. (1999) Metric validation and the receptor-relevant subspace concept. *J. Chem. Inf. Comput. Sci.* **39,** 28–35.
22. Warr, W. A. (1997) Commercial software systems for diversity analysis. *Perspectives in Drug Discovery and Design* **7/8,** 115–130.
23. Martin, E. J., Blaney, J. M., Siani, M. A., Spellmeyer, D. C., Wong, A. K., and Moos, W. H. (1995) Measuring diversity: experimental design of combinatorial libraries for drug discovery. *J. Med. Chem.* **38,** 1431–1436.
24. Higgs, R. E., Bemis, K. G., Watson, I. A., and Wikel, J. H. (1997) Experimental designs for selecting molecules from large chemical databases. *J. Chem. Inf. Comput. Sci.* **37,** 861–870.
25. Martin, E. J. and Critchlow, R. E. (1999) Beyond mere diversity: tailoring combinatorial libraries for drug discovery. *J. Comb. Chem.* **1,** 32–45.
26. Linusson, A., Wold, S., and Norden, B. (1999) Statistical moelcular design of

peptoid libraries. *Molecular Diversity* **4,** 103–114.

27. Linusson, A., Gottfries, J., Lindgren, F., and Wold, S. (2000) Statistical molecular design of building blocks for combinatorial chemistry. *J. Med. Chem.* **43,** 1320–1328.
28. Mount, J., Ruppert, J., Welch, W., and Jain, A. N. (1999) IcePick: A flexible surface-based system for molecular diversity. *J. Med. Chem.* **42,** 60–66.
29. Pickett, S. D., Luttmann, C., Guerin, V., Laoui, A., and James, E. (1998) DIVSEL and COMPLIB—strategies for the design and comparison of combinatorial libraries using pharmacophoric descriptors. *J. Chem. Inf. Comput. Sci.* **38,** 144–150.
30. Clark, R. D. (1997) OptiSim: An extended dissimilarity selection method for finding diverse representative subsets. *J. Chem. Inf. Comput. Sci.* **37,** 1181–1188.
31. Gillet, V. J., Willett, P., Bradshaw, J., and Green, D. V.S. (1999) Selecting combinatorial libraries to optimize diversity and physical properties. *J. Chem. Inf. Comput. Sci.* **39,** 169–177.
32. Zheng, W., Cho, S. H., Waller, C. L., and Tropsha, A. (1999) Rational combinatorial library deisgn. 3. Simulated annealing guided evaluation (SAGE) of molecular diversity: A novel computational tool for universal library design and database mining. *J. Chem. Inf. Comput. Sci.* **39,** 738–746.
33. Good, A. C. and Lewis, R. A. (1997) New methodology for profiling combinatorial libraries and screening sets: cleaning up the design process with HARPick. *J. Med. Chem.* **40,** 3926–3936.
34. Sheridan, R. P., SanFeliciano, S. G., and Kearsley, S. K. (2000) Designing targeted libraries with genetic algorithms. *J. Molecular Graphics and Modelling* **18,** 320–334.
35. Mason, J. S. and Beno, B. R. (2000) Library design using BCUT chemistry-space descriptors and multiple four-point pharmacophore fingerprints: simultaneous optimization and structure-based diversity. *J. Molecular Graphics and Modelling* **18,** 438–451.
36. Hassan, M., Bielawski, J. P., Hempel, J. C., and Waldman, M. (1996) Optimization and visualization of molecular diversity of combinatorial libraries. *Molecular Diversity* **2,** 64–74.
37. Borda, J. C. D. (1781) *Memoire sur les elections au scrutin Histoire de l'Academie Royale des Sciences,* Paris
38. Arrow, K. J. (1963) Social choice and individual values, 2nd ed. Wiley, New York.
39. Czerminski, R. (2001) Evaluating different approaches to consensus scoring, *in preparation.*
40. Saari, D. G. (1995) *Basic geometry of voting.* Springer-Verlag, New York.
41. Meylan, W. M., Howard, P. H., and Boethling, R. S. (1996) Improved method for estimating water solubility from octanol/water partition coefficient. *Environmental Toxicology and Chemistry* **15,** 100–106.

21

Application of Neural Networks to Large Dataset QSAR, Virtual Screening, and Library Design

David A. Winkler and Frank R. Burden

1. Introduction

1.1. Drug Discovery in the Twenty-First Century

In the past decade, it became clear that some fundamental problems were arising in drug discovery. It was becoming harder to find new chemical entities with substantial advantages over existing drugs. Consequently, it became riskier and more expensive to develop new drug entities.

The recent development of combinatorial chemistry has provided a possible solution to this problem. This technology has the potential to improve the efficiency of lead discovery, and provides a new paradigm for drug discovery. Parallel development of high-throughput screening (HTS) methods has enabled biological testing to keep pace with the increased synthesis scale. Assay automation has been greatly assisted by new recombinant technologies that allow ready access to quantities of pure protein gene products such as receptors or enzymes.

The attrition rate of drug lead compounds, as they progress along the development pathway, is high. Pharmaceutical companies are now focusing on building "developability" into chemical libraries at the earliest stage. Their aim is to decrease the percentage of lead compounds failing due to poor ADMET (absorption, distribution, metabolism, excretion, and toxicity) properties.

1.2. Chemical Space Is Vast

Although combinatorial chemistry and HTS have made a dramatic impact on lead discovery, the immense size of the "chemical space" from which libraries can be drawn is not always appreciated. The number of chemical

From: *Methods in Molecular Biology, Combined Library Methods and Protocols*
Edited by: L. B. English © Humana Press Inc., Totowa, NJ

compounds that are drug-like (obeying Lipinski's "rule of five," e.g., *[1]*) and synthetically feasible is vast. For example, simple consideration of the numbers of possible branched chain isomers of alkanes shows that there are more than 4 billion C_{30} isomers alone.

Clearly, more complex compounds with heteroatoms, rings, and unsaturation are capable of being assembled in an almost infinite variety of ways. Estimates of size of the "universe" of chemical compounds range from 10^{60} for drug-like molecules to 10^{400} for molecules obeying the laws of chemical valence. These huge numbers span the range of the estimated number of particles in the universe of 10^{80}.

These numbers are so vast that, even with the most optimistic projections of combinatorial synthesis capabilities and optimum chemical diversity metrics, only a minute fraction of this chemical space could conceivably be explored by real combinatorial libraries.

1.3. New Computational Tools for Data Modeling, ADMET Prediction, and Virtual Screening

Clearly, computational methods for simulating the combinatorial discovery process, predicting ADMET, and exploring chemical space are necessary. Specifically, modeling methods are required for:

- Analyzing large, "dirty" screening data sets to extract structure–activity information for optimization of leads
- Developing mathematical models of targets such as receptors or enzymes (virtual receptors) to allow screening of existing chemical databases
- Using these virtual receptors in conjunction with very large virtual libraries (databases of compounds for which synthesis is possible) to find interesting and unexplored structural motifs, which can form the focus of real combinatorial libraries
- Genetically evolving a focused chemical library enriched in leads by exploring drug-like chemical space
- Exploring chemical space sparsely using methods such as cell-based diversity
- Predicting ADMET properties and using these predictive models to design "developability" into combinatorial libraries

Neural networks are general, model free mapping systems that offer significant advantages in exploring chemical space, extracting information from small and large data sets, and modeling important "developability" properties such as pharmacokinetics, solubility, and toxicity. We present a general review of the application of neural networks to combinatorial discovery. There appear to be no other reviews of this type in the literature. Readers are directed to other reviews of the application of neural nets to bioactive molecule design in the literature cited below.

A major focus of this chapter is a description of several very recent, highly novel neural network techniques and applications of potential importance to combinatorial discovery. Whilst these techniques are so new there are few examples of their applications, we nevertheless feel that they represent important tools for combinatorial chemistry in the immediate future. It is important for readers to gain an appreciation of their methodologies and merits.

2. Review of Applications of Neural Networks in Combinatorial Chemistry

Before reviewing the existing applications of neural networks to combinatorial discovery, we offer brief descriptions of the key concepts in quantitative structure–activity relationships (QSAR), neural networks, and virtual high-throughput screening (VHTS). There are a substantial number of reviews of the applications of QSAR to chemistry and drug design ***(2–13)*** and of applications of neural networks to chemistry ***(13–29)***, to which the reader is referred for more detailed discussions of these topics.

2.1. QSAR

The most appropriate and widely used method for extracting information from large data sets is QSAR and its relatives, quantitative structure–property relationships (QSPR) for property modeling, and quantitative structure–toxicity relationships (QSTR) for toxicity modeling. QSAR is a simple, well validated, computationally efficient method of modeling first developed by Hansch and Fujita several decades ago ***(30)***. QSAR has proven to be very effective for discovery and optimization of drug leads as well as prediction of physical properties, toxicity, and several other important parameters. QSAR is capable of accounting for some transport and metabolic (ADMET) processes and is suitable for analysis of in vivo data.

The basic tenet of QSAR is that molecular properties such as lipophilicity, shape, and electronic properties modulate the biological activity of the molecule. This can be captured in a simple relationship...

$$\log \mathrm{BR} = f\,(r_1, r_2, r_3, r_4, \ldots)$$

where BR is a biological response such as IC_{50}, ED_{50}, LD_{50}, K_i, and the r_n are molecular descriptors (mathematical descriptions of molecular properties). The functional relationship f, often complex and nonlinear, can be derived using statistical methods (multiple linear regression [MLR], partial least squares [PLS]) or neural networks. A training data set is used to derive a model and validation or test sets are used to gauge how predictive the model is.

QSAR modeling can be broken down into four essential steps: conversion of structures into mathematical representations (descriptors) that capture the molecular properties important for the process being modeled; descriptor selection to maximize the information available to the modeling process and minimize the risk of chance correlations; deriving the relationship (mapping) between the molecular descriptors and the biological, physicochemical, or toxicological data; and validating the model and assessing its predictivity.

However, classical QSAR has limitations: it cannot deal with isomers; it is prone to overtraining, overfitting, and chance correlations; it cannot implicitly handle nonlinear dependencies and interaction terms between the parameters; its validation methods are time-consuming and lead to multiple models; and QSAR analyses can be difficult to interpret in terms of interactions at the molecular level.

We believe many of these problems are due to a combination of inadequate molecular representations or descriptors and suboptimal SAR mapping methods. Our work has involved breaking down QSAR into its component processes and developing improvements to each process. We have discovered new QSAR methods that overcome the shortcomings of traditional QSAR.

2.2. Neural Networks

Finding structure–activity relationships is essentially a regression or pattern recognition process. Regression is an "ill-posed" problem in statistics, which sometimes results in QSAR models exhibiting instability when trained with noisy data. Recently neural networks have been used as a more flexible modeling paradigm. Artificial neural networks are computer-based mathematical models developed to have functions analogous to idealized simple biological nervous systems. They consist of layers of processing elements (neurodes), considered analogous to the nerve cells (neurons), interconnected to form a network.

2.2.1. Neurodes and Architecture

The basic unit of an artificial neural network (ANN) is the neurode. Typically, each neurode has numerous inputs ($x_1, x_2, \ldots$), each of which is modified by a weight ($w_1, w_2, \ldots$) (**Fig. 1**).

These modified inputs are summed on entry to the neurode. This net input is then modified by an internal transfer function. This might be a step function that passes a signal if a certain threshold is exceeded. More often, a function (such as a sigmoidal or hyperbolic tangent function) that produces a continuous, differentiable nonlinear signal is used (**Fig. 2**). The output is passed on either as the input for other neurodes, or as an output carrying a result.

The architecture used most commonly is the back-propagation artificial neural network. The network consists of a number of layers, with full or partial connections between the neurodes of each layer. Often, the layers have an extra

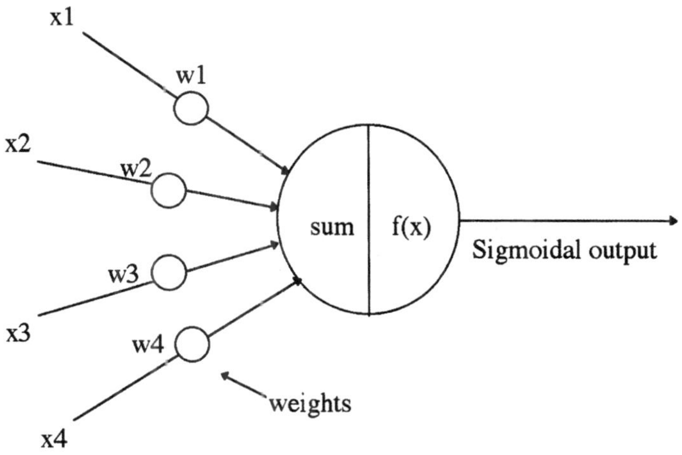

Fig. 1. Artificial neural network neurode.

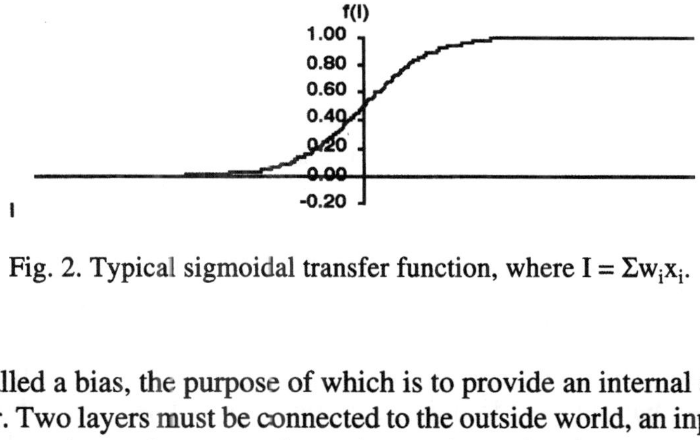

Fig. 2. Typical sigmoidal transfer function, where $I = \Sigma w_i x_i$.

neurode called a bias, the purpose of which is to provide an internal adjustment to the layer. Two layers must be connected to the outside world, an input layer to receive information and an output layer that produces the ANN's response to the information. Frequently, however, there are hidden layers in between—these additional layers allow the network response to be nonlinear. Typically in a multilayered ANN such as the one shown in **Fig. 3**, the transfer functions of the input-layer neurodes are linear functions, whilst those of the hidden and output layers are sigmoidal.

2.2.2. The Process of Learning

The output of an ANN depends on numerous factors—the nature of the neurodes' transfer functions, the architecture of the network, and the weights connecting the neurodes. The weights connecting neurodes are most easily altered. If, in response to a particular input pattern, the network's output is different to the correct output, the weights can be altered so that the network

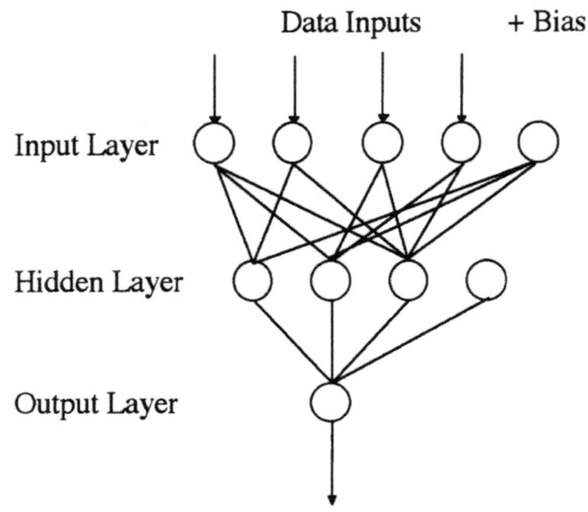

Fig. 3. Example ANN architecture: 3 layered, 4 : 3 : 1 architecture.

more closely approximates the correct result (e.g., biological response of the molecule whose descriptors have been input).

For the neural net to learn rules that relate input patterns to output for a particular data set, it is "trained" on the data—that is, the input patterns are repeatedly shown to the neural network, which modifies the weights connecting neurodes until the error between its outputs and the correct outputs is minimized.

Although a neural network's ability to model the data upon which it is being trained increases with the length of the training procedure, this is not true of a network's predictive capabilities. After a certain number of training cycles, however, the network will begin to "memorize" the training data—that is, it will accurately model the training data but the network's ability to generalize will diminish. This is normally resolved by using a validation set, whose values are predicted but not used in training. Training is stopped when the validation error is minimum and the network's predictive ability is greatest, as **Fig. 4** shows.

Back-propagation is the most common algorithm for adjusting weights. It is a gradient descent algorithm where the network's error is a function of the network's weights. Back-propagation minimizes the average squared error between the network's output and the known data it is approximating by minimizing the error function. The network's weights are altered according to the Delta Rule, in which a proportion of the output error is propagated back through the network to modify the weights.

2.2.3. Pros and Cons of Neural Networks

Regression methods based on neural networks appear to overcome most of the problems with traditional QSAR as they can account for nonlinear structure–activity relationships and can deal with linear dependencies.

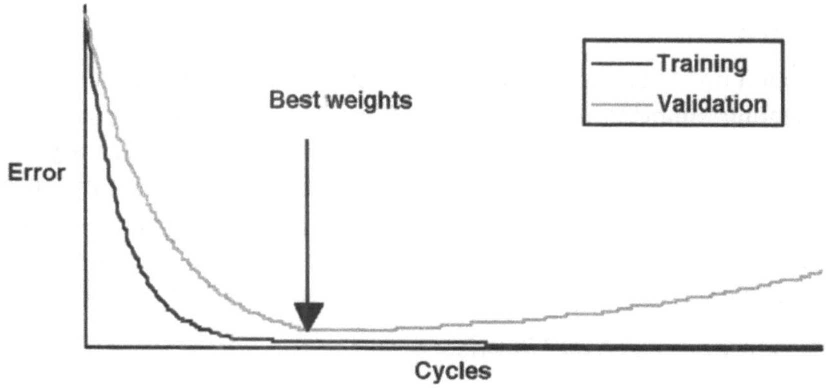

Fig. 4. Typical training and validation error curves.

However, like other regression methods, standard back-propagation neural nets are still prone to overtraining, overfitting, and validation problems. They introduce an additional problem related to overfitting—the need to optimize the neural network architecture. We summarize a number of developments in neural nets, from our work and that of others, which have overcome these shortcomings and allow neural networks to develop very robust models for use in combinatorial discovery.

Neural nets have gained popularity in recent years for extracting information from large data sets provided by combichem/HTS. They are also finding applications in prediction of physicochemical and toxicological properties important for ADMET developability of drugs; database mining, classification, and feature extraction; combinatorial library design and focusing; and simulation of the combinatorial discovery process (virtual HTS [VHTS], *in silico* screening, computational screening).

Pattern recognition, modeling, or regression methods may be classified into two types—unsupervised and supervised learning methods. These differ according to whether a training data set is used to find the model (supervised), or whether the data are classified without prior knowledge of the outcome (unsupervised). Published studies in combinatorial chemistry have mainly described the applications of two types of neural networks: Kohonen, or self-organizing maps (SOMs); and feedforward back-propagation (BP) neural nets. Self-organizing maps are unsupervised learning methods in which the neural net classifies objects into clusters with similar properties. Back-propagation neural nets, the type most often applicable to combinatorial discovery, are supervised learning methods in which a training set of compounds of known properties (e.g., biological activity) are used to train the neural net and obtain a structure–property model.

2.3. VHTS

Recognition of the vastness of chemical space is driving the quest for methods of simulating combinatorial discovery *in silico*. Methods exploring larger regions of chemical space would allow a focusing of combinatorial chemistry libraries into areas with inherent novelty and receptor efficacy. A number of groups are developing neural net-based QSAR models of receptor properties—mathematical receptor surrogates or "virtual receptors." Such virtual receptors are useful screening paradigms for finding leads in large real or virtual databases.

2.3.1. Screening Virtual Libraries

Several computational methods for generating large databases of chemically reasonable structures (virtual libraries) have been developed. They employ strategies such as the mutation of text strings representing chemical structures, the expansion of Markush structural representations, or virtual combinatorial libraries derived by exhaustive enumeration of all substituent variations at specific points on a core scaffold. An example of these large virtual libraries is the ChemSpace™ database, containing approx 10 trillion chemical structures for use in similarity and pharmacophore searches, approx 500,000 times more than all the compounds in Chemical Abstracts.

2.3.2. Target Focused Genetically Evolved Libraries

It is acknowledged that literal simulation of the combinatorial chemistry/HTS process *in silico* still has substantial limitations. Although this will allow searching of chemical space orders of magnitude faster than can real world combinatorial synthesis, it can still explore only a minute fraction of even drug-like chemical space.

An alternative method uses a genetic algorithm to mutate chemical structures, and a fitness function to drive the evolution toward areas of chemical space with desirable properties (**Fig. 5**).

Properties incorporated into the fitness function may include: ease of synthesis (e.g., limits on chiral centres), cost, toxicity (e.g., no alkylating agents or mustards), physicochemical properties (Lipinski's rules), dissimilarity to patented compounds, chemical stability (e.g., no peroxides), etc. Neural net models of structure–activity relationships are clearly important fitness functions in deriving virtual libraries enriched in hits.

3. Review of the Application of Neural Networks to Combinatorial Discovery

The literature contains a substantial number of general reviews on the applications of neural networks to chemistry, QSAR, descriptor selection, toxicity

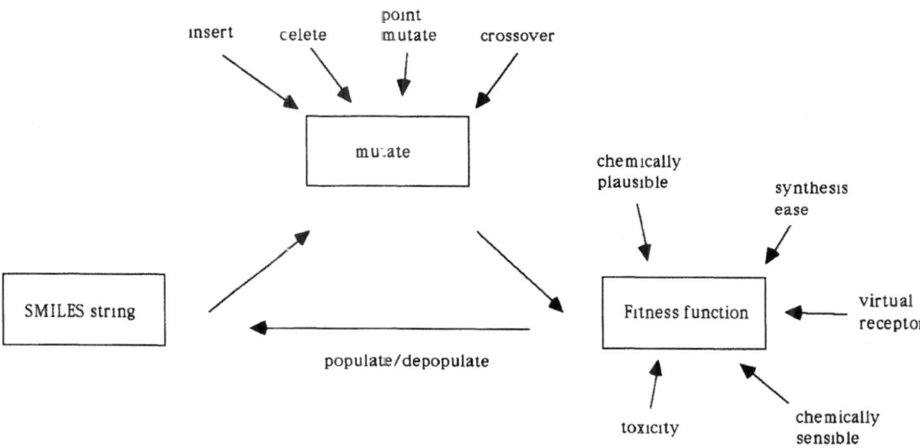

Fig. 5. Overview of a genetically evolved lead generation system.

prediction, and combichem/HTS. Winkler and Madellena reviewed new molecular design tools, such as similarity measures, neural networks, and three-dimensional (3D) QSAR methods in life sciences including drug design *(13,31)*. Jackson summarized recent advances in computational methods for drug design including developments in quantitative structure–activity relationships *(10)*. Wrotnowski reviewed the use of computational intelligence software in modern drug discovery *(32)*. Strengths and weaknesses of computational intelligence methods (neural networks, fuzzy logic, genetic algorithms, fractals, wavelets, and intelligent agents) were described and their current use in drug discovery outlined.

Maggiora et al. reviewed neural networks as a set of computational paradigms with the power to address a wide range of problems from pattern recognition to system identification *(33)*. They discussed issues related to network optimization, data representation, error analysis, and generalization. They particularly emphasized the critical issues of small data sets and noisy data that plague many chemical problems. Burns and Whitesides summarized neural networks, including a description of how they work, applications in pattern recognition, and classification *(34)*.

Balaban and Basak reviewed the advantages of topological indices as molecular descriptors *(35)*. Topological indices represent an efficient tool for molecular modeling using multiparametric linear correlations or nonlinear approaches using neural networks and/or genetic algorithms. Sutter and Jurs reviewed the application of simulated annealing or generalized simulated annealing optimal descriptor selection for neural network models *(36)*. Winkler and Burden reviewed the advantages and disadvantages of traditional QSAR, and that based on neural nets *(37)*. QSAR has advantages of speed and simplicity and of being capable of accounting for some transport and metabolic processes that occur once the compound is administered. Regression methods based on neural networks offer some advantages over MLR methods as they can account for nonlinear SARs and can deal with linear dependencies that sometimes appear in real SAR problems. However, some problems still exist in the development of SAR models using conventional back-propagation neural networks. These authors used to advantage a neural net with Bayesian regularization to produce robust models.

Maddalena has reviewed the role of soft computing methods such as ANNs, genetic algorithms (GAs), fuzzy logic (FL), chaos, fractals, and cellular automata (CA) and their hybrids in molecular design *(38)*. These methods have found application in a wide variety of areas including QSAR, QSPR, variable selection, conformation searching, receptor docking, pharmacophore development, molecular design, combinatorial libraries, surface phenomena, and kinetics and complex system studies. Apostolakis and Caflisch recently reviewed a variety of computational tools that are used to assist drug design, with particular emphasis on QSAR, clustering techniques, and scoring of molecular docking *(39)*.

3.1. Applications to QSAR and VHTS

Combinatorial discovery of new bioactives using high-throughput screening of enormous numbers of congeneric small molecules produced by combinatorial chemistry has stimulated development of theoretical constructs such as virtual combinatorial libraries. High-throughput screening yields more positive responses (hits) than can easily be analyzed manually. In addition, information is contained in the large number of negative responses from inactive molecules. Robust tools are required to extract useful information from these data.

Muskal reviewed the use of neural networks in structure–activity learning and their utility as "electronic assays," capable of rapidly surveying large molecular populations *(40)*. Once a neural network demonstrates adequate predictive performance for a given structure–activity series, it can be used to "electronically screen" compound databases, prospective libraries, and/or virtual libraries for probably active compounds. In a similar way, Keseru et al. developed a VHTS test to identify potentially central nervous system-active (CNS-active) drugs *(41)*. Molecular structures were represented using 2D Unity fingerprints and a feedforward neural network was trained to classify molecules regarding their CNS activity. The neural net recognized at least 89% of CNS-active compounds and so constitutes a useful virtual screen. Winkler and Burden, recognizing a need for a VHTS method, used a specific type of neural network, the Bayesian Regularized Artificial Neural Network (BRANN), to develop SAR models for VHTS *(37)*. This type of network was more robust than standard back-propagation nets and reduced or eliminated the need for lengthy cross-validation. Burden and Winkler also developed a number of computationally efficient molecular indices for use in screening of very large virtual data sets of molecules *(18)*. They discussed the concept of a comprehensive QSAR model as a "virtual receptor" and illustrated this by the results of screening a database of 40,000 molecules.

Lobanov recently presented a novel approach that allowed dramatically increased throughput of *in silico* screening of virtual libraries by evaluating properties of combinatorial products based on the features of the corresponding reagents using neural networks *(42)*. Gasteiger et al. also used neural networks to derive, from quite weak hits, a virtual screening system that can differentiate between active and nonactive compounds and thus allow the design of more focused libraries *(43)*. The success of this endeavor depended critically on the representation of the chemical structure.

Walters et al. have reviewed approaches to the problem of recognizing drug-like molecules *(44)*. It now appears possible to design libraries that are enriched in compounds that have desirable or drug-like properties. Sadowski and Kubinyi also studied this important problem and devised a scoring scheme for the rapid and automatic classification of molecules into drugs and non-drugs *(45)*. Their method used atom type descriptors for encoding the molecular structures and a feedforward neural network for classifying the molecules. The method revealed features in the molecular descriptors that either qualify or disqualify a molecule as being a drug and classified 83% of the Available Chemicals Database (ACD) and 77% of the World Drug Index (WDI) adequately.

In bioinformatics, which is complementary to combinatorial discovery, Huang et al. designed a low order neural network-based filter as a rapid screening agent for identifying single-spanning transmembrane regions from cDNA (complementary DNA) *(46)*. The filter was applied to a library of 2123 anonymous cDNA sequences, which resulted in 61 detections. Evaluation of the detections with two other dissimilar computer prediction algorithms yielded strong transmembrane prediction for 15 of the detections, while 8 of the detections resulted in a definitive negative result. Homologue searches performed on the sequences with detection reports yielded 13 homologues in the predicted reading frame, four of which are membrane associated. Schwabe et al. used similar techniques to identify transcribed sequences from large genomic regions and directly screen genomic libraries *(47)*. Small fragments identified in Southern blots were sequenced and these sequences analyzed by a neural network program that detects coding exons from genomic sequence. They demonstrated efficient isolation of expressed sequences from libraries of whole chromosomes, chromosomal regions, and clones.

So and Karplus tackled the challenging problem of finding an accurate method for estimating the affinity of protein ligands *(48)*. They investigated and compared seven different prediction methods for a set of 30 glycogen phosphorylase (GP) inhibitors with known crystal structures. Five of the methods involve QSAR based on the 2D or 3D structures of the GP ligands alone. The other two methods, LUDI™ and a structure-based neural net binding energy predictor (SBEP) system, made use of the structures of the ligand-receptor complexes. All five QSAR-based models had good predictivity and yielded squared, cross-validated correlation coefficient (q^2) values ranging from 0.60 to 0.82. The LUDI scores were only weakly correlated with biological activity. The SBEP system yielded a q^2 value of 0.60. These models can be applied to VHTS for GP inhibitors. Apostolakis and Caflisch reviewed the variety of computational tools that are used to assist drug design and VHTS *(39)*. These authors reviewed three methods for the estimation of binding energies: neural net-based QSAR; empirical energy functions; and molecular dynamics-based free energy calculations. Bassett and Elling developed a suite of automated methods that aid chemists in lead identification and optimization, using structural and physical feature information from the both hits and inactives. They used a variety of computational intelligence techniques, including neural networks and fuzzy reasoning systems *(49)*.

Tronchet et al. studied the anti-HIV and cytotoxic activities of a heterogeneous series of 1-[(2-hydroxyethoxy)methyl]-6-(phenylthio)thymine (HEPT) analogues and established SAR models suitable for VHTS using a Hansch-type approach, a neural network approach, and a pharmacophore search method (CATALYST™) *(50)*. Maddalena and Johnston carried a similar study on compounds acting at benzodiazepine γ-aminobutyric acid A ($GABA_A$) receptor preparations *(51)*. They found that a suitably optimal ANN pharmacophore model represented the internal physicochemical structure of the receptor site.

3.2. ADME, Pharmacokinetics, and Developability

Lipinski summarized the ranges of physicochemical properties that the large majority of drug-like molecules exhibit *(52)*. Application of these Lipinski (or "rule of five") filters ensures that candidate drugs have a degree of developability inherent, as their physicochemical properties more closely match those required for good pharmacokinetics, and so forth. Lipinski has done additional work on modeling "developability" criteria for incorporation into the library design, QSAR, or optimization processes. Recently, he observed that drug properties have changed so that discovering orally active drugs is becoming more difficult *(53)*. Aqueous solubility is better in phase II drug candidates than in HTS leads and in compounds from HTS follow-up. Lipinski noted that tight SAR feedback from in vivo oral testing biases towards good aqueous solubility. Consequently, lead identification strategy affects the relative importance of poor permeability or poor solubility. Leads from structure-based drug design tend toward larger size, increased hydrogen bonding potential, and poor permeability. HTS-based leads tend towards larger size, higher lipophilicity, and poor solubility. Computational filters such as the "rule of five" flag compounds with profiles that are most problematic for oral activity. Lipinski's ideas have recently been extended to develop similar filters for agrochemicals.

Walters et al. reviewed a variety of successful, mainly informatic approaches to the problem of recognizing drug-like molecules. These methods enable design of libraries that are enriched in compounds that have desirable or drug-like properties *(44)*. Sadowski has used the ideas of Lipinski and informatics to develop fast computational filters for ADMET properties. His filters are based on neural networks and large amounts of data from structural databases or HTS *(54)*. By analogy to the drug/non-drug filter, the method generates fast approximate computational filters that can discriminate between bioavailable/nonavailable compounds, toxic/nontoxic compounds *(45)*.

Vedani and Dobler reviewed the role of neural net-based QSAR and 3D- and 4D-QSAR methods to establish a virtual laboratory for the assessment of receptor-mediated toxicity and the prediction of oral bioavailability *(55)*.

Gobburu and Shelver applied neural networks to predict the pharmacokinetic properties of β-adrenoreceptor antagonists in humans *(56)*. A congeneric series of 10 β-blockers, whose critical pharmacokinetic parameters are well established, was selected for the study. An appropriate neural network system was constructed and tested for its ability to predict the pharmacokinetic parameters from the octanol–water partition coefficient, the pK_a, or the fraction bound to plasma proteins. Neural networks predicted pharmacokinetic values that agreed well with the experimental values and showed better agreement than those predicted by multiple regression models. The results indicate neural networks can be powerful tools in exploration of QSPR. Schapper et al. conducted a QSAR analysis of time- and dose-dependent in vivo anticonvulsant activities of triaminobenzenes *(57)*. The analysis derived the dependence of activity and dose–response behavior on molecular properties. Lockwood et al. used electron density-based wavelet coefficient descriptors (WCDs) as well as 3D and traditional topological descriptors to create an SOM classifying structurally diverse datasets into chemical-reasonable groups for rapid property screening and/or data mining *(58)*. Individual QSAR or ADME models were developed for each cluster using a GA feature selection and/or GA-based clustering methods in addition to neural network pattern recognition. These results provided a basis for the use of WCDs and self-organizing maps in the classification and screening of data sets larger than 10,000 compounds.

3.3. Toxicity Modeling

Models of toxicological endpoints can be invaluable in predicting likely toxicities of commercial bioactive agent (e.g., drug) candidates.

3.3.1. Neural Net Models for Gross Toxicity

QSAR methods based on neural nets have also shown promise in modeling gross or in vivo toxicity of compounds to a number of species. Such models may be faster, and cheaper, and constitute useful surrogates for whole organism toxicity tests involving animal sacrifice. Bradbury reviewed the role of QSARs as tools for predicting the toxicity of chemicals when little or no empirical data are available ***(59)***. Bradbury observed that there has been an evolution of QSAR development and application, from that of a chemical-class perspective, to one that is more consistent with assumptions regarding modes of toxic action.

Calleja et al. studied the relationships between acute toxicity toward five aquatic non-vertebrates and humans, and molecular structure for 38 structurally diverse chemicals ***(60)***. These chemicals were from the 50 priority chemicals prescribed by the Multicenter Evaluation of *In Vitro* Cytotoxicity (MEIC) program. Nonlinear models, derived from PLS regression or BP neural networks, appear to be better than linear models for describing the relation between acute toxicity and molecular structure. BP neural net models, in turn, outperformed nonlinear models obtained from PLS regression. They determined that the physicochemical properties most important for human acute toxicity were the *n*-octanol–water partition coefficient (P_{ow}) and heat of formation.

Devillers and co-workers used neural networks to analyze a large data set of Microtox toxicity results to derive a general QSAR model ***(61–63)***. They used both back-propagation and SOM neural nets to derive models. In a similar study, Xu et al. analyzed a set of 50 alkylated and/or halogenated phenols exhibiting polar narcosis toxicity using feedforward neural networks ***(64)***. Basak used hierarchical quantitative structure–activity relationships (H-QSAR) to construct models for estimating physicochemical, biomedicinal, and toxicological properties of interest ***(65)***. This method uses increasingly more complex molecular descriptors in a graduated approach to model building. Basak applied these techniques to the development of H-QSAR models for estimating the acute aquatic toxicity (LC_{50}) of 69 benzene derivatives to *Pimephales promelas* (fathead minnow). Tang and Bai modeled acute toxicity data of 65 chlorinated aliphatic hydrocarbons to rat LD_{50} using a BP algorithm ***(66)***. The model was used to predict the toxicity of eight chlorinated aliphatic hydrocarbons not used to derive the QSTR model.

Eldred and Jurs developed a QSAR model for the acute oral mammalian toxicity (LD_{50}) of a set of 54 organophosphorus compounds ***(67)***. Feature selection was done with a genetic algorithm to find subsets of descriptors that would support a high quality computational neural network (CNN) model . The best nonlinear CNN model found had a root-mean-square (rms) error of <0.25 log units. Johnson and Jurs successfully modeled the acute oral mammalian toxicity (LD_{50}) of a diverse set of substituted anilines using neural net-based QSAR ***(68)***. Computational neural networks gave the best model, yielding a training set rms error of 0.238 log units and a prediction set rms error of 0.254 log units. Feature selection using computational neural networks to evaluate the fitness of subsets of descriptors chosen by the genetic algorithm was also employed. This routine was able to exploit the nonlinear nature of a CNN, resulting in a model with a training set rms error of 0.233 log units and a prediction set rms error of 0.238 log units.

Ivanciuc devised a model for toxicity of 30 *para*-substituted phenols for *Tetrahymena pyriformis* using neural networks ***(69)***. He used the octanol–water partition coefficient and pK_a as structural descriptors. The quality of the neural net model was comparable to multiple linear regression QSAR models. Basak et al. has used molecular similarity, neural networks, and discriminant analysis methods to predict acute toxic modes of action for a set of 283 chemicals ***(70)***. The majority of these molecules had been determined via toxicodynamic studies to be narcotics, electrophiles/proelectrophiles, uncouplers of oxidative phosphorylation, acetylcholinesterase inhibitors, and neurotoxicants. Nonempirical parameters, such as topological indices and atom pairs, were used as structural descriptors for the development of similarity-based, statistical, and neural network models. Rates of correct classification ranged from 65% to 95% for these 283 chemicals. Eldred et al. reported a mathematical link between the molecular structures and the acute fathead minnow toxicity of a set of 375 organic compounds ***(71)***. Molecular structure descriptors encoding information for topological, geometrical, and electronic aspects of the compounds were used. A genetic algorithm was used to select the best subset of descriptors. The best model (an eight-descriptor, nonlinear CNN model) had rms errors of 0.71, 0.77, and 0.74 log units for the training, cross-validation, and test sets of compounds respectively. Zakarya et al. studied structure–toxicity relationships for 120 diverse insecticidal 1,1,1-trichloro-2,2-*bis*(4-chlorophenyl)ethane-type (DDT-type) molecules using a neural network ***(72)***. On the basis of training results, the neural nets proved superior to regression analysis.

Devillers has reviewed tests developed and proposed as alternatives to whole animal toxicity assays ***(73)***. Among them, the Microtox test has been widely used to estimate the toxicity of agricultural, pharmaceutical, and industrial chemicals producing a large amount of valuable toxicity data. They carried out a critical analysis of the literature and derived a large data bank of more than 1000 chemicals for deriving a general QSAR model for the Microtox test. The molecules were described by means of autocorrelation vectors encoding their hydrophobicity and molar refractivity. Back-propagation neural networks were used to derive a highly reliable model with a wide spectrum of applicability. In a later study, Devillers and Domine reported toxicity models from the Microtox test (*Vibrio fischeri*) ***(63)***. They obtained useful models, albeit with some outliers, from a training set of 1068 organic chemicals described by four different autocorrelation vectors. Addition of the time of exposure as variable allowed them to derive a more powerful model from 2795 toxicity results. The predictive power of this neural network model was assessed by a test set. Eldred et al. conducted a QSAR study of acute aqueous toxicity of 375 diverse organic compounds ***(74)***. The best model used a nonlinear CNN model based on eight calculated molecular structure descriptors. Sutter and Jurs have reviewed the application of simulated annealing to selection of optimal descriptor subsets for neural network models ***(36)***. They illustrated these methods by developing a model for toxicity of benzene derivatives.

3.3.2. Neural Net Models for Mutagenicity and Carcinogenicity

As well as gross toxicity, neural networks have been used to derive models for the very important toxicological properties—mutagenicity and carcinogenicity. Several groups have reported the use of neural networks to analyze the National Cancer Institute (NCI) cancer database ***(75–79)***. Activity patterns across the 60 cell lines provide incisive information on the mechanisms of action of screened compounds and also on molecular targets and modulators of activity within the cancer cells. Mining the database provided useful information for the development of anticancer drugs, for a better understanding of the molecular pharmacology of cancer, and for improvement of the drug discovery process. Benigni and Richard attempted to construct general QSAR toxicity models (e.g., for rodent carcinogenicity) not tailored to congeneric series of chemicals ***(80)***. Their work illustrated some fundamental problems of the application of general QSAR approaches to noncongeneric chemicals. They considered two noncongeneric data sets (mutagenicity and carcinogenicity) with mixed mechanisms of action. They concluded that a successful approach to the problem of QSAR modeling of noncongeneric data will need to consider the multidimensional nature of the problem in both a chemical and a biological sense. Since different chemical classes represent largely independent action mechanisms, some means for extracting local QSARs for constituent classes is necessary. The suitability of neural networks to simultaneously extract more than one QSAR model was illustrated by Burden and Winkler, who used a new, robust structure–activity mapping technique to develop a QSAR model for the toxicity of 278 substituted aromatics toward *T. pyriformis* ***(81)***. Comparisons of their Bayesian neural net models with those derived by classical

PLS analysis showed the superiority of this method. The method was clearly capable of modeling diverse chemical classes and more than one mechanism of toxicity.

Recognizing this advantage, Song et al. used neural networks to classify nitro-substituted polycyclic aromatic hydrocarbons according to their mutagenic activity toward *Salmonella typhimurium* (*82*). The network gave a correct classification rate of 94% for two different classes of compounds: weakly active and strongly active ones. Villemin et al. employed similar methods to find mutagenicity models for aromatic and heteroaromatic nitro compounds (*83*). They showed that neural networks prediction is more accurate than regression analysis prediction. Karelson et al. chose a nonlinear QSAR approach using the Chebyshev polynomial expansion and neural networks to predict the mutagenic toxicity of heteroaromatic and aromatic amines (*84*). Ghoshal et al. employed a back-propagation neural network to correlate mutagenicity, in vitro 5-lipoxygenase inhibitory potency, and estrogen binding affinities with structural and physicochemical descriptors of aromatic and heteroaromatic nitrocompounds, arylhydroxamic acids, and different hexestrol derivatives (*85*). The physicochemical descriptors used were the energy of frontier orbitals, hydrophobicity, and Hammett's constant and/or van der Waals volume. Benigni and Giuliani compared the different families of mathematical models (classical regression, multivariate methods, neural networks) used in the QSAR research for their abilities to predict mutagenicity of compounds (*86*). Ghoshal et al. used a back-propagation type neural net for correlating the mutagenic activity of a dataset of 197 compounds, with energy of the lowest unoccupied molecular orbital (LUMO), and hydrophobicity (*87*). They found the network was a good predictor of activities.

Vracko estimated carcinogenic potency (measured in TD_{50} dose) of molecules using artificial neural networks with a counterpropagation learning strategy (*88*). He used three kinds of descriptors: geometrical structures of molecules, geometrical structures in combination with atomic charges, and energy spectra of occupied orbitals. A set of 45 benzene derivatives was used in this study. The models were able to recognize structures of the training set, and a weak correlation between descriptors and carcinogenic potency was found. In a more wide-ranging study Benigni and Pino used rodent carcinogenicity bioassay data over several decades to evaluate hundreds of chemicals, to better understand the etiology of cancer, and to assess the hazard posed by environmental and industrial chemicals (*89*). They chose a database of 536 rodent carcinogens, and investigated the profiles of tumors induced in the four experimental systems employed: rat and mouse, male and female. They used an unsupervised Kohonen SOM to find the associations among the individual tumor types, and among the tumor profiles induced by the chemicals. They observed that species specificity was generally more important than organ specificity, except for a few tumors in which the species specificity was much more significant than the cross-species sex specificity. For three chemical classes (aromatic amines, electrophilic/alkylating agents, nitroarenes) most represented in the database, they deduced that the important factor in differences among tumor profile were the events that surround the ultimate mechanism of interaction with DNA.

3.3.3. Models for Skin and Eye Irritation

Barratt and co-workers used neural net-based QSAR/QSTR methods to relate the severity of skin corrosivity of acids to parameters that model their skin permeability and cytotoxicity ***(90,91)***. The classification predictions were in agreement with those in the training set for 26 of the 27 acids. The methods provided useful procedures for the prediction of the skin corrosivity potentials of severely corrosive acids, which avoid the use of experimental animals and demonstrate the value of in vitro cytotoxicity parameters as inputs for QSAR analysis.

Barratt also used these methods to develop QSAR models for eye irritation potential of a set of neutral organic chemicals ***(92)***. Patlewicz et al. developed a QSAR model for the eye irritation of cationic surfactants using a data set of 29 in vivo rabbit eye irritation tests on 19 different cationic surfactants ***(93)***. The parameters used were log *P* and molecular volume, log critical micelle concentration (CMC) together with surfactant concentration. The model, constructed using neural network analysis, showed strong positive, nonlinear correlations with surfactant concentration and log CMC and a strongly negative, nonlinear correlation with log *P*. The model explained about 70% of the variance in the data set, consistent with levels of biological variability reported historically for the Draize rabbit eye test. These neural net QSAR methods are useful for interrelating sets of in vivo data in which the biological response parameters are expressed in quite different formats, providing a means of utilizing historical data and thereby extending the availability of in vivo data suitable for the validation of in vitro alternative methods. Such models can be used to screen large databases of chemicals (e.g., national industrial chemical inventories) for hazardous agents, and serve as a validation of QSAR methods for developing useful models of various toxicological endpoints relevant to drug or agrochemical development.

3.4. Library Design

As we showed in the Introduction, according to the rules of combinatorics there are astronomically large number of compounds accessible by combinatorial chemistry. These numbers make even virtual screening of all possible compounds prohibitive if each structure has to be enumerated and evaluated for similarity or particular properties. However, structural diversity of combinatorial products stems from a relatively small number of reagents. This fact may be used to advantage when virtual combinatorial libraries are subjected to virtual screening.

Neural networks may be employed to design, optimize, or focus combinatorial libraries in a number of ways that have been reviewed recently by Sadowski ***(15)***. For example, Lobanov has developed neural network methodologies that dramatically increase throughput of *in silico* screening of virtual libraries by evaluating properties of combinatorial products based on the features of the corresponding reagents ***(42)***. In a similar vein, Agrafiotis has found that it is possible to train a small subset of a library to obtain a neural net model for a nonlinear mapping transformation (NLM) or property prediction, then rapidly transform, or predict the properties of, the rest of the very large library using the neural net model in "readout" mode ***(94,95)***.

McGregor and Muskal have patented a method of using pharmacophore fingerprints for identifying, representing, and productively using high activity regions of chemical space ***(96)***. Pharmacophore fingerprint models may be derived using neural networks. As discussed above, Walters et al. have adopted filters such as the Lipinski "rule of five" and neural network learning systems to design libraries that are enriched in lead compounds with desirable properties ***(44)***. Muskal has proposed applying neural networks to combinatorial library reagent selection and overall library assessment ***(40)***.

Ajay and his colleagues have applied neural networks to designing libraries with CNS activity ***(97)***. CNS-actives and -inactives were selected from the Comprehensive Medicinal Chemistry (CMC) and the MACCS Drug Data Report (MDDR) databases based on whether they were described as having some kind of CNS activity in the databases. This classification scheme resulted in over 15,000 actives and over 50,000 inactives. Each molecule was described by 7 ID descriptors (e.g., number of hydrogen bond donors or acceptors, etc.) and 166 2D descriptors (e.g., presence/absence of functional groups such as NH_2). The models were used to "filter" a large virtual library of 1 million compounds to generate a smaller library of enriched CNS-active leads that would be amenable to combinatorial synthesis. The large virtual library was constructed from scaffolds and side chains frequently found in drug molecules. Sadowski investigated both Kohonen (self-organizing) maps, and feedforward networks as library design tools ***(15)***. Wrede et al. developed a technique for rational sequence-oriented peptide library design based on artificial neural networks and evolutionary optimization strategies ***(98)***. They employed "simulated molecular evolution," where selection of optimized peptides is performed by a trained artificial neural network. New sequence variants are generated by mutations, taking into account the similarities in physicochemical and structural properties between amino acids. As this review shows, neural networks are finding increasing applications in the design of combinatorial libraries.

3.5. Physicochemical Property Models

The prediction of physicochemical properties for molecules is extremely useful for providing data for ADME prediction, chemical scale up, and formulation. Properties such as log *P* are important in pharmacokinetic behavior of compounds and in their oral bioavailability. Knowledge of water solubility aids development of assay systems (ensuring that compounds will dissolve when screened) and for commercial synthesis. Predicted properties, like other computationally derived descriptors, are very useful in deriving QSAR models of biological activity and toxicity. Accurate prediction, rather than direct measurement, of molecular properties is clearly preferable in terms of time and cost-efficiency. Ivanciuc has reviewed the role of neural networks as efficient, general, nonlinear models for computing physicochemical properties of various classes of compounds ***(99)***. He observed that success of neural networks in structure–property models depends mainly on the numerical representation of the structure of the compounds in network calibration and prediction. Three new neural networks were defined to encode into their topological the chemical structure of each compound presented to the network: the Baskin–Palyulin–Zefirov neural device ***(100)***, ChemNet ***(101)***, and Ivanciuc's variant, MolNet. All three neural models use information from the molecular graph only to generate the neural network model.

3.5.1 Octanol–water Partition Coefficient (log P)

As neural networks are parsimonious universal approximators of nonlinear functions, they are excellent candidates for performing the nonlinear regression tasks involved in modeling log *P* and other QSPRs ***(102)***. Huuskonen et al. reported a method for predicting log octanol–water partition coefficients (log *P* values) for a diverse set of 1870 organic molecules based on atom-type electrotopological-state (E-state) indices and neural network modeling ***(103)***. For a test set of 35 nucleosides, 12 nucleoside bases, 19 drug compounds, and 50 general organic compounds not included in the training set, a predictive $r^2 = 0.94$ and rms = 0.41 were calculated by artificial neural networks. Huuskonen attributed the improved prediction ability of artificial neural networks to the nonlinear properties of this method that allowed the detection of high-order relationships between E-state indices and the *n*-octanol–water partition coefficient. Devillers et al. carried out an even larger study using a training set of 7200 chemicals and a back-propagation neural network to predict log *P* of molecules containing nitrogen, oxygen, halogen, phosphorus, and/or sulfur atoms ***(104)***. Chemicals were described by autocorrelation vectors encoding hydrophobicity, molar refractivity, H-bonding acceptor ability, and H-bonding donor ability. The final model achieved excellent results (rms = 0.39, $r = 0.98$) on an external test set of 519 molecules. Analogously, Beck et al. ***(105)***, Bodor et al. ***(106)***, and Breindl et al. ***(107)*** successfully developed neural net-based QSPR models for log *P* of organic compounds based on descriptors from semiempirical molecular orbital (MO) calculations. The models based on these descriptors were able to accurately reproduce the partition coefficients of compounds in the training set and to predict partition coefficients of compounds in test sets. The results compared favorably with those given by the regression analysis approach.

Gakh and co-workers also employed neural networks to predict log *P* and other physical properties of organic compounds ***(108)*** using graph theory to encode the structural information. Huuskonen et al. studied the efficacy of atom-type electrotopological state indices for estimation of the octanol–water partition coefficient (log *P*) values in a set of 345 drug compounds or related complex chemical structures ***(109)***. For the same set of parameters, application of neural networks provided superior prediction ability for training and test sets. Atom-type electrotopological state indices are rapidly calculable from structure so are valuable parameters for fast evaluation of octanol–water partition coefficients that can be applied to screen large databases of chemical compounds, such as combinatorial libraries.

3.5.2. Aqueous Solubility

Aqueous solubility is a desirable property for most drug candidates as it facilitates efficient and unambiguous screening, and is an important factor is designing oral dosage forms. Huuskonen devised an accurate and generally applicable method for estimating aqueous solubilities for a diverse set of 1297 organic compounds based on multilinear regression and artificial neural network modeling ***(110)***. Molecular connectivity, shape, and atom-type E-state indices were used as structural parameters. A predictive $r^2 = 0.92$ and $s = 0.60$ were achieved.

3.5.3. Other Physicochemical Properties

The acidity or basicity of compounds, as expressed by the pK_a, is another property of bioactive compounds. This can have substantial influence on aqueous solubility and pharmacokinetics. Models for pK_a employing neural nets are still emerging although prior experiences with other properties and the recent work by Clark ***(111)*** suggest that neural networks have a useful role in modeling this important property.

Other physicochemical properties are often less important for drug and agrochemical lead development, but factor in the prediction of environmental or occupational safety assessments of databases of, for instance, industrial chemicals. Goll and Jurs correlated the molecular structures of diverse, industrially important, organic compounds to their normal boiling points ***(112)***. Genetic algorithms were used to select the best subsets of molecular descriptors. Computational neural networks were employed to create the models best suited for the prediction of normal boiling points. Cherqaoui and Villemin used back-propagation neural networks to successfully model relationships between structure and boiling point of 150 alkanes ***(113)***. Tetteh and co-workers used a special type of neural net, a radial basis function (RBF) neural network, to simultaneous estimate flash point (T_f) and boiling point (T_b) ***(114)***. Analysis of a database of 400 compounds, divided into training ***(134)***, validation ***(133)***, and testing ***(133)*** sets, yielded a model with the average absolute errors obtained for the validation and testing sets ranges from 10° to 12°C and 11° to 14°C for T_f and T_b, respectively, consistent with experimental error. The neural model was superior to one produced by PLS, strongly suggesting that a nonlinear relationship exists between structure and boiling point/flash point.

4. The Future: New Neural Network Architectures and Novel Applications in Combinatorial Discovery and Development of Bioactive Leads

A major focus of this review is the role of new types and architectures of neural networks and other novel methodologies that have been recently applied to the design and discovery of bioactive agents, especially as they relate to combinatorial discovery. Wrotnowski recently reviewed the application of computational intelligence in modern drug discovery. He compared strengths and weaknesses of neural networks, fuzzy logic, genetic algorithms, fractals, wavelets, and intelligent agents as paradigms for drug discovery ***(32)***. Wrotnowski also provided an assessment of the potential of intelligent computing methods to bring new drugs to the marketplace. Maddalena recently conducted a complementary review of the role of soft computing methods and their hybrids in molecular design ***(38)***. Maddalena summarized their use in QSAR, QSPR, variable selection, conformation searching, receptor docking, pharmacophore development, molecular design, combinatorial libraries, surface phenomena, kinetics, and complex system studies. Both reviewers expect the use of soft, intelligent computing techniques to grow significantly in the future.

Research in neural methods, and their applications to chemistry is an active area. Techniques have been devised that overcome the weaknesses of standard back-propagation neural nets, and novel neural net architectures have been devised that have not yet been applied to combinatorial discovery and bioactive lead development. Another area of active research is in the discovery of better molecular representations that more accurately capture molecular properties important for biological activity. There is a proliferation of possible descriptors currently (more than 1000) and much research is now aimed at finding better, rather than more, representations. This will broaden the applicability of descriptors and may approach the ideal-descriptors of general applicability.

4.1. Improved, Computationally Efficient Molecular Representations

We believe that much more efficient descriptors can be discovered and this is a focus of our work. Until recently QSAR analyses have used relatively simple molecular descriptors based on substituent constants (e.g., Hammett constants, π, or molar refractivities), physicochemical properties (e.g., partition coefficients), or topological indices (e.g., Randic, Weiner, or Kier and Hall indices). Recently we and others have developed several new information rich, computationally efficient representations. The most interesting of these are the molecular eigenvalue indices.

We dissected the QSAR method into its elementary processes: conversion of molecules to relevant descriptors, intelligent descriptor selection, SAR mapping, and validation ***(37)***. We assessed the strengths and weaknesses of current methods of performing each of the component operations and devised new, more efficient methods for each. We derived several new molecular representations: atomistic, functional group, and molecular eigenvalue indices ***(18,115,116)***. Burden developed a method of generating a unique molecular index using eigenvalues of a modified adjacency matrix—a matrix with off-diagonal elements of unity if the atoms are connected and zero if not, for use in database indexing ***(117,118)***. The original form of this index was developed further by Pearlman to become the BCUT (Burden, CAS, University of Texas) index widely used by the pharmaceutical industry as a diversity measure. Our recent work involves finding the eigenvalues of novel molecular matrices in which the diagonal elements represent a range of steric, electrostatic, and lipophilic properties of the corresponding atoms. This type of descriptor appears to capture information relevant to structure–activity relationships not captured by other descriptors ***(119)***. There is increasing evidence in the literature that descriptors of this type have quite general applicability ***(120)***.

These representations have advantages of computational speed, more accurate description of molecular properties relevant to activity, or more generally applicability to diverse chemical classes acting at a common receptor than traditional representations.

4.2. Novel Variable Selection Methods

Many molecular representations have been proposed for use in QSAR. The choice of representation is often a subjective decision, and skillful choices depend on the experience of the researcher and the nature of the modeling problem. Clearly, if too many descriptors are used to find a model, there is a high probability that a chance correlation will occur. If incorrect descriptors, containing irrelevant information are used, no useful model will result.

In QSAR studies of large data sets particularly, variable selection and model building is a difficult, time-consuming, and ambiguous procedure. Kubinyi has reviewed methods such as stepwise regression procedures, neural networks, cluster significance analysis, or genetic algorithms for variable selection *(121)*. He also described a simple and efficient evolutionary strategy, MUSEUM (Mutation and Selection Uncover Models), for variable selection. Random mutation (first by addition or elimination of very few variables, then by simultaneous random additions, eliminations, and/or exchanges of several variables at a time) leads to new models that can be evaluated by appropriate fitness functions. In contrast to common genetic algorithm procedures, only the "fittest" model is stored and used for further mutation and selection, leading to better and better models.

So and Karplus stressed that the selection of appropriate descriptors is an important step in the successful formulation of QSARs *(122–125)*. They compared a number of feature selection routines and mapping methods including forward stepping regression (FSR), genetic function approximation (GFA), generalized simulated annealing (GSA), and genetic neural network (GNN). The GNN method uses a neural network to correlate activity with descriptors that are preselected by a genetic algorithm. A comparison of the predictive qualities for both training and test compounds showed that the GNN protocol achieves the best results and the choice of descriptors by the GNN method are consistent with established SARs on this series of compounds. Luke compared the method of evolutionary programming (EP) to GFA and described how EP can also generate multiple predictors *(126)*. Evolutionary programming, as it is applied here, is able to very quickly generate a series of different predictors and, in direct comparisons, finds good QSARs that were missed by the GFA. Waller and Bradley employed a simple random selection strategy to rapidly identify from a pool of allowable variables those that are most closely associated with a given response variable *(127)*. The FRED (Fast Random Elimination of Descriptors) algorithm starts with a population of offspring models composed of either a fixed or variable number of randomly selected variables. Iterative elimination of descriptors leads to subsequent generations of more fit offspring models. In contrast to common genetic and evolutionary algorithms, only those descriptors contributing to the genetic makeup of less fit offspring models are eliminated from the descriptor pool. A comparison of the results of a FRED analysis with alternative algorithms reveals that this technique is capable of efficiently identifying the same "optimal" solutions. Zupan and Novic optimized a spectrum-like structure representation via genetic algorithm (GA) to determine relevant variables for modeling *(128)*. The fitness function in the variable reduction of the GA procedure was a neural network model. Because the spectrum-like structure representation is reversible, each representation's variable can be back-traced to the structural feature. Hence the variables selected by the GA optimization can elucidate the structural features most relevant for biological activity. The method is robust and, in principle, can be applied to almost any system in which there is a set of molecules that exhibit a specific type of biological, physical, or physicochemical property. It is clear that evolutionary algorithms are being used increasingly in variable selection/feature selection methods for SAR.

Recently we adapted a method called automatic relevance determination (ARD) *(129)*, that uses Bayesian principles, to automatically and objectively select relevant variables from a larger set of descriptors *(130)*. Automatic Relevance Determination allows the network to "estimate" the importance of each input, effectively turning off those that are not relevant. This allows all variables, including those that have little impact on the output, to be included in the analysis without ill effect. The ARD method ensures that irrelevant or highly correlated indices used in the modeling are neglected as well as showing which are the most important variables in modeling the activity data.

Kovalishyn et al. used the cascade correlation neural net to select variables in QSAR studies *(131)*. Their results suggest that these pruning methods can be successfully used to optimize the set of variables for the cascade-correlation learning algorithm neural networks. The use of variables selected by the elaborated methods provides an improvement of neural network prediction ability compared to that calculated using the unpruned sets of variables.

Sutter and co-workers reported a method for automated descriptor selection for quantitative structure–activity relationships using generalized simulated annealing *(36,132)*. The cost function used to evaluate the effectiveness of the descriptors was based on a neural network. The result is an automated descriptor selection algorithm that is an optimization inside of an optimization. Application of the method to QSAR shows that effective descriptor subsets are found, and they support models that are as good or better than those obtained using traditional linear regression methods.

4.3. Improved Structure–Activity and Structure–Property Mapping

Improved methods of structure–property mapping must overcome the current problems of lack of robustness due to instability, overtraining, overfitting, and the validation burden. Improvements to existing neural net architectures (e.g., BP) by regularization and other procedures and application of other architectures not currently exploited to combinatorial discovery show considerable promise in eliminating these problems.

4.3.1. Novel Types of Back-Propagation Neural Net

The current applications of standard feedforward back-propagation neural nets have been summarized in the first part of this work. Several novel variants of back-propagation neural nets have been reported in the literature recently. Winkler and Burden adapted a specific type of back-propagation neural network, the BRANN ***(133,134)***, to development of SAR models ***(135)***. These neural networks incorporate Bayesian regularization, a mathematical process that converts the regression into a "well-posed" statistical problem. Bayesian methods are optimal methods for solving such learning problems. The advantage of BRANN is that the models are robust and the validation process, which scales as $O(N^2)$ in normal regression methods, is unnecessary. These networks automatically solve a number of problems that arise in QSAR modeling such as choice of model, robustness of model, choice of validation set, size of validation effort, and optimization of network architecture. Bayesian regularized neural networks have additional advantages. They are difficult to overtrain as an evidence procedure provides an objective criterion for stopping training. They are difficult to overfit because the BRANN calculates and trains on the number of effective parameters (essentially the number of nontrivial weights in the trained neural network). This is considerably smaller than the number of weights in a standard fully connected back-propagation neural net. These more parsimonious networks are much less likely to be overfitted. Bayesian neural nets are inherently insensitive to the architecture of the network, provided a minimal architecture has been provided. This is a property of Bayesian neural nets that incorporate Occam's razor, making excessively complex models self penalizing. As the architecture is made more complex (e.g., by increasing the number of hidden layer nodes) the number of effective parameters converges to a constant. It has been shown mathematically that they do not strictly require a validation or test set, as they produce the best possible model most consistent with the data. This has the advantage that a single, best model is provided, all data is available for the model, and the validation effort is removed.

Niculescu et al. have reported the application of a related probabilistic neural net to bioactive prediction ***(136)***. These authors investigated the connection between the data preprocessing strategy and kernel choice on the quality of the derived models. Ajay et al. also employed Bayesian methods to design a CNS-active library ***(97)***. A neural network trained using Bayesian methods was trained on CNS-active and CNS-inactive data and correctly predicted up to 92% and 71% accuracy on the actives and inactives. They used the method to generate a small library of potentially CNS-active molecules amenable to combinatorial synthesis.

4.3.2. Novel Types of Neural Net

Although back-propagation neural networks are the most widely used neural net architectures in combinatorial chemistry and bioactive design and discovery, there are a number of other architectures with properties worthy of investigation. Here we consider both novelty of the net architecture and novelty of application. Doucet and Panaye have reported the application of a Hopfield network to molecular design ***(137)***. The basic aim of a Hopfield network is to learn and store a set of patterns so it can recall them later. It is a simple artificial network that is able to store certain memories or patterns in a manner rather similar to the brain in which a full pattern can be recovered if the network is presented with only partial information. The nodes in the network can exist in only one of two possible "states"—on or off—and every node is connected to every other node with some weight. At any instant a node will change its state depending on the inputs it receives from the other nodes. The training and evolution of Hopfield nets is therefore rather different to that of back-propagation neural networks. Doucet and Panaye's results compared favorably with the previous approaches and their method was simpler. Feuilleaubois and co-workers also took advantage of the capability of Hopfield-like neural networks to carry out combinatorial optimization of an objective function ***(138)***. They applied this method to the 3D-pattern search problem. Initial tests indicate that this approach not only successfully retrieved a given pattern, but also suggested partial matches having one or two atoms less than the given pattern. The distributed representation of the problem on Hopfield-like neural networks offered good prospects for parallel implementation.

Campbell and Johnson described how it is possible to reduce numerical data to a mathematical representation called an abductive network (AN) ***(139)***. ANs offer the advantage that correlations may be drawn between variables that are not easily related within a mathematical context. Abductive networks are used for modeling complex relationships among the variables where the functional forms of the relationships are unknown. Abductive modeling is often superior to neural networks and regression because it combines features of each into a more efficient and accurate modeling technique. Given a data set, abductive modeling involves synthesizing an abductive network as a mathematical model of the relationships in the data. An abductive network is a network of functional nodes in which each node contains a mathematical function to compute outputs from a number of inputs. Unlike neural networks, the functions used to compute outputs from inputs may vary throughout the network. In typical applications, polynomials up to degree 3 are used. Campbell and Johnson described several simple examples that illustrate the interesting, and potentially useful properties of abductive networks. They showed that abductive networks more accurately predict some properties than do back-propagation neural networks or linear regression.

Tetteh and co-workers described the application of radial basis function (RBF) neural network models for property prediction and screening ***(114)***. They employed a network optimization strategy based on biharmonic spline interpolation for the selection of an optimum number of RBF neurons in the hidden layer and their associated spread parameter. Comparisons with the performance of a PLS regression model showed the superior predictive ability of the RBF neural model.

Burden investigated the discriminant analysis properties of holographic neural nets (HNNs) ***(140)***. The mapping achieved by HNNs is by a totally different algorithm than BP neural nets. The basis of the holographic method relies on transforming the data to vectors in the complex plane. The vectors have both direction and magnitude, which correspond, respectively, to magnitude and weighting of the original data. The use of the term "holographic" is suggested by the similarity to a class of mathematics found within electromagnetic field theory and in the apparent *enfolding* of information within optical holograms. They rely on the overlaying of a large number of terms in this holographic sense, with various weightings, from which the pattern is reproduced. The training of an HNN is accomplished with very few iterations, often two or three, and the final model is expressed as a set of term coefficients together with associated weights. It should be noted that the holographic neural theory is fundamentally different from the standard connectionist models in artificial neural system theory. In the QSAR context, the method has application as a highly nonlinear discriminant in two- to four-dimensional space. This architecture appears to have advantages over back-propagation neural nets when dealing with category data or classification. It was found that the holographic neural network method gave comparable, and in some cases superior, results to the other discriminant methods. The holographic neural network method was simple to apply and has the advantage that it can be easily refined when new data become available without disturbing the original mapping.

Gobburu et al. described generalized regression neural networks (GRNN) as a special class of artificial neural networks that are memory-based and one-pass learning paradigms ***(141)***. A generalized regression neural network has the advantage over other methods in that it can accurately emulate multidimensional surfaces even with sparse data sets. Gobburu applied GRNN to a series of carboquinones and investigated the effect of various training conditions and influence of the predictor vectors on the performance of the networks.

Domine and co-workers utilized the family of Adaptive Resonance Theory (ART and ART 2-A) based artificial neural networks for unsupervised and supervised pattern recognition ***(142,143)***. The simplest ART network is a vector classifier—it accepts as input a vector and classifies it into a category depending on the stored pattern it most closely resembles. Once a pattern is found, it is modified (trained) to resemble the input vector. If the input vector does not match any stored pattern within a certain tolerance, then a new category is created by storing a new pattern similar to the input vector. Consequently, no stored pattern is ever modified unless it matches the input vector within a certain tolerance. This means that an ART network has both plasticity and stability; new categories can be formed when the environment does not match any of the stored patterns, but the environment cannot change stored patterns unless they are sufficiently similar. Domine analyzed different data sets and identified the advantages and drawbacks of the ART networks. They are also compared ART and ART 2-A with other multivariate techniques such as hierarchical cluster analysis and nonlinear mapping. They concluded that ART 2-A represents a new useful nonlinear statistical tool for QSAR and drug design.

Kovalishyn et al. described the cascade correlation (CC) neural network architecture ***(131)***. The cascade-correlation algorithm starts with a small network and dynamically adds new nodes until the analyzed problem has been solved. This feature of the algorithm obviates predefining the architecture of the neural network prior to network training. Cascade correlation is a supervised learning architecture that builds a near minimal multilayer network topology. The two advantages of this architecture are that there is no need for a user to worry about the topology of the network and that CC learns much faster than the usual learning algorithms. Cascade correlation combines two ideas. The first is the cascade architecture, in which hidden units are added only one at a time and do not change after they have been added. The second is the learning algorithm, which creates and installs the new hidden units. For each new hidden unit, the algorithm tries to maximize the magnitude of the correlation between the new unit's output and the residual error signal of the net. Micheli et al. also used a neural network model recently proposed for the processing of structured data, the recursive cascade correlation (RCC) neural networks model ***(144)***. It allows input of structures as labeled ordered directed graphs and constitutes a novel approach to QSAR. The internal representations developed by the neural networks facilitate discovery of relevant structural features just on the basis of the association between the structure and the target affinity.

Kyngas and Valjakka have developed an evolutionary neural network (ENN) for modeling multifactor data ***(145)***. ENNs can remove insignificant descriptors, choose the size of the hidden layer, and fine tune the parameters needed in training the network. They found that evolutionary neural networks gave more accurate predictions than statistical methods and standard back-propagation neural networks.

4.3.3. Self-Organizing Maps (SOMs)

SOM, also known as a Kohonen neural network, is a projection technique that reduces the descriptor multidimensional space into a space of any given dimensionality. It is a topological-preserving function obtained by unsupervised learning that nonlinearly projects the high-dimensional activity patterns into (usually) two dimensions. The SOM partitions the 2D array into distinct regions, each of which is principally occupied by agents having the same broadly defined properties. Sadowski has reviewed the application of neural methods, including SOM methods, to library design ***(15)***. Anzali et al. have written a review of the applications of self-organizing networks, such as the Kohonen neural network, in drug design ***(19)***. Kireev et al. have assessed nonsupervised neural networks as a new classification tool to process large databases ***(146)***. They assessed SOMs using a database containing more than 2000 organophosphorous compounds with various pesticide activities.

Van Osdol and co-workers applied SOMs to identify potentially useful agents in the National Cancer Institute screening data ***(77)***. The SOM partitions the 2D map into distinct regions and subregions that conform to plausible mechanisms and submechanisms. The results indicate that the activity patterns obtained from the screen contain detailed information about mechanism of action and its basis in chemical structure. The SOMs, unlike the previously applied neural networks, preserved and revealed the relationships among compounds acting by similar mechanisms and therefore have the potential to identify compounds that act by novel mechanisms.

Bernard et al. tested SOM as a nonsupervised procedure for comparing molecular databases ***(147)***. Each chemical compound being represented by a point in the hyperspace of the molecular descriptors. SOMs were used to reflect the multidimensional hyperspace onto a 2D map while preserving the order of distances between the points, but in a nonlinear way. The aim of this work was to apply SOMs to study the overlapping of two databases to obtain information about the extent of their differences in regard to their molecular diversity. The results obtained indicate that SOMs can be used for the search of new leads among available databases and the exploration of new structural domains for a given biological activity. Lockwood et al. used electron density-based wavelet coefficient descriptors (WCDs) as descriptors to create a SOM designed to facilitate the classification of structurally diverse data sets of 20–130 compounds into chemically reasonable groups for rapid property screening and/or data mining ***(58)***. The results of these experiments provide a basis for the use of WCDs and SOMs in the classification and screening of much larger datasets (>10,000 compounds).

Bienfait successfully applied high-resolution Kohonen maps to activity and structure clustering ***(148)***. Gasteiger et al. applied autocorrelation vectors as descriptors and Kohonen neural networks to retrosynthetic analysis, efficient planning of compounds to be synthesized, and ways to split them into sublibraries ***(149)***.

Hanke and Reich used Kohonen nets as a visualization tool for the analysis of protein sequence similarity ***(150)***. The procedure converts sequence (domains, aligned sequences, and segments of secondary structure) into a characteristic signal matrix. This conversion depends on the property or replacement score vector selected by the user. The trained Kohonen network is functionally equivalent to an unsupervised nonlinear cluster analyzer. Protein families, or aligned sequences, or segments of similar secondary structure aggregate as clusters and their proximity may be inspected.

Rose et al. applied Kohonen mapping to a wide matrix of physicochemical property data for a set of antifilarial antimycin analogs containing structural outliers ***(151,152)***. Kohonen mapping compared favorably with nonlinear unsupervised statistical pattern recognition methods for 2D representation of compound similarity and for classification based on antifilarial activity. It may prove a valuable technique for QSAR in situations where a linear method does not model the data well and a high-throughput of test compounds is indicated.

4.3.4. Methods Based on Genetic Properties

Shi et al. ***(76,153)*** employed a genetic function approximation (GFA) to classify the mechanism of action of compounds in the NCI database on the basis of their pattern of growth inhibitory activity against the 60 cell lines ***(78,79)***. They formulated a general "information-intensive" strategy for drug discovery that integrates data on a compound's molecular structure, pattern of growth inhibitory activity, and possible molecular targets in the cell.

Wrede and co-workers described a technique for rational sequence-oriented peptide design in machina ***(98)***. It is based on artificial neural networks and evolutionary optimization strategies. In the natural molecular evolution, amino acid sequences are developed mainly by mutation and selection. Following this principle a biocybernetical design cycle termed "simulated molecular evolution" has been developed in which selection of optimized peptides is performed by a neural network fitness function. New sequence variants are generated by analogy to the rules of natural mutations taking into account the similarities among the individual amino acids in their genetic encoding and their physicochemical and structural properties. They concluded that this technique for peptide design is a useful addition to established peptide-screening technologies.

Burden, Rosewarne, and Winkler used genetic algorithms to attempt to solve the interesting "inverse QSAR" problem of predicting the required molecular properties of a more active molecule from a model *(154)*. They proposed solving this problem by using genetic algorithms to effectively invert the trained neural net model. Devillers developed an intercommunicating hybrid system using a genetic algorithm and a back-propagation neural network model for solving the general problem of designing molecules with specific properties *(155)*. Jiang et al. used a genetic-based recursive algorithm for optimizing the architecture of feedforward neural networks by the stepwise addition of a range of hidden nodes *(156)*. Some new genetic operators, competition and self-reproduction, were introduced and used together with some substantially modified genetic operators, crossover and mutation, to form a modified genetic algorithm (MGA) that ensures asymptotic convergence to the global optima with relatively high efficiency. The proposed methods were applied successfully to chemical analysis and QSAR studies.

4.4. Novel Doesn't Always Mean Better

A number of novel neural net architectures have been shown to offer no significant advantages over standard back-propagation neural nets. Mager and Reinhardt compared the results of the back-propagation and new generalized-regression genetic-neural (GRGN) network methodologies using a series of nonpeptide arginine vasopressin V1 antagonists *(157)*. They showed that both approaches are equivalent with respect to the recognition process while the BP network is superior over GRGN if the sample sizes are reduced by cross-validation.

Liu et al. described the "functional-link net" architecture (FUNCLINK) *(158)*. They claimed that, compared with methods such as adaptive least squares and back-propagation neural nets, FUNCLINK exhibited good recognition and high predictivity. However, Manallack and Livingstone found two disadvantages to the FUNCLINK technique *(159)*. First, the natural ability of neural networks to develop nonlinear relationships is removed with FUNCLINK as these must be specified. Second, the large number of enhanced parameters produced by FUNCLINK increases the possibility of chance effects. They concluded that FUNCLINK adds little to the field of QSAR data analysis.

Livingstone et al. have employed a particular neural net architecture called a reversible nonlinear dimensionality reduction (ReNDeR) net for a low-dimensional display of multivariate data sets *(160)*. The method makes use of the activity values of the hidden neurons in a trained three-layer feedforward network to produce the low-dimensional display. It was claimed that, in contrast to conventional techniques such as principal components analysis or nonlinear mapping, the technique could reconstruct from a point in the low-dimensional display the corresponding multivariate input vector via the weight matrices of the trained network. However, Reibnegger showed that this claim is unjustified in this general form ***(161)***. This method is a useful dimension reduction procedure in spite of loss of some information in the mapping process.

Tetko et al. presented a self-organizing multilayered iterative algorithm that produced linear and nonlinear polynomial regression models that allowed control of the number and the power of the terms in the models ***(162)***. The accuracy of the algorithm was compared to the PLS algorithm in 14 QSAR studies. The results showed that the method is able to select simple models characterized by a high prediction ability and thus provides a considerable interest in QSAR studies. The software was developed using client-server protocol (Java and C++ languages) and is applicable for Web users.

5. Conclusions

An important issue in the application of neural computing methods to the design of bioactive agents and simulation of HTS is whether the methods have lived up to their promise. Manallack and Livingstone (163–165) and Livingstone and Salt ***(11,166)*** reviewed the application of neural methods to QSAR data. They summarized how networks are able to perform the equivalent of discriminant and regression analysis and highlighted their initial susceptibility to overtraining and overfitting, resulting in poor prediction abilities. They reviewed other network algorithms and training regimens emerging in the literature that address these particular problems. More recently, these authors reviewed the work on neural networks in drug design over the last decade ***(16)***. In their review they showed how the overfitting and overtraining problems have been addressed resulting in a technique that surpasses traditional statistical methods. Neural networks have thus largely lived up to their promise, especially when robust regularization methods are used ***(37)***. They predict that the next revolution in QSAR will involve research into producing better descriptors for relating chemical structure to biological activity. Kovesdi et al. also reviewed the methodology and application of neural networks in structure-activity relationships and compared their performance with that of common 3D QSAR methods such as comparative molecular field analysis (CoMFA) and a classical QSAR ***(167)***.

Computational methods of extracting information from screening data sets, discovering comprehensive receptor property models, prediction of ADMET properties, and simulation of combinatorial discovery will play an increasingly important part in drug discovery this century. There is growing evidence that evolutionary algorithms such as GAs will undergo a similar growth in interest and application.

References

1. Lipinski, C. A., Lombardo, F., Dominy, B. W., and Feeney, P. J. (2001) Experimental and computational approaches to estimate solubility and permeability in drug discovery and development settings. *Adv. Drug Deliv. Rev.* **46,** 3–26.
2. Chu, K. C. (1980) The quantitative analysis of structure-activity relationships. *Burger's Medicinal Chemistry,* 4th edit, Ed. M. E. Wolff, Wiley and Son, New York, vol. 1, pp. 393–418.
3. Martin, Y. C. (1981) A practitioner's perspective of the role of quantitative structure-activity analysis in medicinal chemistry. *J. Med. Chem.* **24,** 229–237.
4. Mager, P. P. (1984) Biometrics in medicinal chemistry: a difficult road ahead. *QSAR Des. Bioact. Compd.* 433–442.
5. Trinajstic, N., Randic, M., and Klein, D. J. (1986) On the quantitative structure-activity relationship in drug research. *Acta Pharm. Jugosl.* **36,** 267–279.
6. Martin, Y. C. (1981) A practitioner's perspective of the role of quantitative structure-activity analysis in medicinal chemistry. *J. Med. Chem.* **24,** 229–237.
7. Guo, Z. (1995) Structure–activity relationships in medicinal chemistry: development of drug candidates from lead compounds. *Pharmacochem. Libr.* **23,** 299–320.
8. Karelson, M., Lobanov, V. S., and Katritzky, A. R. (1996) Quantum-chemical descriptors in QSAR/QSPR studies. *Chem. Rev.* **96,** 1027–1043.
9. Kaiser, K. L. E. (1999) Quantitative structure-activity relationships in chemistry. *Can. Chem. News* **51,** 23–24.
10. Jackson, R. C. (1995) Update on computer-aided drug design. *Curr. Opin. Biotech.* **6,** 646–651.
11. Salt, D. W., Yildiz, N., Livingstone, D. J., and Tinsley, C. J. (1992) The use of artificial neural networks in QSAR. *Pestic. Sci.* **36,** 161–170.
12. Hansch, C. (1976) On the structure of medicinal chemistry. *J. Med. Chem.* **19,** 1–6.
13. Winkler, D. A. and Madellena, D. J. (1995) QSAR and neural networks in life sciences. *Ser. Math. Biol. Med.* **5,** 126–163.
14. Peterson, K. L. (2000) Artificial neural networks and their use in chemistry. *Rev. Comput. Chem.* **16,** 53–140.
15. Sadowski, J. (2000) Optimization of chemical libraries by neural networks. *Curr. Opin. Chem. Biol.* **4,** 280–282.
16. Manallack, D. T. and Livingstone, D. J. (1999) Neural networks in drug discovery: have they lived up to their promise? *Eur. J. Med. Chem.* **34,** 195–208.
17. Schneider, G. and Wrede, P. (1998) Artificial neural networks for computer-based molecular design. *Prog. Biophys. Mol. Biol.* **70,** 175–222.
18. Burden, F. R. and Winkler, D. A. (1999) New QSAR methods applied to structure–activity mapping and combinatorial chemistry. *J. Chem. Inf. Comput. Sci.* **39,** 236–242.
19. Anzali, S., Gasteiger, J., Holzgrabe, U., Polanski, J., Sadowski, J., Teckentrup, A., and Wagener, M. (1998) The use of self-organizing neural networks in drug design. *Perspect. Drug Discov. Des.* **9,** 273–299.
20. Svozil, D., Kvasnicka, V., and Pospichal, J. (1997) Introduction to multi-layer feedforward neural networks. *Chemom. Intell. Lab. Syst.* **39,** 43–62.

21. Zupan, J. (1994) Introduction to artificial neural network (ANN) methods: what they are and how to use them. *Acta Chim. Slov.* **41,** 327–352.
22. Sumpter, B. G., Getino, C., and Noid, D. W. (1994) Theory and applications of neural computing in chemical science. *Annu. Rev. Phys. Chem.* **45,** 439–481.
23. Melssen, W. J., Smits, J. R. M., Buydens, L. M. C., and Kateman, G. (1994) Using artificial neural networks for solving chemical problems Part II. Kohonen self-organizing feature maps and Hopfield networks. *Chemom. Intell. Lab. Syst.* **23,** 267–291.
24. Smits, J. R. M., Melssen, W. J., Buydens, L. M. C., and Kateman, G. (1994) Using artificial neural networks for solving chemical problems. Part I. Multilayer feedforward networks. *Chemom. Intell. Lab. Syst.* **22,** 165–189.
25. Lohninger, H. (1993) Neural networks in the chemistry. *Oesterr. Chem. Z.* **94,** 90–92.
26. Gasteiger, J. and Zupan, J. (1993) Neural networks in chemistry. *Angew. Chem.* **105,** 510–536 (See also *Angew. Chem., Int. Ed. Engl.* [1993]), **32,** 503–527.
27. Wythoff, B. J. (1993) Back-propagation neural networks. A tutorial. *Chemom. Intell. Lab. Syst.* **18,** 115–155.
28. Tusar, M., Zupan, J., and Gasteiger, J. (1992) Neural networks and modeling in chemistry. *J. Chim. Phys. Phys.-Chim. Biol.* **89,** 1517–1529.
29. Schneider, G. and Wrede, P. (1998) Artificial neural networks for computer-based molecular design. *Prog. Biophys. Mol. Biol.* **70,** 175–222.
30. Hansch, C. and Fujita, T. (1964) ρ-σ-π Analysis. A method for correlation of biological activity and chemical structure. *J. Am. Chem. Soc.* **86,** 1616.
31. Maddalena, D. J. (1996) Applications of artificial neural networks to quantitative structure-activity relationships. *Expert Opin. Ther. Pat.* **6,** 239–251.
32. Wrotnowski, C. (1999) Counting on computational intelligence. *Mod. Drug Discov.* **2,** 46–48, 51–52, 55.
33. Maggiora, G. M., Elrod, D. W., and Trenary, R. G. (1992) Computational neural networks as model-free mapping devices. *J. Chem. Inf. Comput. Sci.* **32,** 732–741.
34. Burns, J. A. and Whitesides, G. M. (1993) Feedforward neural networks in chemistry: mathematical systems for classification and pattern recognition. *Chem. Rev.* **93,** 2583–2601.
35. Balaban, A. T. and Basak, S. C. (2000) Trends and possibilities for future developments of topological indices. *Abstr. Pap. Am. Chem. Soc.* **220th,** COMP-048.
36. Sutter, J. M. and Jurs, P. C. (1995) Selection of molecular descriptors for quantitative structure-activity relationships. *Data Handl. Sci. Technol.* **15,** 111–132.
37. Winkler, D. A. and Burden, F. R. (2000) Robust QSAR models from novel descriptors and Bayesian regularized neural networks. *Mol. Simul.* **24,** 243–258.
38. Maddalena, D. J. (1998) Applications of soft computing in drug design. *Expert Opin. Ther. Pat.* **8,** 249–258.
39. Apostolakis, J., and Caflisch, A. (1999) Computational ligand deign. *Comb. Chem. High-Throughput Screen.* **2,** 91–104.
40. Muskal, S. M. (1995) Exploiting data from combinatorial synthesis and screening. *Abstr. Pap. Am. Chem. Soc.* **210th,** CINF-023.

41. Keseru, G. M., Molnar, L., and Greiner, I. (2000) A neural network based virtual high-throughput screening test for the prediction of CNS activity. *Comb. Chem. High-Throughput Screen.* **3,** 535–540.
42. Lobanov, V. S. (2000) High-throughput screening of virtual combinatorial libraries with neural networks. *Abstr. Pap. Am. Chem. Soc.* **220th,** CINF-062.
43. Gasteiger, J., Teckentrup, A., and Briem, H. (2000) Analyzing high-throughput screening data by neural networks. *Abstr. Pap. Am. Chem. Soc.* **219th,** COMP-124.
44. Walters, W. P., Ajay, and Murcko, M. A. (1999) Recognizing molecules with drug-like properties. *Curr. Opin. Chem. Biol.* **3,** 384–387.
45. Sadowski, J. and Kubinyi, H. (1998) A scoring scheme for discriminating between drugs and nondrugs. *J. Med. Chem.* **41,** 3325–3329.
46. Huang, G. M., Farkas, J., and Hood, L. (1996) High-throughput cDNA screening utilizing a low order neural network filter. *BioTechniques* **21,** 1110–1114.
47. Schwabe, W., Lawrence, B. J., Robb, A. S., Hopfinger, R. M., Hochgeschwender, U., and Brennan, M. B. (1994) Direct cDNA screening of genomic reference libraries—a rapid method for the identification of transcribed sequences in large genomic regions. *Identif. Transcrib. Sequences [Proc. Int. Workshop],* 3rd, pp. 139–155.
48. So, S.-S. and Karplus, M. (1999) A comparative study of ligand-receptor complex binding affinity prediction methods based on glycogen phosphorylase inhibitors. *J. Comput.-Aided Mol. Des.* **13,** 243–258.
49. Bassett, S. I. and Elling, J. W. (1998) Automating data analysis for high-throughput screening. *Abstr. Pap. Am. Chem. Soc.* **216th,** CINF-009.
50. Tronchet, J. M. J., Grigorov, M., Dolatshahi, N., Moriaud, F., and Weber, J. (1997) A QSAR study confirming the heterogeneity of the HEPT derivative series regarding their interaction with HIV reverse transcriptase. *Eur. J. Med. Chem.* **32,** 279–299.
51. Maddalena, D. J. and Johnston, G. A. R. (1995) Prediction of receptor properties and binding affinity of ligands to benzodiazepine/$GABA_A$ receptors using artificial neural networks. *J. Med. Chem.* **38,** 715–724.
52. Lipinski, C. A. (2001) Drug-like properties and the causes of poor solubility and poor permeability. *J. Pharmacol. Toxicol. Methods* **44,** 235–249.
53. Lipinski, C. A. (2000) Changes in the profiles of drug properties: an experimental, computational, and informatics perspective. *Abstr. Pap. Am. Chem. Soc.* **219th,** CINF-020.
54. Sadowski, J. (2000) Fast computational filters for predicting ADME/Tox. *Abstr. Pap. Am. Chem. Soc.* **219th,** CINF-006.
55. Vedani, A. and Dobler, M. (2000) Multi-dimensional QSAR in drug research: predicting binding affinities, toxicity and pharmacokinetic parameters. *Prog. Drug Res.* **55,** 105–135.
56. Gobburu, J. V. S. and Shelver, W. H. (1995) Quantitative structure-pharmacokinetic relationships (QSPR) of beta blockers derived using neural networks. *J. Pharmaceut. Sci.* **84,** 862–865.
57. Schapper, K.-J., Wiese, M., Dieter, R., Emig, P., Engel, J., Kutscher, B., and

Polymeropoulos, E. E. (1993) QSAR analysis of time- and dose-dependent *in vivo* drug effects using artificial neural networks. *Trends QSAR Mol. Modell.* ***92,*** *Proc. Eur. Symp. Structure-Activity Relationships: QSAR and Molecular Modeling* 9th, pp. 546–549., Strasbourg, France.
58. Lockwood, L., Jr., Breneman, C. M., Embrechts, M. J., Bennett, K. P., and Arciniegas, F. (2000) Use of 2D, 3D, TAE and wavelet coefficient descriptors (WCDs) for generating self-organizing Kohonen maps for QSAR, QSPR, and ADME analyses. *Abstr. Pap. Am. Chem. Soc.* **220th,** COMP-134.
59. Bradbury, S. P. (1994) Predicting modes of toxic action from chemical structure: an overview. *SAR QSAR Environ. Res.* **2,** 89–104.
60. Calleja, M. C., Geladi, P., and Persoone, G. (1994) QSAR models for predicting the acute toxicity of selected organic chemicals with diverse structures to aquatic non-vertebrates and humans. *SAR QSAR Environ. Res.* **2,** 193–234.
61. Devillers, J. and Domine, D. (1997) Neural modeling of the Microtox test. *Abstr. Pap. Am. Chem. Soc.* **213th,** COMP-391.
62. Devillers, J., Bintein, S., Domine, D., and Karcher, W. (1995) A general QSAR model for predicting the toxicity of organic chemicals to luminescent bacteria (Microtox test). *SAR QSAR Environ. Res.* **4,** 29–38.
63. Devillers, J. and Domine, D. (1999) A noncongeneric model for predicting toxicity of organic molecules to Vibrio fischeri. *SAR QSAR Environ. Res.* **10,** 61–70.
64. Xu, L., Ball, J. W., Dixon, S. L., and Jurs, P. C. (1994) Quantitative structure–activity relationships for toxicity of phenols using regression analysis and computational neural networks. *Environ. Toxicol. Chem.* **13,** 841–851.
65. Basak, S. C., Grunwald, G. D., Gute, B. D., Balasubramanian, K., and Opitz, D. (2000) Use of statistical and neural net approaches in predicting toxicity of chemicals. *J. Chem. Inf. Comput. Sci.* **40,** 885–890.
66. Tang, G. and Bai, N. (1999) QSAR toxicity study of chlorated aliphatic hydrocarbons using genetic neural network GA-ANN approach. *Zhongguo Huanjing Kexue* **19,** 539–543.
67. Eldred, D. V. and Jurs, P. C. (1999) Prediction of acute mammalian toxicity of organophosphorus pesticide compounds from molecular structure. *SAR QSAR Environ. Res.* **10,** 75–99.
68. Johnson, S. R. and Jurs, P. C. (1997) Prediction of acute mammalian toxicity from molecular structure for a diverse set of substituted anilines using regression analysis and computational neural networks. *Comput. Assist. Lead Find. Optim., [Eur. Symp. Quant. Structure-Activity Relationships: QSAR and Molecular Modeling],* 11th, pp. 31–48, Lausanne, Switzerland.
69. Ivanciuc, O. (1998) Artificial neural networks applications. Part 4. Quantitative structure-activity relationships for the estimation of the relative toxicity of phenols for Tetrahymena. *Rev. Roum. Chim.* **43,** 255–260.
70. Basak, S. C., Grunwald, G. D., Host, G. E., Niemi, G. J., and Bradbury, S. P. (1998) A comparative study of molecular similarity, statistical, and neural methods for predicting toxic modes of action. *Environ. Toxicol. Chem.* **17,** 1056–1064.
71. Eldred, D. V., Weikel, C. L., and Jurs, P. C. (1999) Prediction of acute fathead

minnow toxicity of organic compounds from molecular structure. *Abstr. Pap. Am. Chem. Soc.* **217th,** COMP-054.

72. Zakarya, D., Boulaamail, A., Larfaoui, E. M., and Lakhlifi, T. (1997) QSARs for toxicity of DDT-type analogs using neural network. *SAR QSAR Environ. Res.* **6,** 183–203.
73. Devillers, J., Bintein, S., and Domine, D. (1996) Predicting the toxicity of chemicals to luminescent bacteria (Microtox test) from linear and non-linear multivariate analyses. *Abstr. Pap. Am. Chem. Soc.* **211th,** COMP-062.
74. Eldred, D. V., Weikel, C. L., Jurs, P. C., and Kaiser, K. L. (1999) Prediction of fathead minnow acute toxicity of organic compounds from molecular structure. *Chem. Res. Toxicol.* **12,** 670–678.
75. Shi, L. M., Fan, Y., Lee, J. K., Waltham, M., Andrews, D. T., Scherf, U., et al. (2000) Mining and visualizing large anticancer drug discovery databases. *J. Chem. Inf. Comput. Sci.* **40,** 367–379.
76. Shi, L. M., Fan, Y., Myers, T. G., and Weinstein, J. N. (1997) Genetic function approximation in the molecular pharmacology of cancer. *Proc. Int. Conf. Neural Networks* **4,** 2490–2493.
77. van Osdol, W. W., Myers, T. G., Paull, K. D., Kohn, K. W., and Weinstein, J. N. (1994) Use of the Kohonen self-organizing map to study the mechanisms of action of chemotherapeutic agents. *J. Natl. Cancer Inst.* **86,** 1853–1859.
78. Weinstein, J. N., Kohn, K. W., Grever, M. R., Viswanadhan, V. N., Rubinstein, L. V., Monks, A. P., et al. (1992) Neural computing in cancer drug development: predicting mechanism of action. *Science* **258,** 447–451.
79. Weinstein, J. N., Myers, T., Buolamwini, J., Raghavan, K., van Osdol, W., Licht, J., et al. (1994) Predictive statistics and artificial intelligence in the U. S. National Cancer Institute's Drug Discovery Program for Cancer and AIDS. *Stem Cells* **12,** 13–22.
80. Benigni, R. and Richard, A. M. (1996) QSARS of mutagens and carcinogens: two case studies illustrating problems in the construction of models for noncongeneric chemicals. *Mutat. Res.* **371,** 29–46.
81. Burden, F. R. and Winkler, D. A. (2000) A quantitative structure-activity relationships model for the acute toxicity of substituted benzenes to *Tetrahymena pyriformis* using Bayesian-regularized neural networks. *Chem. Res. Toxicol.* **13,** 436–440.
82. Song, X.-H., Xiao, M., and Yu, R.-Q. (1994) Artificial neural networks applied to classification of mutagenic activity of nitro-substituted polycyclic aromatic hydrocarbons. *Comput. Chem.* **18,** 391–396.
83. Villemin, D., Cherqaoui, D., and Cense, J. M. (1993) Neural networks studies: quantitative structure–activity relationship of mutagenic aromatic nitro compounds. *J. Chim. Phys. Phys. Chim. Biol.* **90,** 1505–1519.
84. Karelson, M., Sild, S., and Maran, U. (2000) Non-linear QSAR treatment of genotoxicity. *Mol. Simul.* **24,** 229–242.
85. Ghoshal, N., Mukhopadhayay, S. N., Ghoshal, T. K., and Achari, B. (1993) Quantitative structure-activity relationship studies using artificial neural networks. *Indian J. Chem., Sect. B* **32B,** 1045–1050.
86. Benigni, R. and Giuliani, A. (1994) Quantitative structure–activity relationship

(QSAR) studies in genetic toxicology: mathematical models and the "biological activity" term of the relationship. *Mutat. Res.* **306,** 181–186.
87. Ghoshal, N., Mukhopadhyay, S. N., Ghoshal, T. K., and Achari, B. (1993) Quantitative structure-activity relationship studies of aromatic and heteroaromatic nitro compounds using neural network. *Bioorg. Med. Chem. Lett.* **3,** 329–332.
88. Vracko, M. (1997) A study of structure-carcinogenic potency relationship with rtificial neural networks. The use of descriptors related to geometrical and electronic structures. *J. Chem. Inf. Comput. Sci.* **37,** 1037–1043.
89. Benigni, R. and Pino, A. (1998) Profiles of chemically-induced tumors in rodents: quantitative relationships. *Mutat. Res.* **421,** 93–107.
90. Barratt, M. D. (1996) Quantitative structure–activity relationships (QSARs) for skin corrosivity of organic acids, bases and phenols: principal components and neural network analysis of extended datasets. *Toxicol. In Vitro* **10,** 85–94.
91. Barratt, M. D., Dixit, M. B., and Jones, P. A. (1996) The use of *in vitro* cytotoxicity measurements in QSAR methods for the prediction of the skin corrosivity potential of acids. *Toxicol. In Vitro* **10,** 283–290.
92. Barratt, M. D. (1997) QSARS for the eye irritation potential of neutral organic chemicals. *Toxicol. In Vitro* **11,** 1–8.
93. Patlewicz, G. Y., Rodford, R. A., Ellis, G., and Barratt, M. D. (2000) A QSAR model for the eye irritation of cationic surfactants. *Toxicol. In Vitro* **14,** 79–84.
94. Agrafiotis, D. K. and Lobanov, V. S. (2000) Non-linear mapping networks. *J. Chem. Inf. Comput. Sci.* **40,** 1356–1362.
95. Rassokhin, D. N., Lobanov, V. S., and Agrafiotis, D. K. (2001) Non-linear mapping of massive data sets by fuzzy clustering and neural networks. *J. Comput. Chem.* **22,** 373–386.
96. McGregor, M. J. and Muskal, S. M. (2000) Pharmacophore fingerprinting in QSAR and primary library design. Patent WO 99-US25460, US 98-106007.
97 Ajay, Bemis, G. W. and Murcko, M. A. (1999) Designing libraries with CNS activity. *J. Med. Chem.* **42,** 4942–4951.
98. Wrede, P., Schneider, G., and Mueller, G. (1995) Molecular bioinformatics: Evolutionary peptide design in machina. *Bioforum* **18,** 296–297, 300–303.
99. Ivanciuc, O. (1999) Molecular graph descriptors used in neural network models. Department of Organic Chemistry, University "Politehnica" of Bucharest, Rom. Editor(s): Devillers, James, Balaban, Alexandru T. *Topol Indices Relat. Descriptors QSAR QSPR* (1999), pp. 697-777. Publisher: Gordon and Breach Science Publishers, Amsterdam, Neth.
100. Bacherikov, V. A. and Shapiro, Y. E. (1998) Zefirov-Palyulin analysis of the form of the cyclohexanone fragment of 2-arylidene- and 2-(*O*-aroyl)oxymethylene-p-menthan-3-ones. *J. Struct. Chem.* **38,** 947–953.
101. Kireev, D. B. (1995) ChemNet: a novel neural network-based method for graph/property mapping. *J. Chem. Inf. Comput. Sci.* **35,** 175–80.
102. Duprat, A. F., Huynh, T., and Dreyfus, G. (1998) Toward a principled methodology for neural network design and performance evaluation in QSAR. Application to the prediction of LogP. *J. Chem. Inf. Comput. Sci.* **38,** 586–594.

103. Huuskonen, J. J., Livingstone, D. J., and Tetko, I. V. (2000) Neural network modeling for estimation of partition coefficient based on atom-type electrotopological state indices. *J. Chem. Inf. Comput. Sci.* **40,** 947–955.
104. Devillers, J., Domine, D., Guillon, C., and Karcher, W. (1998) Simulating lipophilicity of organic molecules with a back-propagation neural network. *J. Pharmaceut. Sci.* **87,** 1086–1090.
105. Beck, B., Breindl, A., and Clark, T. (2000) QM/NN QSPR models with error estimation: vapor pressure and logP. *J. Chem. Inf. Comput. Sci.* **40,** 1046–1051.
106. Bodor, N., Huang, M.-J., and Harget, A. (1994) Neural network studies. Part 3. Prediction of partition coefficients. *Theochem* **115,** 259–266.
107. Breindl, A., Beck, B., Clark, T., and Glen, R. C. (1997) Prediction of the *n*-octanol/water partition coefficient, logP, using a combination of semiempirical MO-calculations and a neural network. *J. Mol. Model.* **3,** 142–155.
108. Gakh, A. A., Gakh, E. G., Sumpter, B. G., and Noid, D. W. (1994) Neural network-graph theory approach to the prediction of the physical properties of organic compounds. *J. Chem. Inf. Comput. Sci.* **34,** 832–839.
109. Huuskonen, J. J., Villa, A. E. P., and Tetko, I. V. (1999) Prediction of partition coefficient based on atom-type electrotopological state indices. *J. Pharmaceut. Sci.* **88,** 229–233.
110. Huuskonen, J. (2000) Estimation of aqueous solubility for a diverse set of organic compounds based on molecular topology. *J. Chem. Inf. Comput. Sci.* **40,** 773–777.
111. Clark, T., Beck, B., Breindl, A., and Hennemann, M. (1999) The use of quantum mechanics for QSPR and QSAR models with error estimation. *Abstr. Pap. Am. Chem. Soc.* **217th,** COMP-224.
112. Goll, E. S. and Jurs, P. C. (1999) Prediction of the normal boiling points of organic compounds from molecular structures with a computational neural network model. *J. Chem. Inf. Comput. Sci.* **39,** 974–983.
113. Cherqaoui, D. and Villemin, D. (1994) Use of a neural network to determine the boiling point of alkanes. *J. Chem. Soc. Faraday Trans.* **90,** 97–102.
114. Tetteh, J., Suzuki, T., Metcalfe, E., and Howells, S. (1999) Quantitative structure–property relationships for the estimation of boiling point and flash point using a radial basis function neural network. *J. Chem. Inf. Comput. Sci.* **39,** 491–507.
115. Burden, F. R. (1996) Using artificial neural networks to predict biological activity from simple molecular structural considerations. *Quant. Struct. Act. Relat.* **15,** 7–11.
116. Winkler, D. A., Burden, F. R., and Watkins, A. J. R. (1998) Atomistic topological indices applied to benzodiazepines using various regression methods. *Quant. Struct. Act. Relat.* **17,** 14–19.
117. Burden, F. R. (1989) Molecular identification number for substructure searches. *J. Chem. Inf. Comput. Sci.* **29,** 225–227.
118. Burden, F. R. (1997) A chemically intuitive molecular index based on the eigenvalues of a modified adjacency matrix. *Quant. Struct. Act. Relat.* **16,** 309–314.
119. Stanton, D. T. (1999) Evaluation and use of BCUT descriptors in QSAR and QSPR studies. *J. Chem. Inf. Comput. Sci.* **39,** 11–20.
120. Pirard, B. and Pickett, S. D. (2000) Classification of kinase inhibitors using

BCUT descriptors. *J. Chem. Inf. Comput. Sci.* **40,** 1431–1440.

121. Kubinyi, H. (1994) Variable selection in QSAR studies. I. An evolutionary algorithm. *Quant. Struct. Act. Relat.* **13,** 285–294.
122. So, S.-S. and Karplus, M. (1997) Three-dimensional quantitative structure-activity relationships from molecular similarity matrixes and genetic neural networks. 1. Method and validations. *J. Med. Chem.* **40,** 4347–4359.
123. So, S.-S. and Karplus, M. (1997) Three-dimensional quantitative structure-activity relationships from molecular similarity matrixes and genetic neural networks. 2. Applications. *J. Med. Chem.* **40,** 4360–4371.
124. So, S.-S. and Karplus, M. (1996) Genetic neural networks for quantitative structure-activity relationships: improvements and application of benzodiazepine affinity for benzodiazepine/$GABA_A$ receptors. *J. Med. Chem.* **39,** 5246–5256.
125. So, S.-S. and Karplus, M. (1996) Evolutionary optimization in quantitative structure-activity relationship: an application of genetic neural networks. *J. Med. Chem.* **39,** 1521–1530.
126. Luke, B. T. (1994) Evolutionary programming applied to the development of quantitative structure–activity relationships and quantitative structure-property relationships. *J. Chem. Inf. Comput. Sci.* **34,** 1279–1287.
127. Waller, C. L. and Bradley, M. P. (1999) Development and validation of a novel variable selection technique with application to multidimensional quantitative structure-activity relationship studies. *J. Chem. Inf. Comput. Sci.* **39,** 345–355.
128. Zupan, J. and Novic, M. (1999) Optimization of structure representation for QSAR studies. *Analyt. Chim. Acta* **388,** 243–250.
129. Neal, R. M. (1996) Bayesian learning for neural networks, vol. 118, Springer-Verlag, Berlin.
130. Burden, F. R., Ford, M. G., Whitley, D. C., and Winkler, D. A. (2000) Use of automatic relevance determination in QSAR studies using Bayesian neural networks. *J. Chem. Inf. Comput. Sci.* **40,** 1423–1430.
131. Kovalishyn, V. V., Tetko, I. V., Luik, A. I., Kholodovych, V. V., Villa, A. E. P., and Livingstone, D. J. (1998) Neural network studies. 3. Variable selection in the cascade-correlation learning architecture. *J. Chem. Inf. Comput. Sci.* **38,** 651–659.
132. Sutter, J. M., Dixon, S. L., and Jurs, P. C. (1995) Automated descriptor selection for quantitative structure–activity relationships using generalized simulated annealing. *J. Chem. Inf. Comput. Sci.* **35,** 77–84.
133. MacKay, D. J. C. (1992) Bayesian interpolation. *Neural Computat.* **4,** 415–447.
134. MacKay, D. J. C. (1992) A practical Bayesian framework for back-propagation networks. *Neural Computat.* **4,** 448–472.
135. Burden, F. R. and Winkler, D. A. (1999) Robust QSAR models using Bayesian regularized neural networks. *J. Med. Chem.* **42,** 3183–3187.
136. Niculescu, S. P., Kaiser, K. L. E., and Schuurmann, G. (1998) Influence of data preprocessing and kernel selection on probabilistic neural network modeling of the acute toxicity of chemicals to the fathead minnow and *Vibrio fischeri* bacteria. *Water Qual. Res. J. Can.* **33,** 153–165.
137. Doucet, J. P. and Panaye, A. (1998) 3D Structural information: from property

prediction to substructure recognition with neural networks. *SAR QSAR Environ. Res.* **8,** 249–272.

138. Feuilleaubois, E., Fabart, V., and Doucet, J. P. (1993) Implementation of the three-dimensional-pattern search problem on Hopfield-like neural networks. *SAR QSAR Environ. Res.* **1,** 97–114.
139. Campbell, J. L. E. and Johnson, K. E. (1993) Abductive networks: generalization, pattern recognition, and prediction of chemical behavior. *Can. J. Chem.* **71,** 1800–1804.
140. Burden, F. R. (1998) Holographic neural networks as non-linear discriminants for chemical applications. *J. Chem. Inf. Comput. Sci.* **38,** 47–53.
141. Gobburu, J. V. S., Shelver, W. H., and Chen, E. P. (1996) Initial assessment of generalized regression neural networks (GRNN) in QSAR. *Abstr. Pap. Am. Chem. Soc.* **212th,** MEDI-004.
142. Domine, D., Devillers, J., Wienke, D., and Buydens, L. (1996) The heuristic potency of art networks for QSAR data visualization and interpretation. *Abstr. Pap. Am. Chem. Soc.* **211th,** COMP-108.
143. Domine, D., Devillers, J., Wienke, D., and Buydens, L. (1997) ART 2-A for Optimal Test Series Design in QSAR. *J. Chem. Inf. Comput. Sci.* **37,** 10–17.
144. Micheli, A., Sperduti, A., Starita, A., and Bianucci, A. M. (2001) Analysis of the internal representations developed by neural networks for structures applied to quantitative structure-activity relationship studies of benzodiazepines. *J. Chem. Inf. Comput. Sci.* **41,** 202–218.
145. Kyngas, J. and Valjakka, J. (1996) Evolutionary neural networks in quantitative structure–activity relationships of dihydrofolate reductase inhibitors. *Quant. Struct. Act. Relat.* **15,** 296–301.
146. Kireev, D. B., Ros, F., Bernard, P., Chretien, J. R., and Rozhkova, N. (1997) Non-supervised neural networks: a new classification tool to process large databases. *Comput.-Assisted Lead Find. Optim. [Eur. Symp. Quant. Struct. Act. Relat.],* 11th, pp. 255–264.
147. Bernard, P., Golbraikh, A., Kireev, D., Chretien, J. R,. and Rozhkova, N. (1998) Comparison of chemical databases: analysis of molecular diversity with self-organizing maps (SOM). *Analysis* **26,** 333–341.
148. Bienfait, B. (1994) Applications of high-resolution self-organizing maps to retrosynthetic and QSAR analysis. *J. Chem. Inf. Comput. Sci.* **34,** 890–898.
149. Gasteiger, J., Wagener, M., and Sadowski, J. (1996) Assessment of the diversity of combinatorial libraries by an encoding of molecular surface properties. *Abstr. Pap. Am. Chem. Soc.* **211th,** CINF-070.
150. Hanke, J. and Reich, J. G. (1996) Kohonen map as a visualization tool for the analysis of protein sequences: multiple alignments, domains and segments of secondary structures. *Comput. Appl. Biosci.* **12,** 447–454.
151. Rose, V. S., Macfie, H. J. H., and Croall, I. F. (1991) Kohonen topology-preserving mapping: an unsupervised artificial neural network method for use in QSAR analysis. *Pharmacochem. Libr.* **16,** 213–216.

152. Rose, V. S., Croall, I. F., and MacFie, H. J. H. (1991) An application of unsupervised neural network methodology (Kohonen topology-preserving mapping) to QSAR analysis. *Quant. Struct. Act. Relat.* **10,** 6–15.
153. Shi, L. M., Myers, T. G., Fan, Y., and Weinstein, J. N. (1997) Mining the NCI anticancer drug screen database using genetic function approximation (GFA) and cluster analysis. *Abstr. Pap. Am. Chem. Soc.* **213th,** COMP-214.
154. Burden, F. R., Rosewarne, B. S., and Winkler, D. A. (1997) Predicting maximum bioactivity by effective inversion of neural networks using genetic algorithms. *Chemom. Intell. Lab. Syst.* **38,** 127–137.
155. Devillers, J. (1996) Designing molecules with specific properties from intercommunicating hybrid systems. *J. Chem. Inf. Comput. Sci.* **36,** 1061–1066.
156. Jiang, J.-H., Wang, J.-H., Song, X.-H., and Yu, R.-Q. (1996) Network training and architecture optimization by a recursive approach and a modified genetic algorithm. *J. Chemom.* **10,** 253–267.
157. Mager, P. P. and Reinhardt, R. (1999) A comparison of back-propagation and generalized-regression genetic-neural network models. *Drug Des. Discov.* **16,** 49–53.
158. Liu, Q., Hirono, S., and Moriguchi, I. (1992) Application of functional-link net in QSAR. 2. QSAR for activity data given by ratings. *Quant. Struct. Act. Relat.* **11,** 318–324.
159. Manallack, D. T. and Livingstone, D. J. (1994) Limitations of functional-link nets as applied to QSAR data analysis. *Quant. Struct. Act. Relat.* **13,** 18–21.
160. Livingstone, D. J., Hesketh, G., and Clayworth, D. (1991) Novel method for display of multivariate data using neural networks. *J. Mol. Graphics* **9,** 115–118.
161. Reibnegger, G., Werner-Felmayer, G., and Wachter, H. (1993) A note on the low-dimensional display of multivariate data using neural networks. *J. Mol. Graphics* **11,** 129–133.
162. Tetko, I. V., Aksenova, T. I., Volkovich, V. V., Kasheva, T. N., Filipov, D. V., Welsh, W. J., et al. (2000) Polynomial neural network for linear and non-linear model selection in quantitative–structure activity relationship studies on the Internet. *SAR QSAR Environ. Res.* **11,** 263–280.
163. Manallack, D. T. and Livingstone, D. J. (1995) Relating biological activity to chemical structure using neural networks. *Pestic. Sci.* **45,** 167–170.
164. Manallack, D. T. and Livingstone, D. J. (1995) Neural networks and expert systems in molecular design. *Methods Princ. Med. Chem.* **3,** 293–318.
165. Manallack, D. T., Ellis, D. D., and Livingstone, D. J. (1994) Analysis of linear and non-linear QSAR data using neural networks. *J. Med. Chem.* **37,** 3758–3767.
166. Livingstone, D. J. and Salt, D. W. (1992) Regression analysis for QSAR using neural networks. *Bioorg. Med. Chem. Lett.* **2,** 213–218.
167. Kovesdi, I., Dominguez-Rodriguez, M. F., Orfi, L., Naray-Szabo, G., Varro, A., Papp, J. G., and Matyus, P. (1999) Application of neural networks in structure–activity relationships. *Med. Res. Rev.* **19,** 249–269.

Index

I

K

L

Q

X

About the Editor

Lisa B. English graduated *cum laude* with a B.S. in Biology from Drake University, where she began her career as a research scientist investigating cellular transduction in vascular smooth muscle cells. She earned her Ph.D. in Neuroscience from the University of Minnesota, St. Paul, MN, publishing her thesis on brainstem mechanisms of pain and analgesia. After her graduate work, Dr. English was a postdoctoral fellow at Washington University Medical School, St. Louis, MO and The Scripps Research Institute, La Jolla, CA. She was also a postdoctoral associate at the University of Mississippi Medical Center, Jackson, MS. Throughout her post graduate work, Dr. English designed and conducted research studies on the development of the trigeminal pain system, the effects of stress on the hypothalamic-pituitary-adrenal axis, and the role of norepinephrine in depression.

A member of several scientific and professional organizations, including Phi Beta Phi, Beta Beta Beta, the American Medical Writers Association, the Medical Marketing Association, and the Public Relations Society of America, Dr. English is currently the strategic communications manager for BD Biosciences, a business segment of BD (Becton, Dickinson and Company). From genes to proteins to cells, BD Biosciences provides the most comprehensive portfolio of reagents, systems, and technical expertise to support the life sciences and accelerate the pace of discovery and diagnosis.

Prior to joining BD Biosciences, Dr. English worked in medical education and biotechnology marketing communications. During this time, she published several feature articles on combinatorial chemistry, including Miniaturization of HTS Assays, *Innovations in Pharmaceutical Technology,* 2000; Using ADME Data to Improve the Success Rate of Lead Discovery, *Chemistry in Australia,* March 2000, Combinatorial Discovery in Materials Science, *Chemistry in Australia,* November 1999, Keeping Pace with the Drug Discovery Revolution, *Chemistry in Australia,* November 1999, and Combinatorial Chemistry Comes of Age, *Chemical Engineering,* September 1999.

MIX
Papier aus verantwortungsvollen Quellen
Paper from responsible sources
FSC® C105338

If you have any concerns about our products,
you can contact us on
ProductSafety@springernature.com

In case Publisher is established outside the EU,
the EU authorized representative is:
Springer Nature Customer Service Center GmbH
Europaplatz 3, 69115 Heidelberg, Germany

Printed by Libri Plureos GmbH
in Hamburg, Germany